Measuring Statistical Evidence Using Relative Belief

MONOGRAPHS ON STATISTICS AND APPLIED PROBABILITY

General Editors

F. Bunea, V. Isham, N. Keiding, T. Louis, R. L. Smith, and H. Tong

1. Stochastic Population Models in Ecology and Epidemiology *M.S. Barlett* (1960)
2. Queues *D.R. Cox and W.L. Smith* (1961)
3. Monte Carlo Methods *J.M. Hammersley and D.C. Handscomb* (1964)
4. The Statistical Analysis of Series of Events *D.R. Cox and P.A.W. Lewis* (1966)
5. Population Genetics *W.J. Ewens* (1969)
6. Probability, Statistics and Time *M.S. Barlett* (1975)
7. Statistical Inference *S.D. Silvey* (1975)
8. The Analysis of Contingency Tables *B.S. Everitt* (1977)
9. Multivariate Analysis in Behavioural Research *A.E. Maxwell* (1977)
10. Stochastic Abundance Models *S. Engen* (1978)
11. Some Basic Theory for Statistical Inference *E.J.G. Pitman* (1979)
12. Point Processes *D.R. Cox and V. Isham* (1980)
13. Identification of Outliers *D.M. Hawkins* (1980)
14. Optimal Design *S.D. Silvey* (1980)
15. Finite Mixture Distributions *B.S. Everitt and D.J. Hand* (1981)
16. Classification *A.D. Gordon* (1981)
17. Distribution-Free Statistical Methods, 2nd edition *J.S. Maritz* (1995)
18. Residuals and Influence in Regression *R.D. Cook and S. Weisberg* (1982)
19. Applications of Queueing Theory, 2nd edition *G.F. Newell* (1982)
20. Risk Theory, 3rd edition *R.E. Beard, T. Pentikäinen and E. Pesonen* (1984)
21. Analysis of Survival Data *D.R. Cox and D. Oakes* (1984)
22. An Introduction to Latent Variable Models *B.S. Everitt* (1984)
23. Bandit Problems *D.A. Berry and B. Fristedt* (1985)
24. Stochastic Modelling and Control *M.H.A. Davis and R. Vinter* (1985)
25. The Statistical Analysis of Composition Data *J. Aitchison* (1986)
26. Density Estimation for Statistics and Data Analysis *B.W. Silverman* (1986)
27. Regression Analysis with Applications *G.B. Wetherill* (1986)
28. Sequential Methods in Statistics, 3rd edition *G.B. Wetherill and K.D. Glazebrook* (1986)
29. Tensor Methods in Statistics *P. McCullagh* (1987)
30. Transformation and Weighting in Regression *R.J. Carroll and D. Ruppert* (1988)
31. Asymptotic Techniques for Use in Statistics *O.E. Bandorff-Nielsen and D.R. Cox* (1989)
32. Analysis of Binary Data, 2nd edition *D.R. Cox and E.J. Snell* (1989)
33. Analysis of Infectious Disease Data *N.G. Becker* (1989)
34. Design and Analysis of Cross-Over Trials *B. Jones and M.G. Kenward* (1989)
35. Empirical Bayes Methods, 2nd edition *J.S. Maritz and T. Lwin* (1989)
36. Symmetric Multivariate and Related Distributions *K.T. Fang, S. Kotz and K.W. Ng* (1990)
37. Generalized Linear Models, 2nd edition *P. McCullagh and J.A. Nelder* (1989)
38. Cyclic and Computer Generated Designs, 2nd edition *J.A. John and E.R. Williams* (1995)
39. Analog Estimation Methods in Econometrics *C.F. Manski* (1988)
40. Subset Selection in Regression *A.J. Miller* (1990)
41. Analysis of Repeated Measures *M.J. Crowder and D.J. Hand* (1990)
42. Statistical Reasoning with Imprecise Probabilities *P. Walley* (1991)
43. Generalized Additive Models *T.J. Hastie and R.J. Tibshirani* (1990)
44. Inspection Errors for Attributes in Quality Control *N.L. Johnson, S. Kotz and X. Wu* (1991)
45. The Analysis of Contingency Tables, 2nd edition *B.S. Everitt* (1992)
46. The Analysis of Quantal Response Data *B.J.T. Morgan* (1992)
47. Longitudinal Data with Serial Correlation—A State-Space Approach *R.H. Jones* (1993)

Monographs on Statistics and Applied Probability 144

Measuring Statistical Evidence Using Relative Belief

Michael Evans

University of Toronto

Canada

CRC Press
Taylor & Francis Group
Boca Raton London New York

CRC Press is an imprint of the
Taylor & Francis Group, an **informa** business

A CHAPMAN & HALL BOOK

CRC Press
Taylor & Francis Group
6000 Broken Sound Parkway NW, Suite 300
Boca Raton, FL 33487-2742

First issued in paperback 2021

ISBN 13: 978-1-03-209856-2 (pbk)
ISBN 13: 978-1-4822-4279-9 (hbk)

Publisher's Note
The publisher has gone to great lengths to ensure the quality of this reprint but points out that some imperfections in the original copies may be apparent.

Library of Congress Cataloging-in-Publication Data

Evans, Michael, 1949-
  Measuring statistical evidence using relative belief / Michael Evans.
    pages cm. -- (Monographs on statistics and applied probability ; 144)
  Includes bibliographical references and index.
    ISBN 978-1-4822-4279-9 (alk. paper)
  1. Observed confidence levels (Statistics) 2. Measurement uncertainty (Statistics) 3. Mathematical analysis. 4. Error analysis (Mathematics) 5. Statistics. I. Title.

QA277.5.E93 2015
519.5'4--dc23

2015006911

Visit the Taylor & Francis Web site at
http://www.taylorandfrancis.com

and the CRC Press Web site at
http://www.crcpress.com

To my wife Rosemary and daughter Heather.

Contents

Preface

The concept of statistical evidence is somewhat elusive. Virtually all approaches to statistical inference refer to the statistical evidence, or the evidence, for something being true or false. But, to our knowledge, no existing theory defines explicitly what this evidence is or at least prescribes how it is to be measured. It is argued here that not to define how to measure evidence is a significant failure for any proposed theory of statistical inference. After all, the purpose of any statistical analysis is to summarize what the statistical evidence is saying about questions of interest. It seems paradoxical that we should hope to do this without being explicit about how to measure statistical evidence.

It is the purpose of this text to provide an overview of recent work on developing a theory of statistical inference that is based on measuring statistical evidence. Of course, one might ask why there is any need to do this beyond perhaps the satisfaction of having a theory that is honest about such a basic concept.

There is a range of approaches to statistical theory from Bayesian theories at one end of the spectrum, to pure likelihood theory and various frequency-based approaches at the other. Many statisticians feel comfortable fitting themselves somewhere along this scale and ignore the failure to adequately deal with statistical evidence. Others even see virtue in adopting different approaches for different problems, as in wearing a Bayesian hat today and a frequentist hat tomorrow. To an extent, these attitudes are based on issues of practicality as, in the end, a practicing statistician has to get on with the business of doing statistical analyses. While this is understandable, this ignores answering the basic question of statistics: what is a correct statistical analysis? The failure to answer this question is a profound and unacceptable gap in the subject of statistics. It almost certainly undermines confidence in the subject and, to an extent, promotes an "almost anything goes" attitude.

Part of the purpose of this book is to show that being explicit about how to measure statistical evidence allows us to answer the basic question of when a statistical analysis is correct. In fact, such a definition prescribes how the inference step should proceed. As one might expect, however, there is more to the story than simply providing a definition. The approach advocated needs to be judged in its entirety. The theory must provide a logical, coherent framework for conducting statistical analyses that can be implemented in problems of practical importance. Furthermore, the theory must be seen to possess properties that add conviction concerning its suitability.

There are some basic issues that underlie many of the controversies and disagreements in statistics. These include things like the meaning of probability, the role of subjectivity, the meaning of objectivity, the role of infinity and continuity and the relevance of the concept of utility. A fairly strong position is taken in this text on all of these. For example, it is argued that statistics is essentially subjective simply because statisticians always make choices in carrying out a statistical analysis. However, objectivity plays a key role through the data, in assessing the relevance of these choices. In essence, subjectivity can never be avoided but its effects can be assessed and to a certain extent controlled. Undoubtedly, the author's position on subjectivity and objectivity is in disagreement with those who advocate pure subjectivity and with those who believe that there is an objective theory of statistics. It is our contention that inferences derived via relative belief, together with checking the model and prior against the data, place statistics on a much firmer logical foundation with greater relevance for scientific applications.

One caveat needs to be stated for what is being proposed. The developments in this text represent an attempt to establish a gold standard for how a statistical analysis should proceed. A gold standard is something to strive to attain in an application, but, for various reasons, we may fall short. The result of such a failure does not entirely invalidate the analysis but it does suggest that the results have to be suitably qualified. Statisticians are already familiar with making such compromises. Consider the first, and perhaps most important, part of a statistical investigation, namely, the collection of the data. The gold standard here is random sampling from populations and the controlled allocation of values of predictor variables, but this is often not realized. Yet statistical analyses are conducted and useful information is acquired in spite of the deficiencies. Any conclusions drawn, however, must be suitably qualified when there is a failure to attain the highest standard in data collection. The difficulties entailed in guaranteeing that all the necessary ingredients hold for an application of a theory do not justify an attitude that the existence of such a theory is irrelevant.

Chapter 1 discusses some basic features of our overall vision, such as the roles of subjectivity, objectivity, infinity and utility in statistical analyses. In developing a theory of statistical inference, it is necessary to carefully delineate the problems to which the theory is to be applied. As such, the domain of application of the theory is provided, namely, what constitutes a statistical problem and what are the ingredients that a statistician needs to specify to conduct a statistical analysis. In this chapter a simple example is presented of a statistical analysis that satisfies our criteria. Chapter 2 considers the meaning of probability and the various positions taken on probability. This topic lies at the heart of many disagreements in statistics and the author contends that there are a number of reasonable ways to think about probability. Indeed, there is no claim concerning the correctness of one approach to probability over others. The theory of statistical inference presented here is basically independent of these interpretations, although, however probabilities are assigned, they are considered to be measuring belief. Chapter 3 begins the discussion of the heart of the matter, namely, attempts to deal with the concept of statistical evidence. This chapter demonstrates that, while many theories of inference make mention of statistical evidence, they don't adequately define what it is or, more important, how it is to be measured. Furthermore, this failure leads to anomalies for these theories. In Chapter 4 a method is provided for measuring statistical evidence and, based on this method, a theory of inference is developed. This theory is based on the assumption that the ingredients chosen for a statistical analysis are *correct*. Chapter 5 discusses how a statistician is to go about choosing the ingredients for a statistical problem and how these choices are to be checked for their *correctness* in an application. Of course, the meaning of correct in this context requires considerable discussion. It is a key point in our development that the theory and application of inference are logically separate from the checking phase, and these shouldn't be confounded, as the problems are quite different. For practical applications Chapter 5 should precede Chapter 4 but the focus in this text is on measuring statistical evidence. Chapter 6 summarizes the text and points to further possible developments.

Certainly what is being advocated here is not completely divorced from what has been discussed by others. In fact, the author would describe the essence of the relative belief approach to statistical inference as simply being careful about the definition and usage of the Bayes factor. Furthermore, there are many similarities with pure likelihood theory, and relative belief could be described as filling in, from the author's point of view, the logical gaps in that theory. Also, the frequentist approach plays a key role, not in determining inferences, but in checking the suitability of the ingredients as well as providing optimality properties of the inferences.

The author is solely responsible for all errors and omissions. Thanks are owed to many. In particular, Luai Al-Labadi, Zeynep Baskurt, Shelly Cao, Zvi Gilula, Irwin Guttman, Gun Ho Jang, Shaocheng Liu, Hadas Moshonov, Saman Muthuku-marana, Mohammed Shakhatreh, Tim Swartz and Tianli Zou were all co-authors on publications connected with the contents of this text and made key contributions. Irwin Guttman introduced me to Bayesian inference and this has had a major influence. Gun Ho Jang contributed numerous ingenious solutions to technical problems. Many readers of the manuscript provided valuable input, including students Stephen Marsh and Yang Guan Jian Guo. Keith O'Rourke made many useful suggestions. The reviewers also provided valuable feedback and the author would especially like to thank Jay Kadane and David Nott. My interest in problems concerned with the foundations of statistical inference was stimulated by Professor D. A. S. Fraser and I am grateful for that and also for instilling in me the belief that these problems are resolvable.

Chapter 1

Statistical Problems

1.1 Introduction

This book is about measuring statistical evidence. More precisely, a definition of statistical evidence is proposed based on the ingredients of a statistical problem as specified by the statistician. A direct consequence of this definition is a theory of statistical inference that has some unique and appealing features.

It may come as a surprise to the lay reader that exactly how one is to measure statistical evidence is not well-resolved in the scientific literature. Most reasonably numerate individuals have encountered the notions of p-values, standard errors, etc., and understand that these concepts are central to how to reason in statistical problems and that they have something to do with characterizing statistical evidence. Yet it is a fact that experienced, professional statisticians can disagree quite dramatically about the right way to reason in statistical contexts.

On examining the various approaches to inference, it will be seen that they commonly fail to precisely define what statistical evidence is. This can be regarded as a significant omission. In fact, it is our view that any valid theory of statistical inference must specify exactly what is meant by statistical evidence. A definition of statistical evidence should serve as a core of the theory of statistical inference and basically dictate how statistical problems are to be solved, namely, the statistical evidence should tell us what the solution to a problem is.

Our proposal for a definition of statistical evidence is provided in Chapter 4. Any such definition is based upon the ingredients of a statistical problem as specified by a statistician. So the current chapter is concerned with discussing exactly what is meant by a statistical problem and what ingredients need to be specified by the statistician. This leads us to exclude certain problems and ingredients that others might prefer to be included. Our defence for this is that our approach covers the vast majority of practically meaningful statistical problems and that by being exclusionary, a lot of unnecessary complexity and ambiguity is eliminated. Above all, our goal is a logical and complete theory of statistical inference that has practical relevance rather than some kind of mathematical generality. In fact, our view is that attempts at mathematical generality often mislead as to what is appropriate statistical reasoning.

Probability is a key concept in any theory of statistical inference. This is of such importance that the entirety of Chapter 2 is devoted to this topic.

1.2 Statistical Problems

The first question to be answered is: what is a statistical problem? The following example characterizes what could be called the archetypal statistical problem. The discussion in this text is restricted to the consideration of such problems and close relatives. We argue that such restrictions are necessary and moreover apply in the vast majority of applications.

Example 1.2.1 *The Archetypal Statistical Problem.*

Suppose there is a *population* Ω with $\#(\Omega) < \infty$, where $\#(A)$ denotes the cardinality of the set A. So Ω is just a finite set of objects. Further suppose that there is a *measurement* $X : \Omega \to \mathcal{X}$. A measurement X is a function defined on Ω taking values in the set $\mathcal{X}$. So $X(\omega) \in \mathcal{X}$ is the measurement of object $\omega \in \Omega$. For example, Ω could be the set of all students enrolled at a particular school and $X(\omega)$ the height in centimeters of student ω. So, in this case, $\mathcal{X}$ is a subset of R^1. As another example, Ω could be the set of all students enrolled at a particular school and $X(\omega) = (X_1(\omega), X_2(\omega))$, where $X_1(\omega)$ is the height in centimeters of student ω and $X_2(\omega)$ is the gender of student ω, and so, in this case, $\mathcal{X}$ can be taken to be a subset of $R^1 \times \{M, F\}$.

When considering a variable like height, it is common to treat this as possibly taking on a continuous range of values and to allow the set of possible values to be unbounded. While this may seem innocuous, as argued in Section 1.4, we need to be careful when using infinities in discussing statistical problems. As such, because of the finiteness of Ω, the finite accuracy with which our height measurements are made, and the fact that the measurements will occur within known bounds, $\mathcal{X}$ can be taken to be the set of all the possible values of $X(\omega)$ and this is a finite set. Infinite sets are introduced in Section 1.4 when considering approximations to statistical problems.

The fundamental object of interest in a statistical problem is then the *relative frequency distribution* of X over Ω. For a subset $A \subset \mathcal{X}$ the relative frequency of A is given by

$$r_X(A) = \frac{\#(\{\omega : X(\omega) \in A\})}{\#(\Omega)}.$$

So $r_X(A)$ is just the proportion of elements in Ω whose X measurement is in A. Clearly, knowing the relative frequency distribution is equivalent to knowing the *relative frequency function*

$$f_X(x) = \frac{\#(\{\omega : X(\omega) = x\})}{\#(\Omega)},$$

for $x \in \mathcal{X}$, as one can be obtained from the other. Notice that the frequency distribution is defined no matter what the set $\mathcal{X}$ is.

If we can conduct a *census*, where we obtain $X(\omega)$ for each $\omega \in \Omega$, then f_X is known exactly and there is nothing left to do from a statistical perspective. Of course, statistics exists as a subject precisely because it is, at the very least, generally uneconomical to conduct a census, and typically it is impossible to do so. Sometimes statistical problems are expressed as wanting to know about some aspect of f_X rather than f_X itself. For example, if X is real-valued one may be interested to know the *mean*

$$\mu_X = \sum_{x \in \mathscr{X}} x f_X(x)$$

and *variance*

$$\sigma_X^2 = \sum_{x \in \mathscr{X}} (x - \mu_X)^2 f_X(x)$$

of the relative frequency distribution or perhaps some other characteristic of f_X. Certainly there is nothing wrong with focusing on some characteristics of f_X, but knowing f_X represents full statistical information. As such, we will express our discussion in terms of knowing the true f_X.

The fundamental question of statistics is then, based on partial information about the true f_X, how do we make inferences about the true f_X? Of course, it has to be made clear what is meant by partial information, and subsequent sections will do this, but as part of any statistical problem there is the *observed data* $x_1 = X(\omega_1), \ldots, x_n = X(\omega_n)$ where $\{\omega_1, \ldots, \omega_n\} \subset \Omega$ is selected, in some fashion, from the population. ∎

Note that in Example 1.2.1 there are no infinities and everything is defined simply in terms of counting. Also there is no mention of probability. There is, however, a major *uncertainty* in that f_X is unknown without conducting a census. This is the fundamental uncertainty lying at the heart of all of statistical problems.

Undoubtedly Example 1.2.1 seems very restrictive but there are a number of ways in which it can be generalized without violating its basic characteristic of everything being finite and obtainable via counting. For example, we can consider several finite populations $\Omega_1, \ldots, \Omega_m$ with respective measurements $X_1, \ldots, X_m$ and relative frequency functions $f_{X_1}, \ldots, f_{X_m}$ and then discuss making comparisons among them. Also, the relation of so-called *measurement error* problems to Example 1.2.1 and the use of infinities and continuity in statistical modeling are examined in Section 1.4.

Perhaps most important, problems where the interest is with relationships among variables arise as generalizations of Example 1.2.1. The concept of relationship between variables is based on the concept of a *conditional relative frequency distribution*. Suppose there are two measurements X and Y defined on a population Ω with $(X(\omega), Y(\omega)) \in \mathscr{X}$. The *conditional relative frequency function* of Y given $X = x$ is then defined for $(x, y) \in \mathscr{X}$ by

$$f_{Y|X}(y|x) = \frac{\#(\{\omega : X(\omega) = x, Y(\omega) = y\})}{\#(\{\omega : X(\omega) = x\})} = \frac{f_{(X,Y)}(x,y)}{f_X(x)}$$

whenever $\#(\{\omega : X(\omega) = x\}) \neq 0$ or, equivalently, $f_X(x) \neq 0$. Notice that $f_{Y|X}$ is again obtained by simple counting and that it can be obtained from the *joint relative frequency function* $f_{(X,Y)}$ and the *marginal relative frequency function* f_X. The following makes use of conditional relative frequency and represents the most important application of statistics.

Example 1.2.2 *Relationships among Variables.*

Suppose there is a measurement $(X,Y) : \Omega \to \mathscr{X}$ and our interest is in whether or not there is a *relationship* between the variables X and Y. There is a basic definition of what it means for variables to be related.

Definition 1.2.1 *Variables X and Y, defined on population Ω, are related variables if for some x_1, x_2 such that $f_X(x_1) \neq 0$ and $f_X(x_2) \neq 0$, then $f_{Y|X}(\cdot|x_1) \neq f_{Y|X}(\cdot|x_2)$.*

So two variables are related whenever changing the conditioning variable can result in a change in the conditional distribution of the other variable. Note that it is clearly the case that there is no relationship when all the conditional distributions are the same; in fact, this is equivalent to the *statistical independence* of the variables. In general, the *form of the relationship* is given by how $f_{Y|X}(\cdot|x)$ changes as x changes. From a practical viewpoint, formally at least, most variables are related by this definition, as it seems unlikely that these conditional distributions will always be the same. But the relationship between X and Y can be very weak and the changes deemed to be irrelevant for the application at hand.

It is common in statistical applications for various assumptions to be made about the form of $f_{Y|X}(\cdot|x)$. For example, a *regression assumption* says that, for a real-valued Y, at most the *conditional means*

$$\mu_Y(x) = \sum_{y \in \{Y(\omega):X(\omega)=x\}} y f_{Y|X}(y|x))$$

are changing as we change x. Often a *linear regression assumption* is also made where it is assumed that μ_Y is in some finite linear span of functions of x, namely, $\mu_Y \in L\{v_1, \ldots, v_k\}$ where the v_i are real-valued functions defined on $\{X(\omega) : \omega \in \Omega\}$. Of course, these are assumptions and the methods of Chapter 5 are needed to see if these makes sense in an application. Actually, the regression assumptions make the most sense when Y is allowed to take on a continuous range of values, as discussed in Section 1.4. ∎

Although we will often express concepts in terms of the archetypal statistical problem, it will be seen that these apply much more generally. Section 1.3 is particularly relevant in this regard.

1.3   Statistical Models

The fundamental problem in statistics arises because a census cannot be conducted and so relative frequency distributions such as f_X in Example 1.2.1 cannot be known exactly. Note that, because Ω is finite and because each component of X is bounded and measured to finite accuracy, there are only finitely many possibilities for f_X. For example, if $\#(\Omega) = 10^4, X(\omega)$ is height recorded in centimeters and all heights are in $(0, 300]$, then f_X is in a finite set of cardinality considerably less than $10^4 \times 300$. The important point here is that the set of possibilities for f_X is finite.

In many statistical problems, the statistician is willing to *assume* that f_X is in a restricted set of possible relative frequency functions. We index these possible functions by a variable θ, called the *model parameter,* taking values in a set Θ, called the *model parameter space*, to obtain the *statistical model* $\{f_\theta : \theta \in \Theta\}$. So $f_X \in \{f_\theta : \theta \in \Theta\}$ and note that, at this point, Θ is finite. Accordingly, instead of the true relative frequency function, we can speak equivalently about the *true value* of the model parameter since by assumption there is a unique value $\theta_{true} \in \Theta$ such that $f_{\theta_{true}} = f_X$.

More generally, we can define a *parameter ψ of the model* as $\psi = \Psi(\theta)$ where $\Psi : \Theta \overset{onto}{\to} \Psi$, and for convenience we use the Ψ symbol for both the function and its range. So there is a true value for ψ given by $\psi_{true} = \Psi(\theta_{true})$. In general, we want to make inferences about a *parameter of interest ψ* (which could be θ if we take Ψ to be the identity) and refer to all aspects of θ that distinguish values in $\Psi^{-1}\{\psi\}$ as *nuisance parameters*. We provide a simple example.

Example 1.3.1
Consider a population Ω of eligible voters where $\#(\Omega) = 20,000$ and $X(\omega) = 1$, if ω will vote in the next election, and $X(\omega) = 0$, otherwise. So there are in total exactly 20,001 different possibilities for f_X. Suppose further, however, that it is known that at least 5,000 and no more than 15,000 voters will indeed vote, where this information is based on historical records of elections. So, noting that $1/20,000 = 5 \times 10^{-5}$, the statistical model $\{f_\theta : \theta \in \Theta\}$ is given by $\theta \in \Theta = \{0.25000, 0.25005, 0.25010, \ldots, 0.75000\}$ where $f_\theta(1) = \theta$ and $f_\theta(0) = 1 - \theta$.

Rather than the model parameter θ, one might be interested in the odds parameter $\psi = \Psi(\theta) = \theta/(1 - \theta)$ where the range is $\{0.25/(1 - 0.25), 0.25005/(1 - 0.25005), \ldots, 0.75/(1 - 0.75)\}$. In this case Ψ is 1 to 1, so there are no nuisance parameters. ∎

Many attempts at developing theories of inference run into problems when considering inferences for an arbitrary $\psi = \Psi(\theta)$. This is referred to as the *nuisance parameter problem*.

An important point to note about a statistical model is that, unless it includes all the possible distributions, $\{f_\theta : \theta \in \Theta\}$ is an assumption and as such could be incorrect because $f_X \notin \{f_\theta : \theta \in \Theta\}$. As you might expect, when statistical analysis is based on incorrect assumptions, then we have to question the validity of the analysis. If so, why not always take $\{f_\theta : \theta \in \Theta\}$ to be the set of all possible distributions? Certainly in Example 1.3.1 it seems simple to avoid any assumptions.

There are several reasons why model assumptions are made. First and foremost is that there may be definite information about the form of f_X and this information leads to improved inferences when true. Second, and perhaps most common, is that for very complicated situations, model assumptions are made to simplify the analysis and it is felt that the error introduced by these simplifications will not have a material effect on the inferences drawn. For example, in Example 1.2.2 it seems very unlikely that the regression assumption ever holds exactly. But perhaps the deviation from this assumption is so small that it is immaterial, while the added simplicity of only looking at the conditional mean to examine the relationship is of great benefit.

There is the possibility, however, that the model $\{f_\theta : \theta \in \Theta\}$ could be grossly in error. So it is necessary to consider how to assess and deal with this as part of a statistical analysis. This is discussed in part in Section 1.5 and is more thoroughly treated in Chapter 5.

1.4   Infinity and Continuity in Statistics

So far all the sets introduced have been finite. It is our belief that this finiteness holds in any practical application of statistics. Infinite sets can be used as part of a simplifying approximation but with the awareness that this can bring with it problems of interpretation that can lead us astray when we are not careful. Consider the following example.

Example 1.4.1 *Likelihood Functions.*

A probability density f_θ gives rise to a probability measure P_θ on $\mathscr{X}$ via integration of f_θ over relevant sets. Suppose it is assumed that data $x \in \mathscr{X}$ has been generated from one of the probability distributions in the model $\{f_\theta : \theta \in \Theta\}$. Now we wish to make an inference about the true $\theta \in \Theta$. In such a situation, methods based upon the likelihood function are commonly recommended.

Definition 1.4.1 *For observed data x and model $\{f_\theta : \theta \in \Theta\}$, the likelihood function is defined to be the function $L(\cdot \,|\, x) : \Theta \to [0, \infty)$ given by $L(\theta \,|\, x) = k f_\theta(x)$ for any fixed $k > 0$.*

In reality the likelihood function is an equivalence class of functions, as the constant k is arbitrary and can be chosen for convenience. This indeterminacy causes no problems because likelihood inferences only depend on the ratios of likelihood values.

The motivation behind considering the likelihood function lies in saying that θ_1 is at least as preferable (as a guess or inference about the true value of θ) as θ_2 whenever the *likelihood ratio* $L(\theta_1 \,|\, x)/L(\theta_2 \,|\, x) \geq 1$. This imposes a complete preference ordering on Θ. This *likelihood preference ordering* is natural when each distribution is discrete because

$$\frac{L(\theta_1 \,|\, x)}{L(\theta_2 \,|\, x)} = \frac{k f_{\theta_1}(x)}{k f_{\theta_2}(x)} = \frac{P_{\theta_1}(\{x\})}{P_{\theta_2}(\{x\})}$$

is the ratio of the probability of observing x when θ_1 is true to the probability of observing x when θ_2 is true. Given that we have observed x, it is natural to prefer those θ values which give a higher probability to the observed data.

For the situation where continuous probability distributions are employed, let us suppose, for the moment, that $\mathscr{X}$ is Euclidean, $N_\varepsilon(x)$ is an open ball about x of radius ε and f_θ is continuous and positive at x for each θ. Letting $\mathrm{Vol}(A)$ denote the Euclidean volume of $A \subset \mathscr{X}$, we have that

$$P_\theta(N_\varepsilon(x)) = \int_{N_\varepsilon(x)} f_\theta(z)\,dz \sim f_\theta(x)\mathrm{Vol}(N_\varepsilon(x))$$

as $\varepsilon \to 0$, since $P_\theta(N_\varepsilon(x))/f_\theta(x)\mathrm{Vol}(N_\varepsilon(x)) \to 1$ as $\varepsilon \to 0$. So for small ε

$$\frac{L(\theta_1 \,|\, x)}{L(\theta_2 \,|\, x)} = \frac{k f_{\theta_1}(x)}{k f_{\theta_2}(x)} \approx \frac{P_{\theta_1}(N_\varepsilon(x))}{P_{\theta_2}(N_\varepsilon(x))}$$

and again the likelihood ratio can be seen as comparing the probabilities of observing x. As such, the likelihood preference ordering makes sense in the continuous context too.

But notice a key assumption in this argument, namely, that f_θ is continuous at x for each θ. If g_θ is another integrable function that differs from f_θ at most on a set of volume measure 0, then $P_\theta(A) = \int_A f_\theta(z)\,dz = \int_A g_\theta(z)\,dz$ for every Borel set A. So g_θ could just as easily serve as a density for P_θ. As is well known, f_θ can be modified at countably many points to obtain a valid density g_θ and it is not necessary to use a density that is continuous at each x, at least for the computation of probabilities.

Now this anomaly may be considered a minor irritation, but consider a practical context. In such a situation the value of X is measured to a finite accuracy and as such every coordinate in x is a rational number. The set of $x \in \mathcal{X}$ with rational coordinates is necessarily countable and so a density g_θ for P_θ can be chosen such that $g_\theta(z) = 0$ (or some other constant) for every z with rational coordinates. Therefore, if we use such a g_θ in the definition of the likelihood function, for any actually observed data x the likelihood is identically 0 and the likelihood preference ordering doesn't distinguish among the θ. Clearly this is absurd, unless you don't believe in the relevance of the likelihood preference ordering to inference. ∎

One way out of the dilemma posed with continuous models in Example 1.4.1 is to simply demand that the densities in the definition of the likelihood be continuous at each $x \in \mathcal{X}$. But what aspect of an application implies such a restriction? For us this restriction is imposed by the fact that all sets in a statistical application are finite and, when an infinite set is used, this is as an approximation to a finite object. If this approximation aspect is ignored, then absurdities can arise as in the discussion in Example 1.4.1. If $f_\theta(x)$ can be arbitrarily defined on a set of measure 0, as is certainly mathematically acceptable when considering densities just as mathematical objects, then the notion of an approximation is lost.

Various treatments of statistical theory treat infinite sets as basic ingredients that represent reality. For the developments here, however, while we want to make use of the simplicities available with infinite sets, conditions must be placed on such objects to ensure that they behave appropriately as approximations to entities that are in fact finite. As such, it is required that a density f_θ be defined as a limit. For example, in Example 1.4.1, for each $x \in \mathcal{X}$, it is required that

$$f_\theta(x) = \lim_{\varepsilon \to 0} \frac{P_\theta(N_\varepsilon(x))}{\mathrm{Vol}(N_\varepsilon(x))} \tag{1.1}$$

as this ensures that $P_\theta(N_\varepsilon(x)) \approx f_\theta(x)\mathrm{Vol}(N_\varepsilon(x))$ for small ε. The definition of densities is discussed more generally in Appendix A but it is noted that, if a version of f_θ exists that is continuous at x, then it is given by (1.1). This leads to the usual representative densities and in general the density calculated by differentiating a distribution function will satisfy (1.1). Any time density is used it is assumed that it is a density with respect to the volume measure on the respective space (counting measure is volume measure on discrete sets) and that the density arises as a limit as in (1.1); see Appendix A.

It is necessary to be vigilant to ensure that attempts to be mathematically general do not lead us to introduce absurdities into discussions of inference. Mathematics should serve our developments by allowing us to be precise about concepts rather than imposing some abstract set-theoretic conception of what statistics is as a subject. As such, if a counterexample is presented that depends intrinsically on infinity, then the relevance of the counterexample seems questionable. The following is a notorious example that is commonly cited as a paradox for Bayesian inference methodology.

Example 1.4.2 *Uniform Priors.*

Suppose there is a population Ω, where $\#(\Omega) = N$ is very large, and a measurement X such that $X(\omega) = 1$ when ω possesses some characteristic of interest and $X(\omega) = 0$ otherwise. Then $\mathscr{X} = \{0,1\}$ and $f_X(1)$ is the proportion of individuals in Ω that have the characteristic. Letting θ denote the unknown proportion, we can write $f_X(x) = f_\theta(x)$ where $f_\theta(x) = \theta^x(1-\theta)^{1-x}$ for $x \in \mathscr{X}$. Now it is clear that θ is some fraction of the form m/N. A statistical model for this situation is thus $\{f_\theta : \theta \in \Theta_N\}$ where $\Theta_N = \{0, 1/N, \ldots, (N-1)/N, 1\}$ and this contains all the possibilities for a single observed $x = X(\omega)$. In Example 1.5.1 we will discuss the model for more than one observation.

As will be discussed, uncertainty about the true value of θ can result in placing a probability distribution on Θ_N that expresses beliefs about what the true value is. For example, this could be a uniform distribution Π_N on Θ_N so that each possible value of θ has probability $1/(N+1)$. As N is large, we can conceive of approximating this distribution by the uniform distribution Π on the interval $\Theta = [0,1]$. So the model is expanded to include distributions that are known to be incorrect but this is not felt to create any significant errors in the analysis when N is large. For $i \in \{0,1,\ldots,N\}$ then

$$\Pi\left(\left[\frac{i}{N+1}, \frac{i+1}{N+1}\right)\right) = \frac{1}{N+1} = \Pi_N\left(\left\{\frac{i}{N}\right\}\right).$$

Now suppose that, instead of indexing the distributions in the model by θ, we use $\psi = \Psi(\theta) = \theta^2$. The probability density of ψ is $p_\Psi(\psi) = 1/(2\sqrt{\psi})$ when $\psi \in [0,1]$ and is 0 otherwise. It is often claimed that this presents a contradiction for the initial assignment of probability because the distribution of ψ is no longer uniform, yet knowing ψ is equivalent to knowing θ. But consider the approximation to the finite distribution. Certainly $\psi \in \Psi_N = \{0, 1/N^2, 2^2/N^2, \ldots, (N-1)^2/N^2, 1\}$ and note that these points are no longer equispaced and are compressed toward 0. But also note that

$$\frac{1}{N+1} = \Pi\left(\left[\frac{i}{N+1}, \frac{i+1}{N+1}\right)\right)$$
$$= \Pi_\Psi\left(\left[\left(\frac{i}{N+1}\right)^2, \left(\frac{i+1}{N+1}\right)^2\right)\right) = \int_{i^2/(N+1)^2}^{(i+1)^2/(N+1)^2} (2\sqrt{\psi})^{-1} d\psi$$

and the factor $1/2\sqrt{\psi}$ in p_Ψ is simply adjusting for the fact that the transformation Ψ is modifying lengths at different rates in the interval $[0,1]$. In other words, the lack of uniformity for the continuous ψ is simply an artefact of approximating a discrete distribution by a continuous one. When the approximation is taken into account, there is no contradiction or paradox.

Of course, this explanation demands that the initial assignment of prior probabilities conforms to real beliefs. Our conclusion is that when the finite nature of the model is taken into account, beliefs are transformed appropriately under transformations. ∎

The concern expressed in Example 1.4.2 with the definition of a uniform prior is simply a misapplication of infinite sets to a statistical context. In essence it arises as a result of considering an infinite object as a fundamental aspect of a statistical application and for our discussion here, this is not correct. Of course, many other objections have been raised concerning the use of prior distributions. Some of these are discussed in the following section.

An argument commonly given in favor of considering models with continuous spaces is based on the following example.

Example 1.4.3 *Measurement Models.*
Consider measuring the length of some object. For this purpose there is a measuring device which provides a measurement X with a prescribed accuracy. Let us suppose that this accuracy is in centimeters. As well, for any measuring device there are bounds on the values it can record, so the set $\mathscr{X}$ of possible values for X is finite.

Now, as a thought experiment, imagine taking N measurements of the object where N is very large, a much larger number of measurements than could ever be conceived of being taken although still finite. It is natural to suppose that, due to a variety of factors, there will be variation in the measurements recorded. If the N measurements were actually observed, then this gives rise to a relative frequency function f_X. Note that there are finitely many possibilities for f_X and conceivably it could be known simply by carrying out the N measurements. Certainly knowing f_X will tells us lots about the measurement process and in many ways it is the object of interest. Furthermore, it may be reasonable to suppose that, if a large number of measurements are taken, then a good approximation to f_X is available.

Let θ represent the *true* length of the object we are measuring. Given the device we are using, this true length is also recorded in centimeters. As such, $\theta \in \mathscr{X}$ and represents some quantile of the distribution given by f_X. It is not clear which quantile θ will represent but it may be reasonable to *assume* that θ is the median and, if we are further willing to *assume* that f_X is symmetric, then θ is also the mean of f_X. If, in addition, it is *assumed* that only the location and not the form of f_X will change when measuring different objects, then a statistical model for this situation is given by $f_X \in \{f_\theta : \theta \in \Theta\}$ where $f_\theta(x) = f_E(x - \theta)$, f_E is the relative frequency function of the *measurement error* $E = X - \theta$ and $\theta \in \Theta \subset \mathscr{X}$. A model of this form is called a *measurement model* and obviously this can be generalized in a number of ways.

Our main point here is that the basic measurement model is a finite object that is again derived via simple counting. A number of assumptions are necessary, however, for this model to hold. As such, it is our preference to take Example 1.2.1 as the basic motivating example on which to build statistical theory. The analysis of measurement models can still be carried out, but the principles on which such an analysis is based arise from the simpler and more concrete context. ∎

It is sometimes argued that a continuous model for a variable makes sense because it is possible to conceive of measuring to an ever finer accuracy. Whether or not this is true, in any particular context measurements are taken to some predetermined accuracy. The fact that we may have been able to measure more precisely cannot have any impact on our analysis.

In the end, with sets of very large cardinality, it will not make much difference if such a set is taken to be infinite and, depending on the variables being measured, we take distributions to be continuous. When to do this depends on judgment and it doesn't seem easy to provide specific guidance. The simplifications produced generally seem worth it. It is necessary to acknowledge, however, their role as approximations and not develop the theory of inference by treating infinite objects as fundamental.

Some will disagree with our treatment of statistical models containing infinities as being approximations rather than representing reality and so will not accept our assertion that this restriction covers all practically meaningful statistical problems. It has to be acknowledged, however, that any general theory of inference must be applicable to the essentially finite contexts considered here. Furthermore, a general theory must resolve the inferential issues that arise within the essentially finite context, as the theory of relative belief will be seen to do. So criticism of the theory of relative belief based on a lack of mathematical generality is not valid, unless a more general theory which also accomplishes this can be shown to exist. It is clearly important to separate statistical from mathematical issues and that is what the restriction to essentially finite contexts does.

1.5   The Principle of Empirical Criticism

Perhaps nothing excites more controversy in discussions about statistics than the word *objectivity*. The Merriam–Webster online dictionary gives the following definition:

> *objectivity*: *lack of favoritism toward one side or another.*

That seems reasonably clear, but careful thought suggests that defining objectivity precisely is something that is difficult and perhaps even impossible. For example, we might ask to whom the "lack of favoritism" applies. If it is the scientist involved in an investigation, then it is not clear how it can ever really be said that a scientific investigation has been conducted objectively and has not been influenced by the scientist's subjective opinions. In fact, it seems to be the case that the very best scientific work is highly influenced by a scientist's personal opinions and insights about what is true. Scientific work is not simply a cold, dispassionate examination of the facts by an objective investigator.

There is no denying that objectivity should serve as a guiding light in determining what is believed to be true. To ignore the role that *subjectivity* plays in the acquisition of new knowledge, however, is at the very least misleading. Somehow an approach is needed that acknowledges the primacy of objectivity in determining new knowledge, but at the same time realizes a role for subjectivity in that process. It is our view that the subject of statistics presents an appropriate framework for the development of such an approach to scientific inquiry. Unfortunately, much of the discussion in statistics on this topic seems to polarize between an unrealistic view of what objectivity is and how it is attained, and an idealistic view of the virtues of subjectivity. A better path forward lies somewhere in between.

It seems that most fields develop their own requirements for the acceptance of new knowledge. For example, in mathematics, deriving conclusions by the application of the rules of deductive reasoning to given hypotheses in such a way that the argument can be verified as correct by others is the standard. No requirement is made that the results obtained have anything to do with physical reality. In physics the standard requires experimental verification but also requires that there exists theory that in some sense explains the phenomena in question. For example, dark matter is inferred to exist because without it the very successful theory of gravitation would fail to explain some observed astrophysical phenomena. Still, acceptance of dark matter will undoubtedly require its observation or perhaps a deeper theory. In many fields, statistics plays a key role in determining what is accepted as knowledge. As such, a clear definition of statistical evidence and the degree to which it can be claimed to be objective are necessary.

Our position is that statistical analyses are inherently subjective in nature and there is no way of avoiding this. In fact, pretending that such analyses can be objective only blinds us to inherent defects as, for example, failing to provide a clear definition of what statistical evidence is. Others, such as Berger and Berry (1988), have raised similar concerns.

At the same time, however, one has to acknowledge the virtues and even necessity of objectivity in many applications. One solution to this apparent dilemma is to acknowledge that the data can be objective in a statistical analysis, and to require that the following principle be adhered to.

Principle of Empirical Criticism: Every ingredient chosen by a statistician as part of a statistical analysis must be checked against the observed data to determine whether or not it makes sense.

For example, if a statistician chooses to make the assumption that the true distribution is in $\{f_\theta : \theta \in \Theta\}$, where this is a proper subset of the set of all the possible distributions for the problem, then this must be checked against the observed data for its reasonableness. This is called *model checking* and is discussed in Chapter 5.

Also, if an ingredient is used in a statistical analysis that cannot be checked against the data, then any inferences based on this must be qualified as depending on ingredients that cannot be empirically checked for their suitability. It is our view that any theory of inference that depends intrinsically upon ingredients that cannot be checked is of limited usefulness and, as such, is flawed. The application of the principle, with all subjective inputs passing their checks, does not guarantee objectivity but it provides some comfort that the subjective choices made are at least not contradicted by the objective aspect of the analysis. We note that the principle of empirical criticism is similar to the *falsifiability* criterion of Popper (1959), namely, a scientific theory is not valid unless it is empirically testable.

1.5.1   The Objectivity of the Data

It seems clear that there is only one aspect of a statistical investigation that can ever truly be claimed to be objective, namely, the observed data. This isn't to say that the data are always objective. It is only when certain conditions apply that the data can be claimed to be objective and this has to do with how it is collected. Suppose that there is some system or process, completely independent of the statistician or others with any self-interest in the outcome of a statistical analysis, that produces the data $x \in \mathscr{X}$. It is this independence that renders the objectivity of the data.

Do such systems exist? This is a difficult question to answer, but statistics makes a major contribution in this regard by postulating the existence of a *random system*. Loosely speaking, a random system is a system that can be repeatedly performed, with each performance producing an outcome in a set $\{1,\ldots,N\}$, that exhibits the stability characteristic that the relative frequency of any i converges to $1/N$ as the number of performances increases, and that behaves randomly in a sense discussed in Section 2.4. It cannot be said categorically that a random system exists, but we are willing to accept that they do based upon empirical observation. For example, if N poker chips, labelled 1 through N, are placed in a bowl, thoroughly mixed and then, without looking, a chip is drawn and its label observed, then this seems to have the characteristics desired. Of course, it can never be guaranteed that in practice there isn't some aspect of the system that produces systematic, non-random behavior.

Such a system can be applied to the archetypal statistical problem, discussed in Example 1.2.1, to select a subset $\{\omega_1,\ldots,\omega_n\} \subset \Omega$, giving data $x = (x_1,\ldots,x_n)$ where $x_i = X(\omega_i)$, so that each subset $\{\omega_1,\ldots,\omega_n\} \subset \Omega$ has equiprobability $1/\binom{\#(\Omega)}{n}$ of being selected. So we could think of labelling the $N = \binom{\#(\Omega)}{n}$ subsets of size n of Ω with the labels $1,\ldots,N$ and then use a random system to select a subset. In practice it is simpler to label the elements of Ω as $1,\ldots,\#(\Omega)$ and then use a random system, like the chips in the bowl, to select n chips from the bowl without replacement to get the subset $\{\omega_1,\ldots,\omega_n\}$.

In these circumstances it can be claimed that the data are objective, as the control over the data selection process is entirely through the random system. This assertion is reinforced by the fact that each possible subset has the same chance of being selected. Notice that, if $n = 1$, then for $x \in \mathscr{X}$ we have $P(X(\omega) = x) = f_X(x)$, and if $n = 2$, then for $x_1, x_2 \in \mathscr{X}$ we have

$$P(X(\omega_1) = x_1, X(\omega_2) = x_2) = \begin{cases} f_X(x_1)\left(\frac{\#(\Omega)f_X(x_2)}{\#(\Omega)-1}\right) & \text{if } x_1 \neq x_2 \\ f_X(x_1)\left(\frac{\#(\Omega)f_X(x_2)-1}{\#(\Omega)-1}\right) & \text{if } x_1 = x_2 \end{cases}$$

and clearly a similar formula is available for arbitrary $n \leq N$. From this, when $n \ll \#(\Omega)$, then $x = (x_1,\ldots,x_n)$ can be considered as an approximate *i.i.d.* (independently and identically distributed) sample from the probability distribution with probability function given by f_X. This argument is the basis for the assumption that the data arise as an i.i.d. sample from a distribution. Of course if n is large relative to $\#(\Omega)$, then this assumption cannot be made and we have to incorporate the lack of independence into the analysis.

Consider now the context of Example 1.4.2.

Example 1.5.1 *Coin Tossing.*
Suppose that $X : \Omega \to \{0,1\}$ and $\#(\Omega) = N$ is very large relative to n. It is then clear that the model for $x = (x_1, \ldots, x_n)$ can be approximated by the model where it is assumed that the x_i are i.i.d. Bernoulli(θ) and so

$$f_\theta(x) = \prod_{i=1}^{n} \theta^{x_i}(1-\theta)^{1-x_i} = \theta^{n\bar{x}}(1-\theta)^{n(1-\bar{x})}$$

where $\bar{x} = n^{-1}\sum_{i=1}^{n} x_i$. Using the continuity approximation for the parameter space discussed in Example 1.4.2 leads to letting $\Theta = [0,1]$.

This is just the model for coin tossing but note a subtle distinction. We do not take coin tossing as a basic example on which to build statistical theory. Rather coin tossing arises as an approximation to a context that is in all respects finite. Furthermore, no claim is made that actual coin tossing corresponds to a random system. ∎

1.5.2 The Subjectivity of Statistical Models

A common complaint with Bayesian methods of inference is that they are not scientific because they involve a prior which is based on the beliefs of an individual statistician. There is no argument with the assertion that the prior is subjective, but we have some difficulty with the claim that this renders such an analysis "unscientific." If this is so, then it seems to us that virtually all statistical analyses are unscientific as they are based on personal choices by the statistician. For example, unless the model $\{f_\theta : \theta \in \Theta\}$ in Example 1.2.1 is chosen to be exactly equal to the set of all the possible relative frequency functions based on $\mathscr{X}$ and $\#(\Omega)$, then a subjective choice has been made. Such choices are very common. Furthermore, as discussed in Section 1.6, other aspects of statistical analyses are also chosen subjectively.

The issue of subjectivity is in our view a red herring as it is inherent any time choices are made as part of a statistical analysis. While such choices can reflect good judgment, the fact that subjectivity plays such a key role in statistical analyses has to be a cautionary factor when interpreting results. Rather than argue about subjectivity versus objectivity, it seems more productive to simply acknowledge that all statistical analyses are subjective in nature and then deal with concerns about subjectivity through the principle of empirical criticism. For statistical models this involves *model checking* — asking whether or not the observed data x is surprising for each f_θ in the model.

Some guidance is provided on choosing models in Section 5.2 and model checking is discussed in Section 5.5. Until then it is assumed that the chosen model has passed its checks and so can be used for inference.

1.5.3   The Subjective Prior

For a statistical model $\{f_\theta : \theta \in \Theta\}$, the prior is a probability distribution π on Θ selected to reflect beliefs about what the true value of $\theta \in \Theta$ is. For this to make sense it must be assumed that $f_X \in \{f_\theta : \theta \in \Theta\}$, so the model is correct. The prior is also a subjective choice.

In some approaches to inference the prior is considered to be a necessary reflection of the coherency of the beliefs of the analyst; see Section 2.3.2. As such, it is to be treated as unchangeable, as to modify the prior based on the data is incoherent. That position is not taken here. For us the prior is subject to the principle of empirical criticism just as with any other choice made by the statistician.

A natural question is to ask how the principle of empirical criticism can be applied to the prior. If the model is accepted, then it is natural to ask if the true value of θ is a surprising value for the prior. If it is, then naturally there are concerns about the effects of the prior on the analysis. Of course, the true value of θ is unknown. Nevertheless, it is possible to do such checking and this is discussed in Chapter 5.

As also discussed in Chapter 5, there is a logical separation of the activities of model checking, checking the prior and inference about θ, based on a factorization of the joint distribution for (θ, x). Furthermore, there is a logical sequence: first check the model; if the model passes its checks, then check the prior and, if both the model and prior pass their checks, proceed to make inference about θ.

Some guidance is provided on choosing priors in Section 5.3 and procedures for checking the prior are discussed in Section 5.6. Until then it is assumed that the prior has been chosen and passed its checks, and so is ready to be used as part of the inference step.

1.6   The Concept of Utility

The concept of *utility* arises quite naturally in economic matters and, as such, in gambling where costs and benefits arise. Perhaps the earliest occurrence arose in connection with the so-called *St. Petersburg paradox*. It is asked, what is the fair price to pay for the gamble where a gambler wins 2^n dollars when the first head occurs on the n-th toss of a fair coin. The expected payoff is $\sum_{n=1}^{\infty}(2^n)(2^{-n}) = \infty$ dollars but nobody can pay an infinite sum to make the game fair. As a resolution, it is pointed out that increasing amounts of money cause less and less satisfaction as the amounts increase, so really 2^n dollars gives $u(2^n)$ units of satisfaction or utility, where u is an increasing function satisfying $u''(x) < 0$. For example, if $u(x) = \log_2 x$, then the expected utility in the St. Petersburg game equals $\sum_{n=1}^{\infty} u(2^n)(2^{-n}) = \sum_{n=1}^{\infty} n(2^{-n}) = 2 = \log_2 4$. So, when adopting this utility function, one would presumably be willing to pay 4 dollars to play the game.

There are various treatments of statistics where utility plays a key role. For example, in a problem to estimate a quantity $\psi = \Psi(\theta)$ associated with statistical model $\{f_\theta : \theta \in \Theta\}$, it is commonly proposed, for some function u, that $u(\theta, a)$ be considered as a measure of the utility of the value a as an estimate of the true value of ψ when θ is true. Since the true value of ψ is unknown, the observed data x is used, via an estimator $d(x)$, to estimate the true value. Naturally, a good estimator makes the utility $u(\theta, d(x))$ large for each θ and x. When there is a prior π on θ, then we have the joint distribution of (θ, x) given by $\pi(\theta) f_\theta(x)$. The *principle of maximizing expected utility* is then invoked to select an estimator d that maximizes the expected utility $E(u(\theta, d(x)))$ where E refers to expectation with respect to the joint distribution of (θ, x). A d which maximizes the expected utility is referred to as a *Bayes rule*. An axiomatic development of this approach to statistical problems is discussed in Section 2.3.5.

In many problems it makes sense to talk in terms of losses rather than utilities and the *loss function* $L(\theta, a)$ represents the loss incurred when estimating $\Psi(\theta)$ by a. For example, if Ψ is real valued, the loss $L(\theta, a) = (\Psi(\theta) - a)^2$, which is called *squared error loss*, is often used. Naturally, it is desirable to incur the smallest losses possible. In this case, the utility function can be considered to be $u(\theta, a) = -L(\theta, a)$ and finding an estimator d that minimizes expected loss is equivalent to maximizing expected utility.

A utility or loss function is not considered here as a necessary or even suitable ingredient for a statistical problem. There are several reasons for this. Perhaps the most important reason is that it is clear that a utility is a choice made by the statistician and, in contrast to the model and prior, it is not obvious how to check this choice against the data to ensure that it makes sense. This violates our principle of empirical criticism. In many examples it is clear that the choice of the utility is critical in determining the final inferences, so this violation is not appropriate, as it cannot be said that the inferences are based on ingredients that have been empirically checked for their reasonableness.

Some might argue that the problem determines the appropriate utility function but this is not our experience. The utility or loss is typically chosen because it is intuitively reasonable and mathematically convenient in that inferences can be derived fairly easily. For example, this seems to be a common justification for the use of squared error. A stronger basis is needed for the development of a statistical theory.

Another reason for not employing a utility function is that it is not appropriate to confound the concept of statistical evidence with utility, which has its own issues. It is well known that there are many paradoxes associated with utility; for example, see Poundstone (2010). The following well-known example illustrates the kind of considerations that need to be taken into account when discussing utility.

Example 1.6.1 *The Allais Paradox.*

The following paradox is due to Allais (1953). Suppose you are presented with two contexts where you choose between (a) and (b).

Context 1
(a) You receive $\$10^6$ with probability 1.00.
(b) You receive $\$10^6$ with probability 0.89, receive $\$2 \times 10^6$ with probability 0.10 and receive nothing with probability 0.01.

Allais asserted that most people would choose (a) to avoid the small chance of getting nothing. Now consider another pair of choices.

Context 2

(a) You receive $\$10^6$ with probability 0.11 and nothing with probability 0.89.
(b) You receive $\$2 \times 10^6$ with probability 0.10 and nothing with probability 0.90.

Allais asserted that most people would choose (b) because the difference in the chance of receiving nothing is small and the payoff is much greater with (b).

The paradox arises as follows. Suppose you can receive a prize consisting of items A and B or you can receive a prize consisting of items A and C. It is fundamental to utility theory that your preference between these prizes only depend on the preference between B and C. Note that this does not depend on the utility function. But this implies that you will either prefer (a) over (b) or (b) over (a) in both Contexts 1 and 2 as they differ only by the fact that in Context 1 you have the extra 0.89 probability of receiving $\$10^6$ for both (a) and (b). ■

While developing a theory of utility that accounts for situations like that presented in Example 1.6.1 is a worthy activity, it is our position that these considerations do not have any relevance to the concept of statistical evidence. Given the importance of statistical evidence to applications, it seems better to separate its treatment entirely from the concept of utility. This is not to deny that in some real-world situations there is a need to state a utility function and proceed to maximize the expected utility. But in such a context, it would seem at least fair to compare the results with those obtained on the basis of the evidence alone, and then justify the difference in the answers when this is necessary. For example, statistical evidence may be found that a drug treatment for a medical condition is efficacious but, for a wide variety of reasons, a decision may be made not to bring it to market.

1.7   The Principle of Frequentism

Many statisticians adhere to the *principle of frequentism*. For this there is data x and model $\{f_\theta : \theta \in \Theta\}$ as ingredients for a statistical analysis.

The Principle of Frequentism: Any statistical procedure for inference must be evaluated against competitors by comparing their behaviors in an infinite sequence of performances where the data values $x_1, x_2, \ldots$ are i.i.d. according to f_θ and this evaluation must be done for each $\theta \in \Theta$. One procedure is preferred over another when its average behavior is better in some specified sense.

Of course, we can never actually observe an infinite sequence of outcomes so what this principle is proposing is just a thought experiment. In some cases, however, the comparison can be done exactly through mathematics or, more commonly, approximately through simulation. Furthermore, if we were to observe the values $x_1, \ldots, x_n$, then it would be appropriate to combine these data sets together to form a larger data set, as then our inferences would be more accurate than just using the single data sets.

The statement of the principle is somewhat vague, as it doesn't say how the various inference procedures are to be compared. This is commonly done by proposing some figure-of-merit such as the mean-squared error of an estimator or the power of a test of a statistical hypothesis. Consider an example.

Example 1.7.1 *Location Normal.*

Suppose $x = (x_1, \ldots, x_n)$ is a sample of n from the $N(\theta, 1)$ distribution with $\theta \in \Theta = R^1$ unknown. Let us suppose that our goal is to obtain an estimate $T(x)$ of θ that does well with respect to squared error $(T(x) - \theta)^2$ in the sense that this is as small as possible. Intuitively, squared error seems like a reasonable basis for comparison but still it is a choice. As such, we have to wonder why absolute error $|T(x) - \theta|$ is not used or even something else. As in our discussion of utility, there doesn't seem to be any way to apply the principle of empirical criticism to this choice and, in our experience, it is never dictated by the application.

In any case, let us suppose that the choice of squared error has been made. The principle of frequentism says that for each $\theta \in R^1$ we need to consider the behavior of our criterion in an infinite sequence, say $(T(x_1) - \theta)^2, (T(x_2) - \theta)^2, \ldots$, where the x_i are now independent samples of n from an $N(\theta, 1)$ distribution. To simplify the comparison, a common recommendation is to look at the long-run average value of the criterion

$$\lim_{n \to \infty} \frac{1}{n} \sum_{i=1}^{n} (T(x_i) - \theta)^2$$

which, by the strong law of large numbers, equals the *mean-squared error*

$$MSE_\theta(T) = E_\theta((T(x) - \theta)^2).$$

Again this is a choice, namely, to average the criterion, and something else could have been done. For example, the median of the distribution of $(T(x) - \theta)^2$ in each sequence could be used instead and this does not equal $MSE_\theta(T)$.

Now using $MSE_\theta(T)$ to compare estimators, the principle of frequentism implies a search for a T satisfying $MSE_\theta(T) \leq MSE_\theta(T')$ for every $\theta \in R^1$, for any other estimator T'. But it is easy to see that this is doomed to failure. For the estimator $T_\theta(x) \equiv \theta$ has $MSE_\theta(T_\theta) = 0$, which implies that an optimal T satisfy $MSE_\theta(T) = 0$ for every θ. This implies $T(x) \equiv \theta$ for every θ and no such T exists.

So now another somewhat arbitrary choice is needed and this is commonly based on the identity

$$MSE_\theta(T) = Var_\theta(T) + (E_\theta(T) - \theta)^2.$$

If $E_\theta(T) = \theta$ for every θ, so T is *unbiased* for θ, then $MSE_\theta(T) = Var_\theta(T)$. So it is left to look for unbiased T with the smallest variance for every θ. As is well known, $T(x) = \bar{x}$ is unbiased and, among unbiased estimators, uniformly minimizes the variance in this example. ∎

There is much one could complain about in Example 1.7.1. For example, there are many choices made that seem intuitively reasonable but lack any firm theoretical grounding. It should also be noted that, while successful in obtaining a reasonable estimator in this example, attempts to apply the same reasoning in another context may very well fail. It is well known that there are many estimation problems for which an unbiased estimator does not exist, let alone one with uniformly smallest variance.

One could argue that the problem in Example 1.7.1 is that the wrong criterion is being used to compare inference procedures. Perhaps there is a criterion where the principle of frequentism leads us unambiguously to the right inference. If so, the author doesn't know what it is. In any case, our complaint with the principle is much deeper than the failure to successfully provide answers to problems without arbitrary choices being made.

In essence, an argument is needed to justify why adherence to the principle of frequentism is required. Perhaps the best answer is connected with the interpretation of probability in terms of long-run probability and that this connection implies some kind of objectivity to frequentist statistical arguments. Leaving aside concerns about always interpreting probabilities in terms of long-run frequency (see Chapter 2), this does not imply objectivity in any sense for frequentist methods, as they are model dependent and so subjective in nature, as discussed in Section 1.5. For example, $T(x) = \bar{x}$ is not an unbiased estimator of θ with uniformly minimum variance for all models $\{f_\theta : \theta \in R^1\}$ where $f_\theta(x) = f(x - \theta)$ for some density f on R^n with mean 0.

As such, the relative belief inference methods proposed in this text are not justified by an appeal to the principle of frequentism. Still, as discussed in Chapter 3, it is possible to consider the frequentist properties of Bayesian inferences, but the infinite sequences for this are somewhat different from those used in the principle of frequentism. Bayesian frequentism does have implications for relative belief inferences.

1.8   Statistical Inferences

There are three sources of information about the true f_X in Example 1.2.1.

The data: $x \in \mathcal{X}$ has been observed.
The model: x arises from one of the distributions in $\{f_\theta : \theta \in \Theta\}$.
The prior: beliefs about the true $\theta \in \Theta$ are described by π.

The order in which these items are listed corresponds to their order of importance in conducting a statistical analysis. If the data is in some sense bad, for example, we have a biased sample, then the quality of the analysis is impaired and adjustments must be made. If the data has been produced appropriately but the model is bad, for example, no f_θ is close to being a reasonable approximation to f_X, then the analysis will generally not be acceptable. Finally, if the data and the model are both acceptable but the prior seriously distorts what the data and the model are saying, then again there is a problem. As such, it is necessary to collect the data properly, check the model against the data to ensure that it is a reasonable choice, and then check the prior to assess whether or not there is a serious conflict.

Collecting the data properly is part of the design aspect of statistics which is not the topic of this text and, as such, it is always assumed that this has been carried out appropriately. Checking the model and prior is the topic of Chapter 5 and for now it is assumed that this has been performed and that $\{f_\theta : \theta \in \Theta\}$ and π are acceptable. This now brings us to the topic of inference, namely, how to combine the data, model and prior together to make inferences about some parameter of interest $\psi = \Psi(\theta)$.

EXAMPLE   19

First, a definition is needed for what is meant by an *inference*. There are two basic forms of inference about the unknown θ.

Estimation: Provide estimate $\psi(x)$ of the true value of $\psi = \Psi(\theta)$ together with an assessment of the accuracy of the estimate.
Hypothesis Assessment: Provide a statement of the evidence that the hypothesis $H_0 = \Psi^{-1}\{\psi_0\}$ is either true or false together with an assessment of the strength of this evidence.

An estimation inference is just like the usual practice of quoting an estimate together with its standard error, although somewhat more general. For this we quote $\psi(x)$ together with a region $C(x)$ containing this estimate and interpret the "size" of $C(x)$ as the assessment of the accuracy of the estimate. The word size is in quotes because this is application dependent and refers to some measure of the set like volume or cardinality, etc. A hypothesis assessment inference is based on the evidence that the true value of ψ is some fixed value ψ_0 and there is the provision for finding evidence either for or against H_0. In addition, a calibration is proposed of how strong this evidence is, either for or against. This involves an assessment of the evidence for ψ_0 in comparison with the evidence for each of the other possible values of ψ. As such, this approach is different from the usual formulation of a hypothesis testing problem where one tests $\Psi(\theta) = \psi_0$ versus $\Psi(\theta) \neq \psi_0$.

Analogous inferences can also be developed for *prediction problems*. In general, a prediction problem involves estimating a future $y \in \mathcal{Y}$ or assessing whether $y_0 \in \mathcal{Y}$ is a plausible value, based on having observed x from model $\{f_\theta : \theta \in \Theta\}$, when y has model $\{g_\delta(\cdot\,|\,x) : \delta \in \Delta\}$ with $\delta = \Delta(\theta)$. The connection between the models arises via Δ with $\delta_{true} = \Delta(\theta_{true})$.

Chapter 4 provides a measure of statistical evidence based on relative belief. Once this definition is given, the forms of estimation and hypothesis assessment inferences are determined so there are no choices to be made by the statistician. As opposed to some other developments, estimation and hypothesis assessment proceed from a common basic idea — a measure of statistical evidence. Before defining this, however, the principle of conditional probability, discussed in Section 2.1.2, and the principle of evidence, discussed in Section 4.2, are both needed.

1.9   Example

This section presents a simple example of what is being advocated for a statistical analysis without providing justifications. This involves specifying a model and a prior, checking the model against the data, checking the prior against the data, assessing the bias in the prior and then, presuming these ingredients pass the checks, implementing inferences based on the definition of statistical evidence. In Chapters 4 and 5 these topics are much more extensively developed. Also, Baskurt and Evans (2013) contains much of the development concerning measuring statistical evidence and assessing the bias in a prior.

Consider the situation where $x = (x_1, \ldots, x_n) \in \{0,1\}^n$ is observed and the x_i are assumed to be i.i.d. Bernoulli(θ) with $\theta \in [0,1]$. This specifies our choice of the sampling model for generation of the data. The prior is taken to be a beta(α_0, β_0) distribution, where α_0 and β_0 are specified, so the prior density is given by

$$\pi(\theta) = \frac{\Gamma(\alpha_0 + \beta_0)}{\Gamma(\alpha_0)\Gamma(\beta_0)} \theta^{\alpha_0 - 1}(1 - \theta)^{\beta_0 - 1}.$$

Let us suppose for this example that, based on considerations as discussed in Section 5.3, the choice $\alpha_0 = \beta_0 = 4$ provides an appropriate prior. For example, asking for symmetry about $\theta = 1/2$ and requiring the interval $(0.000, 0.225)$ to contain 0.05 of the prior probability leads to this choice of prior. Let the parameter of interest be $\Psi(\theta) = \theta$ and, in particular, suppose interest is in assessing the hypothesis H_0 that $\theta = 1/2$.

Note that the parameter is ranging over a continuous set of values in this example. In practice, however, it is necessary to remember that this is an approximation to a situation where there are in reality only a finite set of θ values, perhaps lying on a equispaced grid in $[0, 1]$. This aspect of the problem is ignored here other than to point out where this issue can be significant.

1.9.1   Checking the Model

For this situation, model checking is based on the conditional distribution of the data given the minimal sufficient statistic $T(x) = n\bar{x}$, the number of ones observed. It is known that this conditional distribution is independent of θ. Let $D : \mathscr{X} \to R^1$ be a discrepancy statistic that in some sense measures deviation from model correctness. Letting $f_D(\cdot | T(x))$ denote the conditional probability function of D given T, the check involves computing

$$P_D(\{d : f_D(d | T(x)) \le f_D(D(x) | T(x))\} | T(x)) \tag{1.2}$$

which is the conditional probability of a value from the conditional distribution having a probability of occurrence no greater than the probability of the observed value. If this probability is small, then the observed value of the discrepancy statistic is located in a region of low probability for D and so the observed data is surprising for the model. This in turn leads to doubts about the validity of the model. Note that large values of (1.2) do not suggest model correctness. Model checking only looks for indications that the model is wrong, as it can virtually never be said that it is correct.

The conditional distribution of the data given $T(x) = n\bar{x}$ is uniform on the set of binary sequences $z = (z_1, \ldots, z_n)$ with $T(z) = n\bar{x}$ ones. Values can be generated from this distribution by taking such a binary sequence and randomly permuting the elements.

Suppose the data

$$x = (1, 1, 0, 0, 0, 0, 0, 0, 0, 0, 1, 1, 1, 1, 0, 1, 0, 0, 1, 0) \tag{1.3}$$

is observed so $T(x) = 8$. Actually, this data was generated from a Bernoulli$(1/2)$ distribution so the model and the hypothesis are both correct.

EXAMPLE   21

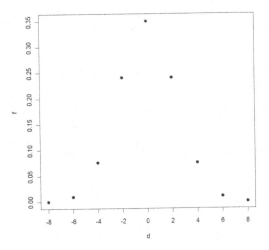

Figure 1.1 *The conditional probability function $f_D(\cdot \mid T(x))$.*

One concern is that there is a possible time trend in the data, namely, the distribution could have changed as we sampled and so the model would be incorrect. One possible way to assess this is via the discrepancy statistic $D(x) = n_1\bar{x}_1 - n_2\bar{x}_2$ where $n_1\bar{x}_1$ is the number of ones in the first n_1 sample values and $n_2\bar{x}_2$ is the number of ones in the last n_2 sample values. As such, $D(x)$ is comparing the number of ones in the two parts of the sample. Here we will take $n_1 = n_2 = 10$ and obtain the observed value $D(x) = 2 - 6 = -4$ for the data (1.3). Figure 1.1 is a plot of the conditional probability function of D computed using simulation. The value of (1.2) is 0.17 and so the observed value is not very surprising and so does not contradict the model.

Of course, other discrepancy statistics could also be employed. Additional issues associated with model checking are discussed in Chapter 5.

1.9.2 Checking for Prior-Data Conflict

Checking for prior-data conflict here requires the computation of the probability

$$M_T\left(m_T(t) \le m_T(T(x))\right) \tag{1.4}$$

where m_T is the prior predictive density of T, namely,

$$
\begin{aligned}
m_T(t) &= \int_0^1 \binom{n}{t} \theta^t (1-\theta)^{n-t} \frac{\Gamma(\alpha_0 + \beta_0)}{\Gamma(\alpha_0)\Gamma(\beta_0)} \theta^{\alpha_0-1}(1-\theta)^{\beta_0-1}\, d\theta \\
&= \frac{\Gamma(n+1)}{\Gamma(t+1)\Gamma(n-t+1)} \frac{\Gamma(\alpha_0+\beta_0)}{\Gamma(\alpha_0)\Gamma(\beta_0)} \frac{\Gamma(t+\alpha_0)\Gamma(n-t+\beta_0)}{\Gamma(n+\alpha_0+\beta_0)}.
\end{aligned}
$$

Essentially m_T gives the prior distribution of the minimal sufficient statistic T. The value of (1.4) serves to locate the value of $T(x)$ in its prior distribution, as this value

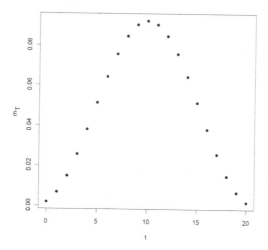

Figure 1.2 *Plot of* m_T.

will be small when $T(x)$ lies in a region of low probability. As demonstrated in Chapter 5, this probability is indicating whether or not the true value of θ is lying in a region of low probability for the prior.

In this example (1.4) becomes

$$M_T\left(\frac{\Gamma(t+\alpha_0)\Gamma(n-t+\beta_0)}{\Gamma(t+1)\Gamma(n-t+1)} \leq \frac{\Gamma(T(x)+\alpha_0)\Gamma(n-T(x)+\beta_0)}{\Gamma(T(x)+1)\Gamma(n-T(x)+1)}\right)$$

and this is easily computed by generating $\theta \sim \text{beta}(\alpha_0, \beta_0)$ and then $t \sim$ binomial(n, θ) many times. When $\alpha_0 = \beta_0 = 1$, then both sides of the inequality are always equal and so (1.4) always equals 1. Therefore, with a uniform prior there is never any prior-data conflict.

Figure 1.2 is a plot of m_T for the prior with $\alpha_0 = \beta_0 = 4$. For the data (1.3), the value of (1.4) is 0.728. Accordingly, the data do not contradict the prior.

There are many additional concepts and issues associated with checking for prior-data conflict. For example, what can be done if it is determined that a prior-data conflict exists? Chapter 5 contains a full development of this topic.

Another aspect of a prior that needs to be checked is whether or not a prior induces bias into an application. This requires the prescription of what statistical evidence is and so is deferred until Section 1.9.4. Logically, an assessment of the bias induced by a prior should precede inference, as extreme bias can render an inference unusable. An assessment of bias does not depend on the observed data and is properly done at the design stage.

EXAMPLE   23

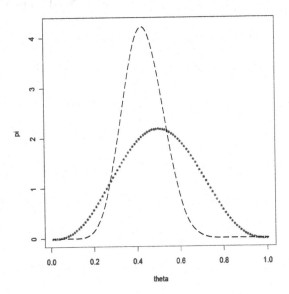

Figure 1.3 *The prior* $\star\star\star$ *and the posterior* $---$ *densities.*

1.9.3 Statistical Inference

First the conditional (posterior) density of θ given $T(x)$ is determined. This is given by the joint for $(\theta, T(x))$ divided by the marginal for $T(x)$, namely,

$$\pi(\theta \,|\, n\bar{x}) = \frac{\binom{n}{n\bar{x}} \theta^{n\bar{x}}(1-\theta)^{n-n\bar{x}} \frac{\Gamma(\alpha_0+\beta_0)}{\Gamma(\alpha_0)\Gamma(\beta_0)} \theta^{\alpha_0-1}(1-\theta)^{\beta_0-1}}{m_T(n\bar{x})}$$

$$= \frac{\Gamma(n+\alpha_0+\beta_0)}{\Gamma(n\bar{x}+\alpha_0)\Gamma(n-n\bar{x}+\beta_0)} \theta^{n\bar{x}+\alpha_0-1}(1-\theta)^{n-n\bar{x}+\beta_0-1}.$$

Therefore, the posterior of θ is a $\text{beta}(n\bar{x}+\alpha_0, n-n\bar{x}+\beta_0)$ distribution, and, when $\alpha_0 = \beta_0 = 4$, this is a $\text{beta}(n\bar{x}+4, n-n\bar{x}+4)$ distribution.

Figure 1.3 is a joint plot of the $\text{beta}(4,4)$ prior and the $\text{beta}(12,16)$ posterior using the data (1.3). Clearly, the data has led to some learning concerning the true value of θ.

As discussed in Chapter 4, the measure of the evidence that θ is the true value is defined to be equal to the relative belief ratio $RB(\theta \,|\, x)$. In this case, $RB(\theta \,|\, x)$ is the ratio of the posterior density to the prior density, namely,

$$RB(\theta \,|\, x) = \frac{\pi(\theta \,|\, n\bar{x})}{\pi(\theta)} = \frac{\Gamma(\alpha_0)\Gamma(\beta_0)}{\Gamma(\alpha_0+\beta_0)} \frac{\Gamma(n+\alpha_0+\beta_0)}{\Gamma(n\bar{x}+\alpha_0)\Gamma(n-n\bar{x}+\beta_0)} \theta^{n\bar{x}}(1-\theta)^{n-n\bar{x}}$$

which measures the change in the probability of θ being the true value from a priori to a posteriori. If $RB(\theta \,|\, x) > 1$, then there is evidence from the data that θ is the true value, evidence against θ being the true value when $RB(\theta \,|\, x) < 1$, and no evidence either way when $RB(\theta \,|\, x) = 1$. While $RB(\theta \,|\, x)$ is proportional to the likelihood in

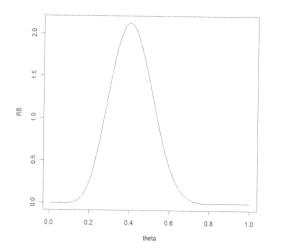

Figure 1.4 *Plot of RB(θ|x)*.

this example, it cannot be multiplied by a positive constant without destroying its interpretation as the evidence that θ is the true value.

The estimate of θ is the value maximizing $RB(\theta|x)$, as this has the maximal evidence in its favor. In Chapter 4 it is proved that this maximal value is always greater than 1. The value where $RB(\theta|x)$ is maximized is easily seen to equal $\bar{x}$ in this case. For the data (1.3) this equals 0.400.

The accuracy of this estimate is assessed via the 0.95-credible region

$$C_{0.95}(x) = \{\theta : RB(\theta|x) \geq c_{0.95}(x)\}$$

where $c_{0.95}(x)$ is determined to be the infimum of the constants k such that the posterior probability content of the set $\{\theta : RB(\theta|x) \leq k\}$ is no less than 0.05. Notice that $\bar{x} \in \{\theta : RB(\theta|x) \geq k\}$ whenever $k \leq \sup_\theta RB(\theta|x)$. The size of $C_{0.95}(x)$, which in this case is length as it is an interval, is then taken as the assessment of the accuracy of $\bar{x}$ as an estimate. The form of $C_{0.95}(x)$ is determined by the definition of evidence as, if $\theta_1 \in C_{0.95}(x)$ and $RB(\theta_2|x) \geq RB(\theta_1|x)$, then logically it is necessary that $\theta_2 \in C_{0.95}(x)$ as well.

Figure 1.4 is a plot of the relative belief ratios based on the data (1.3) and the choice $\alpha_0 = \beta_0 = 4$. The 0.95-credible interval is $C_{0.95}(x) = (0.227, 0.593)$ and its length of $0.593 - 0.227 = 0.366$ indicates that there is still a reasonable degree of uncertainty about the true value of θ.

To assess the hypothesis $H_0 = \{\theta_0\}$, then $RB(\theta_0|x)$ is the evidence that H_0 is true. As subsequently discussed in Chapter 4, it is necessary to calibrate $RB(\theta_0|x)$ and this is done by comparing the value $RB(\theta_0|x)$ with each of the other possible values $RB(\theta|x)$. In particular, it is necessary to see where $RB(\theta_0|x)$ lies with respect

EXAMPLE   25

to the posterior distribution of $RB(\cdot|x)$. For example, one measure of the strength of this evidence as given by the posterior probability that the true value of θ has a relative belief ratio no greater than $RB(\theta_0|x)$, namely,

$$\Pi(RB(\theta|x) \leq RB(\theta_0|x)|T(x)). \tag{1.5}$$

When $RB(\theta_0|x) < 1$ and (1.5) is small, then there is strong evidence against H_0 because there is a large posterior probability that the true value has a larger relative belief ratio. When $RB(\theta_0|x) > 1$ and (1.5) is large, then there is strong evidence in favor of H_0 because there is a small posterior probability that the true value has a larger relative belief ratio and θ_0 has the most evidence in its favor in $\{\theta : RB(\theta|x) \leq RB(\theta_0|x)\}$.

For the data (1.3) and the choice $\alpha_0 = \beta_0 = 4$, the relevant value of the relative belief ratio is $RB(1/2|x) = 1.421$, and since this is greater than 1, there is evidence in favor of H_0. For (1.5) the value

$$\Pi(RB(\theta|x)) \leq RB(1/2|x)|n\bar{x} = 8) = 0.309$$

is obtained and so the evidence in favor of H_0 is only moderate as there is a posterior probability of 0.691 that the true value of θ has a larger relative belief ratio.

There are issues associated with continuity when assessing the strength of evidence and, in particular, with the computation of (1.5). This is further discussed in Sections 4.4.1 and 4.7.1.

1.9.4   Checking the Prior for Bias

As previously mentioned, there is also the need to assess if the chosen prior induces any bias into the problem. In the case of the uniform prior it may seem obvious that it does not but this issue is in general more complicated and subtle. Of use here is the Savage–Dickey ratio result, as presented in Chapter 4, which shows that the relative belief ratio can also be expressed as

$$RB(\theta|x) = \frac{m_T(T(x)|\theta)}{m_T(T(x))}$$

where $m_T(\cdot|\theta)$ is the conditional prior probability function of T given that θ is true.

The bias against $H_0 = \{\theta_0\}$ can be measured by the prior probability, when H_0 is true, that evidence will be obtained against H_0, namely,

$$M_T\left(\frac{m_T(t|\theta_0)}{m_T(t)} < 1 \,\middle|\, \theta_0\right), \tag{1.6}$$

where $M_T(\cdot|\theta_0)$ denotes the prior distribution of T when the true value is θ_0. If (1.6) is large, then there is bias against H_0, as it is known, even before collecting the data, that it is likely to obtain evidence against H_0. For this problem $M_T(\cdot|\theta_0)$ is the binomial(n, θ_0) distribution.

The bias in favor of $H_0 = \{\theta_0\}$ is measured by the prior probability, when $\theta = \theta_* \neq \theta_0$ is true, that evidence in favor of H_0 is obtained, namely,

$$M_T \left(\frac{m_T(t \mid \theta_0)}{m_T(t)} > 1 \,\middle|\, \theta_* \right). \tag{1.7}$$

For this problem $M_T(\cdot \mid \theta_*)$ is the binomial(n, θ_*) distribution. Typically (1.7) is decreasing as θ_* moves away from θ_0, so it makes sense to take the θ_* to be values that just differ from θ_0 by the smallest amount that is practically significant. If (1.7) is large for such choices of θ_*, then there is bias in favor of H_0, as it is known, even before we see the data and when H_0 is false, that evidence will be obtained in favor of H_0. The size of a difference that is practically meaningful depends on the application.

For the data (1.3) and the choice $\alpha_0 = \beta_0 = 4$, the bias in favor of H_0 is

$$M_T(m_T(t \mid 0.50)/m_T(t) < 1 \mid 1/2) = 0.265$$

so there is little bias against $H_0 = \{1/2\}$ induced by the beta$(4,4)$ prior. For $\theta_* = 0.50 \pm 0.05$, the bias in favor of H_0 is

$$M_T(m_T(t \mid 0.50)/m_T(t) > 1 \mid 0.45) = 0.692,$$
$$M_T(m_T(t \mid 0.50)/m_T(t) > 1 \mid 0.55) = 0.692$$

and these are equal because of symmetry. So there is some bias in favor of $H_0 = \{1/2\}$ induced by the beta$(4,4)$ prior.

The bias can be somewhat controlled by the choice of the prior but primarily by the choice of the sample size. Controlling the bias together with checking for prior-data conflict, provide powerful tools for dealing with criticisms often leveled at the use of priors.

1.10   Concluding Comments

This chapter has been about providing background for our later developments of a theory of statistical inference based upon measuring statistical evidence using relative belief. For any theory of inference it must be made clear to what problems it is to be applied and what the ingredients of the theory are. These issues have been dealt with here. Given that relative belief theory is different from other theories of inference, it is necessary to discuss why there is a need to develop something that is different. Chapter 3 is an examination of other theoretical approaches to dealing with the concept of statistical evidence and Chapter 4 develops relative belief theory itself.

Chapter 2

Probability

2.1 Introduction

This chapter discusses the concept of probability. The focus is not with the mathematical aspects of probability but rather with the meaning of probability, although this will sometimes use mathematics. Ultimately, the subject of probability is applied in the real world and so a clear understanding is required of what probability statements mean. It is necessary to be able to explain exactly what is meant by the probability of an event is, for example, 0.735.

This proves to be much harder than one might think at first. Arguments about the meaning of probability have persisted almost since the earliest development of the concept. These disagreements have lead to substantial division in the statistical community and contribute to the lack of a firm foundation for a statistical theory that is generally accepted. This chapter is, in part, a survey of the different schools of thought concerning the meaning of probability.

The author's view, however, is that, while the meaning of probability is important, perhaps much more important is that probability theory gives us the basic rules for reasoning in statistical contexts. These rules are common to all the interpretations of probability. Thus a theory of statistical inference that is invariant to how probabilities are assigned or interpreted is preferable. It is worth noting that this goes a long way toward resolving various disputes within statistics. One is still permitted to question the validity of any probability assignments made, and to demand their justification, but the idea that there are different theories of statistics based on different meanings for probability is not appropriate.

2.1.1 Kolmogorov Axioms

Various concepts have been developed that extend or modify the typical approach to probability as a measure of uncertainty. The discussion in this text is restricted to probabilities that obey the *Kolmogorov axioms*.

Suppose there is a set Ω and a $\omega \in \Omega$ whose value is concealed, perhaps by it being a future response from some system or by some other mechanism. So there is uncertainty about what the true value of ω is. A probability measure P is then an expression of the degrees of belief that ω is in various subsets of Ω. If $A \subset \Omega$, and $P(A) = 1.0$, then this means that there is certainty that $\omega \in A$, while $P(A) = 0.2$ means we are only 20% certain of this and $P(A) = 0$ means it is certain that $\omega \notin A$.

More formally, the domain of P is restricted to be a Boolean σ-algebra.

Definition 2.1.1 *A Boolean σ-algebra $\mathscr{F}$ on a set Ω is a collection of subsets of Ω that satisfy (i) $\Omega \in \mathscr{F}$, (ii) if $A \in \mathscr{F}$ then $A^c \in \mathscr{F}$, (iii) if $A_1, A_2, \ldots \in \mathscr{F}$ then $\cup_{i=1}^{\infty} A_i \in \mathscr{F}$.*

Effectively, $\mathscr{F}$ is closed under countably many Boolean operations. A probability measure is then defined as follows.

Definition 2.1.2 *A probability measure P defined on a Boolean σ-algebra $\mathscr{F}$ on a set Ω satisfies (i) $P : \mathscr{F} \to [0,1]$, (ii) $P(\Omega) = 1$, (iii) if $A_1, A_2, \ldots \in \mathscr{F}$ are mutually disjoint, then $P(\cup_{i=1}^{\infty} A_i) = \sum_{i=1}^{\infty} P(A_i)$.*

The triple $(\Omega, \mathscr{F}, P)$ is referred to as a *probability model*.

Part (iii) of Definition 2.1.2 is known as the *countable additivity axiom* and some would prefer that only *finite additivity* be required for P, namely, for all $n \in \mathbb{N}$, if $A_1, A_2, \ldots, A_n \in \mathscr{F}$ are mutually disjoint, then $P(\cup_{i=1}^{n} A_i) = \sum_{i=1}^{n} P(A_i)$. Note that countable additivity is equivalent to finite additivity whenever Ω is finite. It is easily seen that countable additivity is a necessary restriction if we want probability to behave sensibly on infinite sets. For example, giving up countable additivity may also entail giving up results like the theorem of total probability as in $P(A) = \sum_{i=1}^{\infty} P(A \mid C_i) P(C_i)$ where $\{C_i : i = 1, 2, \ldots\}$ is a partition of Ω. So without countable additivity, conditional probability may no longer behave "properly." Therefore, the discussion here is restricted to the countably additive case any time probability measures are used on infinite sets. Note that it is easy to prove that countable additivity holds if and only if, whenever $A_n \in \mathscr{F}$ for all n and $A_n \downarrow A$ as $n \to \infty$, then $P(A_n) \to P(A)$. From this it is seen that countable additivity is just a continuity requirement. Finite additivity is discussed again in Section 2.3.

Restricting to the Kolmogorov axioms is not to say that other approaches are not reasonable or perhaps even inevitable. But the centrality and importance of the Kolmogorov axiomatization cannot be denied. Furthermore, extending our considerations beyond the conventional view of probability only seems to compound the problem of giving a suitable definition of statistical evidence. Interested readers are referred to books such as Fine (1973), Walley (1991) and Halpern (2003) for discussions of various generalizations of probability beyond Kolmogorov's formulation.

2.1.2 Conditional Probability

There is another axiom of probability that plays a fundamental role, particularly in applications. In fact, it is the main principle of inference.

The Principle of Conditional Probability: For probability model $(\Omega, \mathscr{F}, P)$ and $A, B \in \mathscr{F}$ with $P(B) > 0$, if it is known that the unknown ω is in B, namely, it is known that the event B has occurred, then replace $P(A)$ by $P(A \mid B) = P(A \cap B)/P(B)$.

The principle of conditional probability says that, when told that $\omega \in B$, then our initial degree of belief $P(A)$ that $\omega \in A$ must be replaced by $P(A \mid B)$. Effectively, having learned that B is true, P is replaced by $P(\cdot \mid B)$. While there are many arguments as to why this replacement should be made, there is no mathematical reason that compels us to do so. In fact, other rules for how to modify beliefs, as expressed by P, can also be considered. But our attention here is restricted to this rule. Note that the principle of conditional probability is essentially a rule for telling us how probabilities should be changed after obtaining evidence.

The principle of conditional probability can be generalized to weaken the requirement that $P(B) > 0$. Extensive use will be made of this and the reader is referred to Ash (1972) for the mathematical details.

In an application there is a need to be concerned with how the information $\omega \in B$ arises. There are a number of examples (see Section 2.2) where a lack of care on this point leads to apparent paradoxes. Basically, it is necessary to specify a function Ξ on Ω such that the information obtained is equivalent to being told the value $\Xi(\omega) = \xi_0$ such that $B = \Xi^{-1}\{\xi_0\}$. The function Ξ is referred to hereafter as an *information generator*. Examples 2.2.1 and 2.2.2 illustrate some of the difficulties associated with conditional probabilities when we are not careful about specifying Ξ.

In statistical contexts, the observed information is typically the data and it is necessary to be clear about how this was generated to unambiguously apply the principle of conditional probability. This is the role played by the statistical model. Consider the following example.

Example 2.1.1 *The Posterior Distribution.*

Suppose $x \in \mathscr{X}$ is observed where x was generated from a probability distribution P_θ with $\theta \in \Theta$ unknown. Throughout this text the mathematical structure discussed in Appendix A is assumed. So let f_θ be the density of P_θ with respect to volume measure μ on $\mathscr{X}$, where $f_\theta(x)$ is unambiguously defined, for each $x \in \mathscr{X}$, as

$$f_\theta(x) = \lim_{\varepsilon \to 0} \frac{P_\theta(N_\varepsilon(x))}{\mu(N_\varepsilon(x))},$$

for any sequence $N_\varepsilon(x)$ of neighborhoods shrinking nicely to x as $\varepsilon \to 0$. Therefore,

$$P_\theta(B) = \int_B f_\theta(z) \mu(dz)$$

for each Borel subset $B \subset \mathscr{X}$.

Suppose also that a prior probability measure Π is specified on Θ with density π with respect to volume measure ν on Θ. Again assume the mathematical structure of Appendix A so that

$$\pi(\theta) = \lim_{\varepsilon \to 0} \frac{\Pi(N_\varepsilon(\theta))}{\nu(N_\varepsilon(\theta))}$$

for any sequence $N_\varepsilon(\theta)$ of neighborhoods shrinking nicely to θ as $\varepsilon \to 0$. So

$$\Pi(A) = \int_A \pi(\theta) \nu(d\theta)$$

for each Borel subset $A \subset \Theta$.

Therefore, the joint distribution of (θ, x) is given by $\Pi \times P_\theta$, which has density $\pi(\theta)f_\theta(x)$ with respect to volume measure $\nu \times \mu$ on $\Theta \times \mathcal{X}$. Corresponding to observing x, there is the information generator $\Xi(\theta, x) = x$. The conditional probability of $A \subset \Theta$, given that we have observed x, is

$$\Pi(A \mid x) = \int_A \pi(\theta \mid x) \, \nu(d\theta)$$

where

$$\pi(\theta \mid x) = \frac{\pi(\theta)f_\theta(x)}{m(x)}$$

is the density of the measure $\Pi(\cdot \mid x)$ with respect to ν and

$$m(x) = \int_\Theta \pi(\theta)f_\theta(x) \, \nu(d\theta).$$

The marginal prior probability measure M of x is given by

$$M(B) = \int_\Theta \pi(\theta)P_\theta(B) \, \nu(d\theta)$$

and so

$$M(B) = \int_\Omega \pi(\theta) \left(\int_B f_\theta(z) \, \mu(dz) \right) \nu(d\theta)$$

$$= \int_B \left(\int_\Theta \pi(\theta)f_\theta(z) \, \nu(d\theta) \right) \mu(dz) = \int_B m(x) \, \mu(dz).$$

This implies that m is the density of M with respect to volume measure μ on $\mathcal{X}$.

The measure $\Pi(\cdot \mid x)$ and density $\pi(\cdot \mid x)$ are referred to as the *posterior measure* and the *posterior density*, respectively, of θ. By the principle of conditional probability the posterior $\Pi(\cdot \mid x)$ gives the appropriate probabilities for θ after having observed x. The measure M and density m are referred to as the *prior predictive measure* and the *prior predictive density*, respectively, of x. The prior predictive M gives the appropriate probabilities for x before observing it, just as the prior Π gives the appropriate probabilities for θ before observing x. ∎

A significant issue sometimes arises with conditional probability in the continuous case.

Example 2.1.2 *Borel Paradox.*

The original Borel paradox refers to an example where there is a uniform distribution on a unit sphere and the information is given that a random point lies on a particular great circle. The conditional distribution of the point is different, however, depending on whether the given great circle is treated as a line of longitude or as a line of latitude. So the conditional probability distribution is not determined until specifying the information generator Ξ.

Such ambiguities sometimes arise in statistical problems. Suppose that $(\omega_1, \omega_2) \sim N_2(0, I)$ and we are told that $\omega_1 = \omega_2$ but not the value of (ω_1, ω_2). This could arise when comparing two means. The question then is, what is the appropriate conditional distribution of (ω_1, ω_2) given that $C = \{(\omega_1, \omega_2) : \omega_1 = \omega_2\}$ has been observed? Again there does not appear to be a clear answer to this question. For example, is the information generated by observing the value $\Xi_1(\omega_1, \omega_2) = \omega_1/\omega_2$ and reporting C when $\Xi_1(\omega_1, \omega_2) = 1$, or by observing the value $\Xi_2(\omega_1, \omega_2) = \omega_1 - \omega_2$ and reporting C when $\Xi_2(\omega_1, \omega_2) = 0$? This matters, as different conditional distributions are obtained depending on which information generator is used. In essence, the ambiguity arises because all the ingredients of the problem have not been fully specified.

One could argue that the Borel paradox is just another artefact of continuity. For when (ω_1, ω_2) has discrete distribution P with $P(C) > 0$, the conditional distribution $P(\cdot | C)$ is the same no matter which information generator Ξ is used (although see Examples 2.2.1 and 2.2.2 where randomization may play a role in generating the information). As such, specifying what discretization is appropriate and how it relates to C will resolve this ambiguity. In general, if further ingredients cannot be specified, then we are left with making a somewhat arbitrary choice of an information generator to resolve the issue. ∎

2.2   Principle of Insufficient Reason

This principle corresponds to the earliest applications of probability and is commonly associated with Laplace.

The Principle of Insufficient Reason: If $\omega \in \Omega = \{\omega_1, \ldots, \omega_m\}$ is unknown and we have no reason to distinguish among the ω_i as possibilities for the value of ω, the probability that ω equals ω_i is $1/m$ for each i.

The consequence of applying this principle is that a uniform probability distribution is placed on Ω. While the principle is stated for a finite set, it can obviously be generalized to some infinite sets. For example, the uniform distribution on the interval $(0, 1)$ arises via the principle. It is not at all obvious how to generalize the principle to an arbitrary infinite set and this can be viewed as a defect. For us, however, the essential finiteness of all sets in an application make this criticism a moot point. There are also ambiguities concerning the identification of the appropriate set Ω for the application of the principle in a given context. For example, in tossing a coin twice, the response could be taken to be the sequence of outcomes or just the number of heads recorded. The uniform distribution on the sequences does not lead to a uniform distribution on the number of heads.

Clearly the principle of insufficient reason is using probability to measure an individual's degree of belief. As such, probability doesn't represent something more fundamental and physical and this has led to many criticisms of the principle. For example, Howie (2002) documents the rise of the frequentist interpretation of probability as a reaction against the blind application of the principle. For us probability is always used to measure degree of belief, no matter how it is assigned. For statistical applications the relevance of assigned probabilities can only be judged against the data, as discussed in Chapter 5, and so the principle is not rejected out of hand.

The following, known as *the three prisoners problem* and a variant called the *Monte Hall problem*, illustrate applications of the principle of insufficient reason and show that care needs to be taken when applying the principle of conditional probability.

Example 2.2.1 *The Three Prisoners Problem.*

Suppose there are three prisoners I, II and III; they are told that two will be executed while one will live and that this has already been decided but not revealed to the prisoners. Prisoner I asks the jailer to name one of the other prisoners who will be executed and is told II, as the jailer doesn't think that this information is informative to I. The question is: does this information change I's beliefs about whether or not he will live?

Based on the principle of insufficient reason, I's initial probability of not being executed is $1/3$. After being told that II will be executed, I concludes, based on insufficient reason, that this probability is now $1/2$ and is happier as his degree of belief in the event that he will live has increased. Formally, the set of possible outcomes is $\Omega = \{(I,II,III), (I,III,II), (II,III,I)\}$ where the first two coordinates of $\omega \in \Omega$ indicate those who will be executed. Then $A = \{(II,III,I)\}$ is the event that I will live and $B = \{(I,II,III), (II,III,I)\}$ is the event that II will be executed. Using the principle of insufficient reason $P(A) = 1/3, P(B) = 2/3$ and then, by the principle of conditional probability, $P(A\,|\,B) = P(A \cap B)/P(B) = (1/3)/(2/3) = 1/2$.

Certainly the result seems paradoxical, as there wouldn't appear to be any value in I learning that II will be executed, but given this supposedly useless information, his probability of living has increased. But now I starts having some doubts based upon considering how the jailer generates the information about who will be executed. This is equivalent to specifying an information generator Ξ, as discussed in Section 2.1.2.

It is clear that $\Xi(I,II,III) = II, \Xi(I,III,II) = III$ but the value of $\Xi(II,III,I)$ is not determined. If $\Xi(II,III,I) = II$, then the relevant conditional probability is $1/2$ but, if $\Xi(II,III,I) = III$, then the relevant conditional probability is 0. More generally, it may seem plausible that the the jailer used a random system to generate what is reported to I. So suppose there is an auxiliary variable U such that $P(U = II) = p, P(U = III) = 1 - p$ where $p \in [0,1]$ is unknown to prisoner I. Then consider the information generator $\Xi^*(\omega, U)$ for the enlarged response (ω, U) given by

$$\Xi^*((I,II,III),U) = II \text{ for all } U,$$
$$\Xi^*((I,III,II),U) = III \text{ for all } U,$$
$$\Xi^*((II,III,I),II) = II,$$
$$\Xi^*((II,III,I),III) = III. \tag{2.1}$$

With this information generator for the enlarged problem and, using the theorem of total probability, the unconditional probability that II will be reported by the jailer is $1(1/3) + 0(1/3) + p(1/3) = (1+p)/3$ and the unconditional probability that I will live and II will be reported is $p(1/3)$. Therefore, the conditional probability that I will live is $p/(1+p)$ and this can be any number in $[0, 1/2]$. In fact, the only time that this conditional probability is $1/2$ is when $p = 1$ and it is less than $1/3$ whenever $p < 1/2$. So given that there is ambiguity about the way the information is generated, there is no justification in I believing that the relevant conditional probability is $1/2$.

Another application of the principle of insufficient reason can be considered here. For suppose that $p \sim U(0,1)$ as this value is unknown to I. Then the conditional probability that I will not be executed, having been told that II will be executed, is $\int_0^1 p(1+p)^{-1} dp = 1 - \log 2 = 0.307$, which is a smaller than $1/3$. So in this case I definitely doesn't want to be told who will be executed! ∎

The Monte Hall problem is a notorious example where many analysts ignore the information processing aspect of conditional probability, producing considerable confusion and sometimes erroneous answers.

Example 2.2.2 *The Monte Hall Problem.*

In the game show Let's Make a Deal a contestant is shown three closed doors I, II and III and there is a desirable prize behind one and a goat behind each of the other doors. The contestant is asked to pick a door. By the principle of insufficient reason the contestant concludes that the probability they will win the prize by selecting any door is $1/3$ and chooses I. After the contestant picks door I, the host of the show, Monte Hall, opens door II to reveal a goat. The contestant is then offered the opportunity to switch the door they have selected to door III. Should the contestant switch or not?

The set of possible outcomes is $\Omega = \{(I,II,III),(I,III,II),(II,III,I)\}$ where the first two coordinates of $\omega \in \Omega$ indicate the doors concealing goats. Then $A = \{(II,III,I)\}$ is the event that the contestant will win the prize by not switching and $B = \{(I,II,III),(II,III,I)\}$ is the event that door II conceals a goat. Using the principle of insufficient reason $P(A) = 1/3, P(B) = 2/3$ and then, by the principle of conditional probability, $P(A \mid B) = P(A \cap B)/P(B) = (1/3)/(2/3) = 1/2$ and so there doesn't appear to be any reason to switch.

It is necessary to consider, however, how the information was generated. Again it is clear that $\Xi(I,II,III) = II, \Xi(I,III,II) = III$ and the value of $\Xi(II,III,I)$ is not determined. If $\Xi(II,III,I) = II$, then the relevant conditional probability of winning by not switching is $1/2$ but, if $\Xi(II,III,I) = III$, then the relevant conditional probability is 0 since $\Xi^{-1}\{II\} = \{(I,II,III)\}$. More generally, it may seem plausible that the host used a random system to generate what is reported to the contestant. So suppose there is an auxiliary variable U such that $P(U = II) = p, P(U = III) = 1 - p$ where $p \in [0,1]$ is unknown to the contestant. Then consider the information generator $\Xi^*(\omega, U)$ for the enlarged response (ω, U), which is again given by (2.1). With this information generator the unconditional probability that II will be opened by the host is $1(1/3) + 0(1/3) + p(1/3) = (1+p)/3$ and the unconditional probability that the contestant will win by not switching and II will be opened is $p(1/3)$. Therefore, the conditional probability that the contestant will win by not switching is $p/(1+p)$ and this can be any number in $[0,1/2]$. But this implies that the conditional probability that the contestant will win by switching is always greater than $1/2$ and so the contestant should always switch doors. Note that when $p = 1/2$, then the conditional probability the contestant will win by switching is 2/3. Similarly, when the principle of insufficient reason is applied to p, the probability of winning by switching is $1 - 0.307 = 0.693$. So overall, based on this analysis, the contestant should always switch.

Sometimes the conclusion that the contestant should switch is reached by the following argument: the probability is 2/3 that the initial choice conceals a goat so the probability is 2/3 that switching will result in the contestant winning. But this ignores the way the information is generated and is only appropriate when the host randomly chooses a door based on the value of an auxiliary variable U with $P(U = II) = 1/2$. If the contestant is told how the information was generated, then the probability of winning by switching is not ambiguous but otherwise it is. It can still be concluded, however, that switching is appropriate.

This example has engendered a considerable amount of discussion among both statisticians and non-statisticians. The paper by Morgan, Chaganty, Dahiya and Dovial (1991) presents a correct analysis. ■

In both Examples 2.2.1 and 2.2.2 the problems have not been with the principle of insufficient reason, but rather with the application of the principle of conditional probability. The principle of insufficient reason does mislead us, however, as it seems that this principle lies behind the initial appeal of the uniform conditional probabilities. But when we are careful about the reasoning, the initial probability assignments seem quite sensible.

More generally, the principle of insufficient reason is applied in physics. Nobody really knows why these applications are appropriate but their validity is established by the verification of the predictions the theory makes.

Example 2.2.3 *Statistical Mechanics.*

Consider a perfect gas in a finite volume container. Maxwell idealized this as many hard spheres undergoing perfectly elastic collisions and with their positions and velocities all random. By random he used the principle of insufficient reason to argue that the position vectors were uniformly distributed in the container and independent of the velocity vectors. For the velocity vectors he assumed a uniform distribution for the directions and independence and identical distributions for the components of the velocity vectors. This implies that the distribution of the velocity vector is $N_3(0, \sigma^2 I)$, for some $\sigma^2 > 0$, and so the distribution of the squared magnitude of the velocity vector is a rescaled chi-squared(3) distribution. The value of σ^2 depends on some physical constants. This reasoning led to a model for the behavior of a perfect gas that produced empirically verified facts and so became an accepted model. Boltzmann used similar reasoning to determine the distribution of energies of particles in a perfect gas. A more detailed description and the derivations can be found in Ruhla (1989).

This example doesn't justify the use of the principle of insufficient reason in general, but it does illustrate the value of this type of argument. Although these models produce meaningful results, it can't be said for certain that the assumptions concerning the distributions are correct or even that describing the physical situation by probability distributions is valid. ■

As in Example 2.2.3, when probability distributions are assigned in statistical problems it can only be said that the probability assignments are not contradicted by the data. This lack of full verifiability of the assumptions shouldn't deter us, however, from reasoning via statistical inference. There is a necessity for this reasoning to proceed via a logical, justifiable process and that is the point of a theory of statistical inference.

2.3 Subjective Probability

Subjective probability refers to assignments of probability that are based on an individual's beliefs. These beliefs are presumably based on experience or other observed evidence but the point is that there is no attempt to try and justify the probability assignments as being based on some external, objective source. In fact, as argued in this text, in a sense all probabilities are primarily subjective as their assignment is based on choices made by statisticians. The most that can be said is that the probability assignments made are reasonable rather than correct. Seminal works on subjective probability can be found in Ramsey (1931), de Finetti (1937), with both papers reprinted in Kyburg and Smokler (1964) and Savage (1972).

2.3.1 Comparative or Qualitative Probability

A minimal approach to probability is to simply refer to a *qualitative probability ordering* $\preceq$ on a Boolean algebra (only required to be closed under finite unions) $\mathscr{F}$ of subsets of Ω, where $A \preceq B$ is interpreted as event A is not more probable than event B. The ordering $\preceq$ implies an equivalence $A \equiv B$ whenever $A \preceq B$ and $B \preceq A$, so that A and B are equiprobable, and a strict ordering $A \prec B$ whenever $A \preceq B$ and $B \not\preceq A$, so that A is less probable than B. This is a weak interpretation of probability as no numerical values are specified, only the ordering. This seems sensible when considering subjective beliefs.

It is reasonable to ask, however, what conditions need $\preceq$ satisfy so that there exists a probability measure P *agreeing* with $\preceq$ in the sense that if $A \prec B$, then $P(A) < P(B)$ and if $A \equiv B$, then $P(A) = P(B)$. For our discussion here we suppose that Ω is finite. The paper Kraft, Pratt and Seidenberg (1959) contains such a result for the finite case but we follow the development in Narens (2007).

Note that $\preceq$, as a binary relation on $\mathscr{F}$, can also be considered as a subset of $\mathscr{F} \times \mathscr{F}$. Let I be an arbitrary finite set and $\Gamma = \{\gamma_i : i \in I, \gamma_i \in \preceq\}$ be a listing of elements of $\preceq$ where an element of $\preceq$ can be listed multiple times. So the elements of I are used to label elements of $\preceq$. For such a Γ and $\omega \in \Omega$ define

$$\Gamma_l(\omega) = \#(\{\gamma_i \in \Gamma : \omega \text{ is an element of the left side of } \gamma_i\}),$$
$$\Gamma_r(\omega) = \#(\{\gamma_i \in \Gamma : \omega \text{ is an element of the right side of } \gamma_i\}),$$

so $\Gamma_l(\omega)$ equals the number of times element ω occurs on the left side of an element of Γ and $\Gamma_r(\omega)$ equals the number of times ω occurs on the right side of an element of Γ. The following definition is needed.

Definition 2.3.1 *The ordering $\preceq$ satisfies the finite cancellation axioms if, whenever $\Gamma = \{\gamma_i : i \in I, \gamma_i \in \preceq\}$, with I a finite set, consists only of strict inequalities or equivalences and $\Gamma_l(\omega) = \Gamma_r(\omega)$ for every $\omega \in \Omega$, then Γ must only contain equivalences.*

So the definition says that, if each element of Γ is of the form $A \equiv B$ or $A \prec B$ for some $A, B \in \mathcal{F}$, where the number of times ω occurs on the left side of an element of Γ equals the number of times ω occurs on the right side of an element of Γ, and this holds for all $\omega \in \Omega$, then we must only have equivalences in Γ.

Definition 2.3.1 is a natural requirement for a P existing that agrees with $\preceq$ because of the following result.

Theorem 2.3.1 *If Ω is finite, and the probability model $(\Omega, \mathcal{F}, P)$ agrees with $\preceq$ on $\mathcal{F}$, then $\preceq$ satisfies the finite cancellation axioms.*

Proof. Suppose Γ is as in Definition 2.3.1 and it contains a strict inequality, say $A_j \prec B_j \in \Gamma$. Then we must have $P(A_j) < P(B_j)$ and this implies $\sum_{i \in I} P(A_i) < \sum_{i \in I} P(B_i)$, which in turn implies

$$\sum_{i \in I} \sum_{\omega \in A_i} P(\{\omega\}) < \sum_{i \in I} \sum_{\omega \in B_i} P(\{\omega\}). \tag{2.2}$$

But each term on the left in (2.2) occurs the same number of times on the right and conversely, since $\Gamma_l(\omega) = \Gamma_r(\omega)$. Therefore, the two sides must be equal and there is a contradiction. ∎

Accordingly, suppose that $\preceq$ satisfies the following axioms.

Axiom 1: $\preceq$ is a reflexive binary relation on $\mathcal{F}$.
Axiom 2: $\phi \prec \Omega$.
Axiom 3: $\preceq$ satisfies the finite cancellation axioms.

The following result is due to Scott (1964) and is proved in Narens (2007).

Theorem 2.3.2 *(Scott's Theorem) If $\mathcal{F}$ is a Boolean algebra on finite Ω, and $\preceq$ on $\mathcal{F}$ satisfies Axioms 1–3, then there exists probability model $(\Omega, \mathcal{F}, P)$ agreeing with $\preceq$.*

These results can be extended substantially and Narens (2007) can be consulted for this.

In practice it seems that we need to proceed in a much simpler way than trying to prescribe $\preceq$ so that it satisfies the finite cancellation axioms without referring to numerical probabilities. For example, if $\#(\Omega) = N$ and a qualitative probability on $\mathcal{F} = 2^\Omega$ is required, then it is necessary to specify $\preceq$ for $\binom{2^N}{2}$ pairs and make sure that this satisfies finite cancellation — surely an impossible task when N takes on even a modest value like $N = 20$. Alternatively, it seems much more feasible to order the singleton sets labelling the elements of Ω as ω_i for $i = 1, \ldots, N$ so that $\{\omega_i\} \preceq \{\omega_{i+1}\}$. This ordering doesn't specify an ordering for all elements of 2^Ω but a simple approach then is to specify an increasing function $f : \{1, \ldots, N\} \to R^1$, with $f(i) = 0$ when $\{\omega_i\} \equiv \phi$, and such that $f(i+1)/f(i)$ represents the degree to which it is believed that ω_{i+1} is more probable than ω_i. Now define a probability measure on 2^Ω by $P(A) = \sum_{\omega_i \in A} f(i) / \sum_{i=1}^N f(i)$ and note that P agrees with the ordering on the singleton sets and imposes a complete ordering on 2^Ω. Ultimately, numerical values for probabilities are needed for the definition of statistical evidence, so a qualitative probability alone is not satisfactory for applications.

2.3.2   Probability via Betting

One compelling argument for subjective probability assignments is based on betting and the following principle.

> *Principle of Avoiding Sure Losses: A rational gambler will never price gambles on the value of an unknown $\omega \in \Omega$ so that there is a sure loss.*

A combination of gambles that guarantees a loss for a gambler is known as a *Dutch book*. Of course a rational gambler would accept the possibility of a sure gain.

Suppose that $\omega \in \Omega$ is unknown and there is a market comprised of rational gamblers. A *gamble* is a function $X : \Omega \to R^1$ where the interpretation is that the purchaser of X receives $X(\omega)$ if ω is true. No units are specified so perhaps the easiest interpretation is that the gambler receives $X(\omega)$ of utility. Note that $X(\omega)$ can be negative and this corresponds to a loss for the purchaser. Let $L(\Omega)$ denote the set of all bounded gambles and note that any linear combination of gambles is a gamble. Let $P : L(\Omega) \to R^1$ be a *price function,* or using de Finetti's terminology, a *prevision,* where $P(X)$ is the price a gambler is willing to pay to either offer or buy the gamble X. If you buy X, your gain is $X(\omega) - P(X)$ and, if you sell X, your gain is $P(X) - X(\omega)$, when ω is revealed. Note that negative gains are positive losses. Suppose further that all gamblers are willing to sell or purchase any finite number of gambles and any fraction of gambles. So, for example, a gambler may buy one unit of X and sell $\sqrt{2}$ units of Y for the combination that has payoff $(X(\omega) - P(X)) + \sqrt{2}(P(Y) - Y(\omega))$.

The following result justifies the same buying and selling price for a gamble under the assumption that the gambler is rational and that the principle of no sure losses holds.

Lemma 2.3.1 *The buying and selling price of $X \in L(\Omega)$ must be the same.*

Proof. Suppose a gambler would buy X for p_1 and sell X for p_2 where $p_1 > p_2$. This combination results in the gain $(X - p_1) + (p_2 - X) = p_2 - p_1 < 0$ and the gambler has a sure loss. So by the principle of avoiding sure loss we must have $p_1 \leq p_2$.

If $p_1 < p_2$, then for any $\varepsilon > 0$ the gambler will not pay $p_1 + \varepsilon$ for X and will not sell X for $p_2 - \varepsilon$. But this combination of gambles has gain $(X - p_1 - \varepsilon) + (p_2 - \varepsilon - X) = p_2 - p_1 - 2\varepsilon$ and this is positive when ε is small enough, ensuring a sure gain. Since the gambler is rational, we must have that $p_1 = p_2$. ∎

Another way of expressing the principle of avoiding a sure loss is to say that the prevision P is coherent.

Definition 2.3.2 *The prevision P is **coherent** if for every $m, n \in \mathbb{N}$ and every $X_1, \ldots, X_m, Y_1, \ldots, Y_n \in L(\Omega)$,*

$$\sup_{\omega \in \Omega} \left\{ \sum_{i=1}^{m} (X_i(\omega) - P(X_i)) - \sum_{i=1}^{n} (Y_i(\omega) - P(Y_i)) \right\} \geq 0.$$

A gambler with a coherent prevision will never have a sure loss on any finite combination of gambles. There is an equivalent characterization of a coherent P.

Lemma 2.3.2 *The prevision P is coherent if and only if*

$$\sup_{\omega \in \Omega} \sum_{i=1}^{n} \lambda_i (X_i(\omega) - P(X_i)) \geq 0 \tag{2.3}$$

for all $\lambda_1, \ldots, \lambda_n \in R^1$.

Proof. If (2.3) holds, then choosing the λ_i in (2.3) to be integers establishes the coherence of P.

Now suppose that P is coherent. Then (2.3) holds whenever all the λ_i are rational simply by multiplying through by a common denominator. Suppose, however, that for some $\lambda_1, \ldots, \lambda_n \in R^1$ there exists $\delta > 0$ such that $\sup_{\omega \in \Omega} \sum_{i=1}^{n} \lambda_i (X_i(\omega) - P(X_i)) \leq -\delta$. Let $\alpha_i = \sup_{\omega \in \Omega} |X_i(\omega) - P(X_i)|$ and note that this is real since X_i is bounded. Let $\alpha = \max\{\alpha_1, \ldots, \alpha_n\}$, $\varepsilon = \delta / 2n\alpha$ and $\rho_i \in [\lambda_i - \varepsilon, \lambda_i + \varepsilon] \cap \mathbb{Q}$.

When $X_i(\omega) - P(X_i) \geq 0$, then

$$\rho_i(X_i(\omega) - P(X_i)) \leq (\lambda_i + \varepsilon)(X_i(\omega) - P(X_i)) \leq \lambda_i(X_i(\omega) - P(X_i)) + \varepsilon\alpha.$$

When $X_i(\omega) - P(X_i) < 0$, then

$$\rho_i(X_i(\omega) - P(X_i)) \leq (\lambda_i - \varepsilon)(X_i(\omega) - P(X_i)) \leq \lambda_i(X_i(\omega) - P(X_i)) + \varepsilon\alpha.$$

So for all $\omega \in \Omega$

$$\sum_{i=1}^{n} \rho_i(X_i(\omega) - P(X_i)) \leq \sum_{i=1}^{n} \lambda_i(X_i(\omega) - P(X_i)) + n\varepsilon\alpha \leq -\delta + \delta/2 = \delta/2 < 0$$

and this contradicts the coherence of P. ∎

From Lemma 2.3.2 it can be established that P has the properties of an expectation operator and conversely.

Theorem 2.3.3 *The prevision P is coherent if and only if (i) $P(X+Y) = P(X) + P(Y)$ for every $X, Y \in L(\Omega)$, (ii) $P(\lambda X) = \lambda P(X)$ for every $X \in L(\Omega), \lambda \in R^1$, (iii) $P(X) \geq 0$ whenever $X \in L(\Omega)$ satisfies $X \geq 0$, (iv) $P(1) = 1$.*

Proof. Suppose that P is coherent. Then

$$0 \leq \sup_{\omega \in \Omega} \{(X(\omega) + Y(\omega) - P(X+Y)) - (X(\omega) - P(X)) - (Y(\omega) - P(Y))\}$$
$$= P(X) + P(Y) - P(X+Y)$$

and similarly $0 \leq P(X+Y) - P(X) - P(Y)$, proving (i). Also

$$0 \leq \sup_{\omega \in \Omega} \{(\lambda X(\omega) - P(\lambda X)) - \lambda(X(\omega) - P(X))\} = \lambda P(X) - P(\lambda X)$$

and similarly $0 \leq P(\lambda X) - \lambda P(X)$, proving (ii). Since $0 \leq \sup_{\omega \in \Omega} -(X(\omega) - P(X))$ implies $P(X) \geq -\sup_{\omega \in \Omega} -X(\omega) = \inf_{\omega \in \Omega} X(\omega) \geq 0$, this proves (iii). Finally, $0 \leq 1 - P(1)$ and $0 \leq P(1) - 1$ proves (iv).

For the converse, suppose there exists $X \in L(\Omega)$ such that $\sup_{\omega \in \Omega}(X(\omega) - P(X)) = -\delta < 0$. Define $Y \in L(\Omega)$ by $Y(\omega) = P(X) - \delta - X(\omega)$ and note that $Y \geq 0$. But by (i), (ii) and (iv), $P(Y) = P(X) - \delta - P(X) = -\delta < 0$, which contradicts (iii). Therefore, $\sup_{\omega \in \Omega}(X(\omega) - P(X)) \geq 0$ for all $X \in L(\Omega)$ and this implies (2.3). ∎

From the theorem we see that restricting P to indicator functions implies that P corresponds to a finitely additive probability measure.

Corollary 2.3.1 *A coherent prevision P gives a finitely additive probability measure on 2^{Ω}. If u is paid for the gamble with payoff $u + v$ whenever $\omega \in A$ and 0 otherwise, then $P(A) = u/(u + v)$.*

Proof. For $A \in 2^{\Omega}$ let $P(A)$ be alternative notation for $P(I_A)$ and the result is immediate. For the second statement, $X(\omega) = (u + v)I_A(\omega)$ and $P(X) = u$. Therefore, $P(X) = (u + v)P(A) = u$ and the result follows. ∎

The betting approach to probability can be extended to handle *contingent gambles* where for $B \subset \Omega$ a gambler buys or sells gamble X for $P(X|B)$, with the purchaser receiving $X(\omega)$ when $\omega \in B$ and, when $\omega \notin B$, the bet is *called off* with the price $P(X|B)$ returned to the purchaser. The payoff on this gamble is clearly $I_B(X - P(\cdot|B))$. The function $P(\cdot|B) : L(\Omega) \to R^1$ is called a *conditional pricing function* or *conditional prevision*. Of course it is required that $P(\cdot|B)$ satisfy Theorem 2.3.3 so that it is coherent.

Consider the interaction of $P(\cdot|B)$ with the unconditional coherent prevision P.

Theorem 2.3.4 *It is the case that $P(A \cap B) = P(A|B)P(B)$.*

Proof. Suppose that $P(A \cap B) > P(A|B)P(B)$ and consider the combination of gambles where the gambler sells the gamble I_A conditional on B, buys the unconditional gamble $I_{A \cap B}$ and sells $P(A|B)$ units of the unconditional gamble I_B. The gain for the gambler on this combination of gambles is

$$-I_B(\omega)(I_A(\omega) - P(A|B)) + (I_{A \cap B}(\omega) - P(A \cap B)) - P(A|B)(I_B(\omega) - P(B))$$

$$= \begin{cases} P(A|B)P(B) - P(A \cap B) & \text{if} \quad \omega \in B^c \\ P(A|B)P(B) - P(A \cap B) & \text{if} \quad \omega \in A^c \cap B \\ P(A|B)P(B) - P(A \cap B) & \text{if} \quad \omega \in A \cap B \end{cases}$$

which is always less than 0 since $\Omega = B^c \cup (A^c \cap B) \cup (A \cap B)$ and the gambler has a sure loss. So we must have $P(A \cap B) \leq P(A|B)P(B)$. Now supposing that $P(A \cap B) < P(A|B)P(B)$, reverse all the above gambles (buying instead of selling and selling instead of buying) to get a sure loss for the gambler. Therefore, by the principle of avoiding sure losses, $P(A \cap B) = P(A|B)P(B)$. ∎

So when $P(B) > 0$ the usual formula for a conditional probability applies, namely, $P(A|B) = P(A \cap B)/P(B)$.

Now for a $X \in L(\Omega)$ consider the situation where a gambler buys the gamble X given B for $P(X|B)$ and buys the gamble X given B^c for $P(X|B^c)$. This is clearly equivalent to just buying X so these gambles must have the same payoffs

$$I_B(X - P(X|B)) + I_{B^c}(X - P(X|B^c)) = X - P(X)$$

and $X = I_B X + I_{B^c} X$ implies $P(X) = P(B)P(X|B) + P(B^c)P(X|B^c)$, which is the theorem of total probability in this setting. This can be generalized to a finite partition $\mathscr{B} = \{B_1, \ldots, B_m\}$ where $\Omega = \cup_{i=1}^{m} B_i, B_i \cap B_j = \phi$ for $i \neq j$, and then

$$P(X) = \sum_{i=1}^{m} P(B_i)P(X|B_i). \tag{2.4}$$

Sometimes it is argued, based on the betting formulation for probability, that Kolmogorov's axiom (iii) should be weakened to only require finite additivity. Consider the following example.

Example 2.3.1 *Uniform Probability on $\mathbb{N}$.*

If $\Omega = \mathbb{N}$, then define $P(A) = \lim_{n\to\infty} \#(A \cap \{1, \ldots, n\})/n$ whenever this limit exists. It can be shown that P can be extended to a finitely additive probability measure on 2^{Ω}; see Walley (1991). Clearly $P(A) = 0$ for any finite set A while $P(A) = 1/2$ if A is the subset of even natural numbers and so $\sum_{i=1}^{\infty} P(\{2i\}) = 0 \neq 1/2$. Therefore, P is not countably additive. ∎

Generally, there exist finitely additive uniform probability measures on noncompact sets. While this may seem beneficial, the behavior of finitely additive P is much more complex than the countably additive case. For example, it is not necessary that (2.4) holds for countably infinite $\mathscr{B}$ as it does for a P satisfying Kolmogorov's axioms. It is clear that P in Example 2.3.1 does not satisfy the theorem of total probability when $\mathscr{B}$ is countably infinite since $\Omega = \cup_{i=1}^{\infty} \{i\}, P(\Omega|\{i\}) \equiv 1$, but $1 = P(\Omega) \neq \sum_{i=1}^{\infty} P(\Omega|\{i\})P(\{i\}) = 0$.

Following the discussion in Section 1.4, we prefer to let the case when Ω is finite be our guide as to what is appropriate. As such, the infinite case is to be considered as an approximation to something finite in practice and should reflect the properties of the finite context it is approximating. So the continuity condition on P is imposed here. Further discussion on this point, and its relevance to the theory of statistical inference, can be found in Kadane (2011).

Some authors, such as de Finetti, argue that the betting formulation is the *correct* interpretation for probability and various arguments are advanced for this. Our preference is to say that it is one possible interpretation for probability. In fact, the emergence of probability from such simple rules of rational behavior is just another argument supporting probability as the basis for logical reasoning in situations of uncertainty. A particularly appealing aspect of this approach is that there is no need to introduce the more difficult concept of randomness; see Section 2.4. A negative aspect is that it involves the concept of utility and there is no reason to suppose that this is an essential aspect of the concept of probability.

2.3.3 Probability and No Arbitrage

Probability arises in finance via the no arbitrage principle.

> *No Arbitrage Principle: There are no situations where an investor can invest nothing and for which there is a possibility of a gain with no possibility of a loss.*

The no arbitrage principle is an equilibrium principle in the sense that arbitrage opportunities may momentarily exist in a market but they disappear as investors move in and remove them. The no arbitrage principle is similar to the principle of avoiding sure losses. It relates to how an individual prices or values certain possibilities and so could be thought of as subjective in nature.

A *portfolio A* is some finite collection of investments. Suppose there is a time 0 when investments are made and Cost(portfolio A) refers to the total cost of all the investments in A at time 0. Also suppose that there is a time 1 when investments are liquidated and Payoff(portfolio A) is the amount realized from A. Let $r > 0$ denote the *risk-free rate of return* over this period, namely, if 1 unit is invested at time 0 at the risk-free rate, then at time 1 this has grown to equal $1 + r$. Assume an investor can always borrow and invest at the risk-free rate. Further assume that an investor can always buy (go long) an investment as well as sell investments short (borrow the investment and sell it at time 0, returning the investment at time 1) with no borrowing cost. The following is a consequence of the no arbitrage principle.

Lemma 2.3.3 *(Law of One Price) If Payoff(portfolio A) $\leq$ Payoff(portfolio B), then Cost(portfolio A) $\leq$ Cost(portfolio B).*

Proof. Suppose Cost(portfolio A) > Cost(portfolio B). Then at time 0 sell A and buy B, which entails no investment by the investor. Then the payoff at time 1 is Payoff(portfolio B) – Payoff(portfolio A) + (Cost(portfolio A) – Cost(portfolio B))$(1 + r) > 0$ and this violates the no arbitrage principle. ∎

From this result it follows that, if Payoff(portfolio A) = Payoff(portfolio B), then Cost(portfolio A) = Cost(portfolio B).

The following example illustrates how probability arises from these considerations.

Example 2.3.2 *Option Pricing.*

Suppose there is an *underlying* investment that has value ω_0 at time 0 and value $\omega_1 = \omega_0 u$ or $\omega_1 = \omega_0 d$ at time 1 where $d < u$. Actually, it must be that $d < 1 + r < u$. For if $1 + r < d$, then borrow ω_0 at rate r and buy the underlying for a payoff of at least $\omega_0 d - \omega_0(1 + r) > 0$, which violates no arbitrage. If $1 + r > u$, then sell the underlying short at time 0 and invest the proceeds at the risk-free rate for payoff of at least $\omega_0(1 + r) - \omega_0 u > 0$, which violates no arbitrage. It makes sense to also require $d < 1$, so there is the possibility of a loss of capital, but this is not necessary.

A *call option* is a *financial derivative* based on the underlying. At time 0 the seller of the call receives C from the buyer by granting them the right to purchase the underlying from the seller at time 1 for the *strike price K*. The payoff to the holder of the call at time 1 is

$$(\omega_1 - K)_+ = \begin{cases} (\omega_0 u - K)_+ & \text{if} \quad \omega_1 = \omega_0 u \\ (\omega_0 d - K)_+ & \text{if} \quad \omega_1 = \omega_0 d \end{cases}$$

and the positive part is used because the option will not be exercised if the price at time 1 is less than K. The problem then is to determine C.

For this use the law of one price, namely, determine a portfolio of known cost which has the same payoff as C. Start with x_0 and purchase Δ_0 units of the underlying so the cost of this portfolio is x_0 with payoff

$$(x_0 - \Delta_0\omega_0)(1+r) + \Delta_0\omega_1 = \begin{cases} (x_0 - \Delta_0\omega_0)(1+r) + \Delta_0\omega_0 u & \text{if} \quad \omega_1 = \omega_0 u \\ (x_0 - \Delta_0\omega_0)(1+r) + \Delta_0\omega_0 d & \text{if} \quad \omega_1 = \omega_0 d. \end{cases}$$

If this portfolio is to have the same value at time 1 as the portfolio consisting of the call, then

$$(x_0 - \Delta_0\omega_0)(1+r) + \Delta_0\omega_0 u = (\omega_0 u - K)_+$$
$$(x_0 - \Delta_0\omega_0)(1+r) + \Delta_0\omega_0 d = (\omega_0 d - K)_+. \tag{2.5}$$

Solving (2.5) for the unknowns x_0 and Δ_0 leads to

$$\Delta_0 = \frac{(\omega_0 u - K)_+ - (\omega_0 d - K)_+}{\omega_0(u-d)}$$
$$x_0 = (1+r)^{-1}[(\omega_0 u - K)_+ - \Delta_0\omega_0(u - 1 - r)]$$
$$= \frac{1}{1+r}\left[(\omega_0 u - K)_+ - \frac{(\omega_0 u - K)_+ - (\omega_0 d - K)_+}{u-d}(u - 1 - r)\right]$$
$$= (1+r)^{-1}[\pi(\omega_0 u - K)_+ + (1-\pi)(\omega_0 d - K)_+]$$

where

$$0 \le \pi = \frac{1+r-d}{u-d} \le 1.$$

Notice that π is a probability, called the *risk-neutral probability*. Therefore, by the law of one price,

$$C = (1+r)^{-1}E_{P_\pi}((\omega_0 z - K)_+) \tag{2.6}$$

where z follows the risk-neutral probability measure given by

$$P_\pi(\{u\}) = \pi, P_\pi(\{d\}) = 1 - \pi.$$

A similar argument can be used for obtaining the price P of a *put option*. A put option gives the buyer the right to sell an underlying for strike price K at time 1 to the seller of the put. This is given by

$$P = (1+r)^{-1}E_\pi((K - \omega_0 z)_+). \tag{2.7}$$

Formulas (2.6) and (2.7) are known as the *Black–Scholes formulas* for this situation.
■

Example 2.3.2 can be substantially generalized. For example, consider the model where there are n equidistant time steps with the underlying price changing by the factors u or d at each time step. This is known as the binomial model where, for a call exercised at time n, (2.6) becomes

$$C = (1+r)^{-n}E_{P_\pi^n}((\omega_0 z - K)_+)$$

with $P_\pi^n = P_\pi \times \cdots \times P_\pi$ being the risk-neutral probability measure in this case. If the time intervals are allowed to shrink and increase in number, then under some reasonable conditions, the risk-neutral probability measure converges to a Brownian motion with drift.

Perhaps the most remarkable aspect of this context is that probability appears to arise again from a principle of rational behavior. There is no need to discuss long-run performance characteristics or even randomness, as probability just appears naturally. More discussion on probability and its role in finance can be found in Shafer and Vovk (2001).

2.3.4 Scoring Rules

Another justification for probability arises through assessing the accuracy of forecasters. Suppose $\mathscr{B} = \{B_1, \ldots, B_m\}$ is a finite indexed family of subsets of Ω, and a forecaster provides $r_i \in [0,1]$ as a *probability forecast* that event B_i will occur for each i. For example, suppose $\Omega = \{0,1\}^7$, where $\omega_i = 1$ indicates rain and $\omega_i = 0$ indicates no rain on day i, B_j is an event concerning days in the next week when it will rain and r_j is the forecaster's stated probability that B_j will be true. The forecast is called a *coherent forecast* if there exists probability measure P such that $P(B_i) = r_i$ for $i = 1, \ldots, m$. It would seem that one would want to make coherent forecasts but why is this?

Consider assessing the accuracy of forecasts by a scoring rule.

Definition 2.3.3 *A proper scoring rule is a function* $s : \{0,1\} \times [0,\infty] \to [0,\infty)$ *such that (i) $ps(1,r) + (1-p)s(0,r)$ is uniquely minimized at $r = p$ for all $p, r \in [0,1]$ and (ii) $s(0,\cdot), s(1,\cdot)$ are continuous.*

Here $s(I_A(\omega), r)$ is a measure of the difference between forecast r and $I_A(\omega)$ where $I_A(\omega) = 1$ indicates that event A has occurred. If you have probability p that A will occur and you forecast r, then $ps(1,r) + (1-p)s(0,r)$ is your expected score and this is minimized by choosing $r = p$. For the full forecast, the *penalty*, based on scoring function s, is given by $\sum_{i=1}^m s(I_{B_i}(\omega), r_i)$. To *weakly prefer* forecast r_i over r_i' means that

$$\sum_{i=1}^m s(I_{B_i}(\omega), r_i) \le \sum_{i=1}^m s(I_{B_i}(\omega), r_i')$$

for all $\omega \in \Omega$, while to *strongly prefer* forecast r_i over r_i' means that the inequality is strict for all $\omega \in \Omega$.

The following result is proved in Predd et al. (2009).

Theorem 2.3.5 *Suppose a forecaster's forecasts are assessed using a proper scoring rule. (i) If forecast r_i is coherent, then there is no forecast r_i' weakly preferred to r_i. (ii) If forecast r_i is incoherent, then there is a coherent forecast r_i' strongly preferred to r_i.*

So Theorem 2.3.5 suggests that, if a forecaster wants to make accurate forecasts, at the very least their forecasts should arise via an assignment of a probability measure for the uncertain outcome ω.

If the forecaster has subjective probabilities $P(B_i)$ for the B_i, then the following result suggests what their forecast should be.

Theorem 2.3.6 *For a proper scoring rule, the forecast minimizing the expected penalty with respect to probability measure P is given by $r_i = P(B_i)$ for $i = 1, \ldots, m$.*

Proof. The i-th term in $E_P\left(\sum_{i=1}^m s(I_{B_i}(\omega), r_i)\right)$ is $P(B_i)s(1, r_i) + P(B_i^c)s(0, r_i)$, which is minimized by $r_i = P(B_i)$ and the result follows. ∎

So a well-calibrated forecaster provides the forecast $r_i = P(B_i)$, at least if they measure the accuracy of their forecasts using a proper scoring rule, as this guarantees there is no better forecast and on average they do the best.

A commonly used scoring rule is the *quadratic scoring rule* $s(x, r) = (x - r)^2$. As $ps(1, r) + (1 - p)s(0, r) = p(1 - r)^2 + (1 - p)r^2 = (r - p)^2 + p(1 - p)$ is minimized at $r = p$, this scoring rule is proper. The penalty derived from the quadratic scoring rule is

$$\sum_{i=1}^m s(I_{B_i}(\omega), r_i) = \sum_{i=1}^m (I_{B_i}(\omega) - r_i)^2$$

and, as the scoring rule is proper, $r_i = P(B_i)$ provides the smallest expected penalty under P.

When $\mathscr{B}$ is a partition of Ω, then $\sum_{i=1}^m (I_{B_i}(\omega) - r_i)^2 = 1 - 2\sum_{i=1}^m r_i I_{B_i}(\omega) + \sum_{i=1}^m r_i^2$. In that case, Theorem 2.3.6 implies that $r_i = P(B_i)$ maximizes the expected value of $2\sum_{i=1}^m r_i I_{B_i}(\omega) - \sum_{i=1}^m r_i^2$, which is known as the *Brier score*. This was introduced by Brier (1950) in the context of assessing weather forecasters.

Another proper scoring rule is given by the *logarithmic scoring rule* $s(x, r) = \log(|1 - x - r|)$ with penalty $\sum_{i=1}^m \log(|1 - I_{B_i}(\omega) - r_i|)$. By Theorem 2.3.6 the expected value of the logarithmic penalty is minimized by $r_i = P(B_i)$.

More extensive discussion of scoring rules can be found in Brier (1950), de Finetti (1974), Savage (1971), Lindley (1982) and Predd et al. (2009).

2.3.5 Savage's Axioms

As already discussed, there is no use made of utility in our developments of a theory of statistical evidence. For a theory of statistical evidence should be as free as possible of personal valuations as to what is good or bad. Certainly utility seems implicit in approaches to probability that are based on some form of pricing and so it is often considered as a primary component of a statistical problem. Statistical problems are then cast as part of a more general scenario where a participant is faced with making a decision, namely, selecting an *action* $a \in \mathscr{A}$, from a set $\mathscr{A}$ of possibilities, based on having observed a state $s \in \mathscr{S}$. This choice is effected by a *decision function* $d \in \mathscr{D}$ where $\mathscr{D}$ is the set of all functions $d : \mathscr{S} \to \mathscr{A}$. For example, $\mathscr{A}$ could be the set of all possible values of an unknown $a, s \in \mathscr{S}$ is observed and then a prediction or estimate $d(s)$ of a is quoted for some $d \in \mathscr{D}$. The purpose of decision theory is to select d. Now consider a theory of decision making due to Savage (1972), as presented in Fishburn (1970).

It is *assumed* that there is a preference ordering $\prec_{\mathscr{D}}$ on $\mathscr{D}$. So for $d_1, d_2 \in \mathscr{D}$ write $d_1 \prec_{\mathscr{D}} d_2$ whenever the actions taken by d_2 are preferable to those taken by d_1.

This ordering reflects the personal utilities of the decision maker. The ordering $\prec_{\mathscr{D}}$ induces an equivalence relation $\equiv_{\mathscr{D}}$ by $d_1 \equiv_{\mathscr{D}} d_2$ whenever $d_1 \not\prec_{\mathscr{D}} d_2$ and $d_2 \not\prec_{\mathscr{D}} d_1$ and also the relation $\preccurlyeq_{\mathscr{D}}$ by $d_1 \preccurlyeq_{\mathscr{D}} d_2$ whenever $d_1 \prec_{\mathscr{D}} d_2$ or $d_1 \equiv_{\mathscr{D}} d_2$.

For $a \in \mathscr{A}$ we define d_a by $d_a(s) = a$ for every $s \in \mathscr{S}$. Then the preference ordering on $\mathscr{D}$ induces a preference ordering $\prec_{\mathscr{A}}$ on $\mathscr{A}$ via $a_1 \prec_{\mathscr{A}} a_2$ whenever $d_{a_1} \prec_{\mathscr{D}} d_{a_2}$.

Say that $d_1, d_2 \in \mathscr{D}$ *agree* on A whenever $d_1(s) = d_2(s)$ for all $s \in A$. Define $d_1 \prec_{\mathscr{D}} d_2$ given A whenever $d_1' \prec_{\mathscr{D}} d_2'$ for all $d_1', d_2' \in \mathscr{D}$ such that d_i and d_i' agree on A while d_1' and d_2' agree on A^c. A subset $A \subset \mathscr{S}$ is said to be *null* if and only if $d_1 \equiv_{\mathscr{D}} d_2$ whenever d_1 and d_2 agree on A^c so the states in A are irrelevant for establishing preference among decision functions. If $\{\mathscr{S}_i : i \in I\}$ is a partition of $\mathscr{S}$ and $d \in \mathscr{D}, i \in I$, define $d_{a,i}$ by $d_{a,i}(s) = a$ when $s \in \mathscr{S}_i$ and $d_{a,i}(s) = d(s)$ otherwise.

There is an induced preference ordering $\prec_{\mathscr{S}}$ on subsets of $\mathscr{S}$ given by $A \prec_{\mathscr{S}} B$ whenever for all $a_1, a_2 \in \mathscr{A}$, such that $a_1 \prec_{\mathscr{A}} a_2$, we have $d_{a_1,a_2}^A \prec_{\mathscr{D}} d_{a_1,a_2}^B$ where $d_{a_1,a_2}^A(s) = a_2$ if $s \in A$ and $d_{a_1,a_2}^A(s) = a_1$ if $s \in A^c$. So the set of states A can be viewed as less preferable than the set of states B since d_{a_1,a_2}^A will choose the preferable action a_2 whenever $s \in A$ while d_{a_1,a_2}^B will do this whenever $s \in B$ and d_{a_1,a_2}^A is less preferable than d_{a_1,a_2}^B.

Consider the following axioms for $\prec_{\mathscr{D}}$.

Axiom 1: $\prec_{\mathscr{D}}$ satisfies (i) $d_1 \prec_{\mathscr{D}} d_2$ implies $d_2 \not\prec_{\mathscr{D}} d_1$ and (ii) $d_1 \not\prec_{\mathscr{D}} d_2, d_2 \not\prec_{\mathscr{D}} d_3$ implies $d_1 \not\prec_{\mathscr{D}} d_3$.

Axiom 2: If d_1, d_1' agree on A, d_2, d_2' agree on A and d_1, d_2 agree on A^c, d_1', d_2' agree on A^c, then $d_1 \prec_{\mathscr{D}} d_2$ if and only if $d_1' \prec_{\mathscr{D}} d_2'$.

Axiom 3: If $A \subset \mathscr{S}$ is not null and d_1, d_{a_1} agree on A, d_2, d_{a_2} agree on A, then $d_1 \prec_{\mathscr{D}} d_2$ given A if and only if $d_{a_1} \prec_{\mathscr{D}} d_{a_2}$.

Axiom 4: If (i) $d_{a_1} \prec_{\mathscr{D}} d_{a_2}, d_1$ agrees with d_{a_2} on A and with d_{a_1} on A^c, d_2 agrees with d_{a_2} on B and with d_{a_1} on B^c, and (ii) if $d_{a_1'} \prec_{\mathscr{D}} d_{a_2'}, d_1'$ agrees with $d_{a_2'}$ on A and with $d_{a_1'}$ on A^c, d_2' agrees with $d_{a_2'}$ on B and with $d_{a_1'}$ on B^c, then $d_1 \prec_{\mathscr{D}} d_2$ if and only if $d_1' \prec_{\mathscr{D}} d_2'$.

Axiom 5: The relation $\prec_{\mathscr{A}}$ is not empty.

Axiom 6: For all $a \in \mathscr{A}$, for all $d, d' \in \mathscr{D}$ such that $d \prec_{\mathscr{D}} d'$, there exists a finite partition $\{\mathscr{S}_i : i \in I\}$ of $\mathscr{S}$ such that for all $i \in I$, we have $d_{a,i} \prec_{\mathscr{D}} d'$ and $d \prec_{\mathscr{D}} d_{a,i}'$.

Axiom 7: If for every $s \in A, d \prec_{\mathscr{D}} d_{d'(s)}$ given A, then $d \prec_{\mathscr{D}} d'$ given A, and if for every $s \in A, d_{d'(s)} \prec_{\mathscr{D}} d$ given A, then $d' \preccurlyeq_{\mathscr{D}} d$ given A.

Axiom 1 is a simple consistency requirement. Axiom 2 is an instance of the *sure thing principle*, as it says that our preference for d_2 over d_1 does not depend on the states where these decision functions agree. Recall, however, Example 1.6.1. Similarly, Axiom 3 says that preference for constant act a_2 over constant act a_1 given non-null A, is equivalent to this preference unconditionally. Axioms 4 and 7 seem like natural consistency requirements. Axiom 5 says that there are preferences among the actions. Axiom 6 seems the most difficult to justify.

The following result is proved in Fishburn (1970).

Theorem 2.3.7 *If $\prec_\mathscr{D}$ satisfies Axioms 1–7, then (i) there exists a unique finitely additive probability measure P on $2^\mathscr{S}$ such that $A \prec_\mathscr{S} B$ if and only if $P(A) < P(B)$, (ii) P has the property that if $B \subset \mathscr{S}$ and $\rho \in [0,1]$, then there exists $C \subset \mathscr{S}$ such that $P(C) = \rho P(B)$ and (iii) there exists bounded $u : \mathscr{A} \to R^1$ such that $d_1 \prec_\mathscr{D} d_2$ if and only if $E_P[u(d_1(s))] < E_P[u(d_2(s))]$ and u is unique up to a positive affine transformation.*

So u can be considered as a utility function and the *principle of maximum expected utility* follows as the most preferred decision function maximizes $E_P(u(d))$ over $d \in \mathscr{D}$.

The result can be interpreted as saying that probability is essential to rational decision making and that certainly has significance. Again probability emerges without any reference to randomness but at the cost of confounding probability with preferences and utility. Also, Theorem 2.3.7 (ii) implies that the set of states $\mathscr{S}$ must be an uncountably infinite set and this is contrary to our goals as expressed in Section 1.4. In any case, the formalism as expressed in the axioms is not essential for a theory of statistical evidence.

2.3.6   Cox's Theorem

In Cox (1946, 1961) an attempt is made to derive probability from very elementary considerations independent of relative frequency and of utility or pricing. As such, this has a great deal of appeal and is often cited, as in Jaynes (2003).

Cox's development proceeds along the following lines. For each $U, V \subset \Omega$ where $U \neq \phi$, let $Bel(V \mid U) \in R^1$ and interpret this number as the belief that an unknown $\omega \in V$, given that we know $\omega \in U$. For notational purposes we write $Bel(V) = Bel(V \mid \Omega)$. Cox then assumes that Bel has a few seemingly natural properties and argues that these properties imply the existence of a 1 to 1, onto function $g : R^1 \to R^1$ such that, for $V \in 2^\Omega$, $P(V) = g(Bel(V))$ defines a probability measure on 2^Ω and $g(Bel(V \mid U))g(Bel(U)) = g(Bel(V \cap U))$ when $U \neq \phi$, so $Bel(\cdot \mid U)$ behaves like conditional probability. Therefore, Bel is then *isomorphic* to a probability measure. So this says that any reasonable characterization of how one should reason in contexts of uncertainty should correspond to the rules of probability.

In Halpern (1999a, 1999b) reservations are raised concerning Cox's result. In particular, it is not applicable when Ω is finite and so, considering our discussion in Section 1.4, this seems very unnatural. In Paris (2009), a rigorous proof of a version of Cox's theorem is provided that holds in finite contexts. For this, consider Bel as being defined on a number of sample spaces Ω in a set $\mathscr{A}$. We might consider $\mathscr{A}$ to be the set of all finite domains. The following result is proved in Paris (2009).

Theorem 2.3.8 *(Cox's Theorem) If for all domains Ω in a set $\mathscr{A}$, there is a decreasing function $S : [0,1] \to [0,1]$, a continuous function $F : [0,1]^2 \to [0,1]$ such that F is strictly increasing in each coordinate in $(0,1]^2$ and Bel has the following properties:*
(i) $0 \leq Bel(V \mid U) \leq 1$ for all $U, V \subset \Omega$ where $U \neq \phi$,
(ii) $Bel(\phi \mid U) = 0$ and $Bel(\Omega \mid U) = 1$ $U \subset \Omega$ where $U \neq \phi$,

(iii) $Bel(V \cap V' | U) = F(Bel(V' | V \cap U), Bel(V | U))$ *for all* $U, V, V' \subset \Omega$ *such that* $V \cap U \neq \phi$,

(iv) $Bel(V^c | U) = S(Bel(V | U))$ *for all* $U, V \subset \Omega$ *where* $U \neq \phi$,

(v) for all $0 \leq \alpha, \beta, \gamma \leq 1$ *and* $\varepsilon > 0$ *there is an* $\Omega \in \mathscr{A}$ *and* $U_1, U_2, U_3, U_4 \subset \Omega$ *such that* $U_1 \supset U_2 \supset U_3 \supset U_4$ *with* $U_3 \neq \phi$ *and* $|Bel(U_4 | U_3) - \alpha| < \varepsilon, |Bel(U_3 | U_2) - \beta| < \varepsilon$ *and* $|Bel(U_2 | U_1) - \beta| < \varepsilon$,

then there is an isomorphism g such that $g \circ Bel$ *is a finitely additive probability measure on* 2^{Ω} *for each* Ω, *and* $g(Bel(V | U))g(Bel(U)) = g(Bel(V \cap U))$ *when* $U \neq \phi$.

Restricting attention to finite Ω implies that $Bel(\cdot | U)$ takes only finite many values and so condition (i) certainly seems reasonable, perhaps requiring only a location-scale transformation of Bel. It is certainly reasonable to assume that the impossible event and the sure event always attain the minimum and maximum belief as in condition (ii).

Condition (iii) says that our belief that both V and V' are true, given that U holds, is determined by our belief that V is true given U and our belief that V' is true given that both V and U hold. For example, consider U to be the event it rained yesterday, V the event it rains today and V' the event that it rains tomorrow. Condition (iii) says our belief that it will rain today and tomorrow given that it rained yesterday can be computed from our belief that it will rain today given that it rained yesterday and our belief that it will rain tomorrow given that it rained today and yesterday.

Condition (iv) indicates that our belief that V^c is true given U can always be determined as a function of our belief that V is true given U. Furthermore, our belief in V^c given U decreases as our belief in V given U increases.

Condition (v) seems harder to justify, as it implies a certain complexity to the beliefs one holds. In particular, one needs to be able to make very fine distinctions between the strengths of beliefs.

2.4 Relative Frequency Probability

The treatments of probability in Section 2.3 have been under the heading of subjective probability because they are associated with an individual's beliefs or rational behavior. There is no attempt to claim that probability is some aspect of the physical world that is measurable. If ones beliefs are consistent, and one behaves according to such beliefs, then probability emerges in a very natural way.

One's beliefs, as expressed via a probability measure P, may be seriously in error in the sense that what is believed to be true is substantially different from reality. For the reasoning system used, however, this doesn't matter, as the system of reasoning needs to be independent of whether or not our beliefs are in any sense "correct." As a convenient analogy, think of a theorem based upon mutually consistent hypotheses. The hypotheses correspond to our beliefs and the reasoning process used to deduce the conclusion does not depend on the correctness of the hypotheses. The theorem is true, given the hypotheses, provided the reasoning has proceeded correctly. Similarly, a theory of statistical inference is needed that can be applied unambiguously to any specification of beliefs, as expressed by a probability measure, to produce inferences; see Chapter 4. The inferences are then correct given the beliefs.

Statistics is applied in the real world to reason. So if applied correctly, according to a suitable theory of statistical inference, the inferences are correct and the relevance of the results rests solely on the specification of the beliefs involved. Accordingly, it makes sense that these beliefs be checked for their correspondence with reality. Essentially, there is only one way to ultimately make this assessment, namely, through the data, and this is the subject of Chapter 5. It is virtually never the case, however, where it can be said that our beliefs are categorically correct, only that they are not contradicted by the data. This is all that statistics can say.

A common claim, however, by those who argue for the correctness of basing a theory of statistical inference on the principle of frequentism, is that the probabilities used correspond to real-world entities. This is because the probabilities involved are supposed to correspond to those obtained from relative frequencies, as discussed in Section 1.2. Certainly the relative frequency distribution is the object of interest in a statistical investigation but, in fact, the probabilities quoted in a frequentist analysis do not correspond to these probabilities but rather are based upon the model as chosen by the statistician. So in a real sense they are subjective.

Still, the relative frequency approach to probability is important enough that it bears closer examination.

2.4.1   Long-Run Relative Frequency

Consider a system that can be repeatedly performed and in a single performance gives a value $\omega \in \Omega$ where Ω is finite. Suppose there is a sequence of outcomes $\omega_1, \omega_2, \ldots$ from repeated performances of the system. For the first n performances $\omega_1, \omega_2, \ldots, \omega_n$ define the relative frequency of $A \subset \Omega$ by

$$r_n(A) = \frac{\#\{i : \omega_i \in A, i = 1, \ldots, n\}}{n}.$$

Note that r_n defines a probability measure on 2^Ω for each n. Loosely speaking, a system that can be repeatedly performed is a *random system* if it satisfies:

(i) for any sequence of performances $\omega_1, \omega_2, \ldots$, then $\lim_{n \to \infty} r_n(A)$ exists for each $A \subset \Omega$,

(ii) the sequence of performances $\omega_1, \omega_2, \ldots$ is "random."

It is important to note that (i) and (ii) do not comprise a definition of a random system; rather, they provide a characterization of how it is thought a random system should behave.

If (i) is satisfied, then it is natural to define a probability measure on 2^Ω by $P(A) = \lim_{n \to \infty} r_n(A)$. The randomness requirement is not necessary for this definition but intuitively it seems essential. Whether or not a random system actually exists is a question that cannot be answered categorically. Of course, the finiteness of whatever is observed is a limitation, but the methods of Chapter 5 can be employed to assess whether or not the outcomes observed do have appropriate characteristics, at least to some degree. In the end, however, we are left with the assertion that some systems, such as games of chance performed under appropriate conditions, satisfy our characterization. This is an important point, because underlying our belief in the objectivity of the data is the belief that the data was produced by a random system. In fact, in experimental design, systems that are believed to be random systems are employed as part of the data collection process. It has been argued here that the primary reason for this is to ensure objectivity, at least if the random system possesses appropriate uniform properties.

2.4.2 Randomness

Actually the most difficult aspect of the relative frequency approach to probability lies with characterizing what is meant by randomness. The following example shows that this is not captured by statistical independence.

Example 2.4.1 *Champernowne's Sequence.*

Suppose that $\Omega = \{0, 1, 2, 3, 4, 5, 6, 7, 8, 9\}$. It is clear that the sequence $0, 1, 2, 3, 4, 5, 6, 7, 8, 9, 1, 0, 1, 1, 1, 2, 1, 3, 1, 4, \ldots$ is being constructed deterministically from the natural numbers in their natural order. It can be shown, however, that $\lim_{n \to \infty} r_n(\{i\}) = 1/10$ for all $i \in \Omega$, $\lim_{n \to \infty} r_n(\{(i, j)\}) = 1/100$ for $(i, j) \in \Omega^2$ whenever we take pairs in the sequence some fixed distance apart, etc. In fact, all the independence requirements of a sequence generated as *i.i.d.* from a uniform distribution on Ω will satisfied by this sequence but it is definitely not random. More details on this sequence can be found in Champernowne (1933). ■

So the idea of a random sequence is something beyond what is meant by an *i.i.d.* sequence of observations. A successful theory of randomness has been developed based upon ideas found in Kolmogorov (1963). A close examination of this theory reveals two somewhat startling facts. First, randomness has nothing to do with probability. Properly expressed, it is concerned with the inherent complexity of a sequence or object. Second, there is no statistical test for randomness. Certainly structure can be recognized, and so an assertion made that something is not random, but it is never possible to assert that a sequence is random.

While this is somewhat disturbing for a frequentist theory of statistics, this simply reinforces our claim that probability is about expressing beliefs and nothing else. The notion of randomness is, however, inherent to our characterization of a random system. This is important for statistics, as it ensures the objectivity of the data collected in the sense that the data collection process is outside the control of any interested parties. Probabilities may indeed be assigned based on frequencies that arise through counts, but these probabilities represent beliefs and have nothing to do with randomness.

The theory of randomness and associated concepts is discussed in Fine (1973), Li and Vitanyi (1993) and Shafer and Vovk (2001). While an interesting topic, and certainly relevant for statisticians for the reasons given, we do not see this as being relevant to the development of a successful theory of statistical inference. Further discussion, on an attempt to build a theory of probability based on limiting frequencies, can be found in von Mises (1964).

2.5   Concluding Comments

The logical interpretation of probability where, for sets of propositions A and B, the number $P(A \mid B) \in [0, 1]$ is a measure of the degree to which the truth of B entails the truth of A has not been discussed here. This can be considered as a generalization of logical implication. A concise treatment can be found in Fine (1973). This interpretation, however, is not seen as relevant to our discussion of evidence, where probabilities are always interpreted as degrees of belief no matter how they are assigned.

The main point of this chapter is that there is little point in arguing about what probability is as some measurable quantity in the universe. It is not clear to us that conceiving of precise, true values for probabilities even makes sense, no matter how they are assigned. Such a discussion seems somewhat like trying to decide on what is ultimately meant by the concept of truth itself. Certainly this is interesting to consider, but the concept of truth and the reasoning process associated with it have proven to be profoundly useful without being able to ultimately answer this question. Similarly, it seems that the concept of probability and the associated rules of reasoning in contexts of uncertainty are much more important than trying to ultimately settle on what probability is. It is our view that it is only relevant to determine whether or not the assignment of probabilities makes sense in light of an application. This is the topic of Chapter 5, where probabilities are assessed in light of observed data. It can never be claimed, based on such an analysis, that the probabilities assigned are correct, but when these are not contradicted by the data, there can be a sound basis for drawing appropriate inferences.

Chapter 3

Characterizing Statistical Evidence

3.1 Introduction

This chapter discusses a variety of approaches that have been considered for characterizing statistical evidence and reasons why it is believed that these are not fully successful. Our general point is that any valid theory of statistical inference must at least prescribe how to measure statistical evidence. This seems essential if we are going to speak of "the statistical evidence." Many theories of inference stop somewhat short of stating explicitly how to measure statistical evidence yet still make reference to it. This leads to confusion and ambiguity, especially with respect to how one is supposed to conduct a statistical analysis. While there can be many problems with statistical analyses, particularly when these are carried out on observational data and don't account for inherent biases, the failure to use a tool appropriate to the task of assessing evidence can be a contributing factor in any deficiencies that statistical theory brings to this task.

Chapter 4 is concerned with a theory of statistical inference based upon a measure of the statistical evidence that an unknown takes a certain value. The definition relies on the prescription of a statistical model and a proper prior and, of course, the data. To see why this definition is suitable, however, it is necessary to understand the deficiencies in other methodologies where the concept of statistical evidence plays a key role.

3.2 Pure Likelihood Inference

The pure likelihood approach to inference is perhaps the most focused and logical attempt to characterize the concept of statistical evidence based on the model and the data as the only ingredients for a statistical problem. The books by Edwards (1992) and Royall (1997) are notable in this regard.

3.2.1 Inferences for the Full Parameter

Likelihood inferences are based on the likelihood function as defined in Definition 1.4.1. So for observed data x and model $\{f_\theta : \theta \in \Theta\}$, all inferences about θ are based on $L(\cdot \,|\, x) : \Theta \to [0, \infty)$ given by $L(\theta \,|\, x) = kf_\theta(x)$ for some fixed $k > 0$. The basic motivating idea arises from the discrete case where $f_\theta(x)$ is the probability of observing x given that θ is the true value. This imposes a preference ordering on the elements of Θ.

The *Likelihood Preference Ordering:* θ_1 is not preferred to θ_2, denoted $\theta_1 \preceq \theta_2$, whenever $L(\theta_1 | x) \leq L(\theta_2 | x)$.

The motivation for this preference ordering is that observing x when θ_2 is true is at least as probable as observing x when θ_1 is true. Clearly, bearing in mind our comments in Section 1.4, this interpretation carries over to the case of continuous models as well.

The likelihood preference ordering is an axiom or principle of the pure likelihood approach to inference in the sense that our inferences are required to conform to it. This immediately implies an estimate of θ given by the *maximum likelihood estimate (MLE)* $\theta_{MLE}(x) = \arg\sup_\theta L(\theta | x)$, since $\theta_{MLE}(x)$ is the most preferred value. Typically the MLE exists and is unique, although problems can arise if there isn't enough data. In this section it is assumed that $\theta_{MLE}(x)$ exists and is unique to simplify the discussion.

A natural quantification of the uncertainty in the MLE is given by quoting a region $C(x) \subset \Theta$ containing $\theta_{MLE}(x)$ where the "size" of $C(x)$ gives our assessment of the accuracy of $\theta_{MLE}(x)$. Here "size" could be length of an interval or volume of a set in a Euclidean space or some other measure that has some direct relevance to the application. The form of $C(x)$ is implied by the preference ordering. For, if $\theta_1 \in C(x)$ and $L(\theta_1 | x) \leq L(\theta_2 | x)$, then necessarily $\theta_2 \in C(x)$. So a *likelihood region* $C(x) = \{\theta : L(\theta | x) \geq c(x)\}$, for some function $c : \mathscr{X} \to [0, \infty)$, gives a relevant assessment of the accuracy, as $\theta_{MLE}(x)$ is always an element of any non-null likelihood region.

Completing the specification of a likelihood region requires the provision of $c(x)$, as this is not determined by the preference ordering. This requires another axiom or principle of the pure likelihood approach.

The Law of the Likelihood: the ratio $L(\theta_1 | x)/L(\theta_2 | x)$ measures the strength of the evidence supporting θ_1 over θ_2.

The interpretation of this is as follows. If $L(\theta_1 | x)/L(\theta_2 | x) = 2$, the evidence supports θ_1 twice as much as θ_2 and if $L(\theta_1 | x)/L(\theta_2 | x) = 1/3$, the evidence supports θ_1 one third as much as θ_2.

If $C(x) \neq \phi$ and $\theta \in C(x)$, then

$$L(\theta | x)/L(\theta_{MLE}(x) | x) \geq c(x)/L(\theta_{MLE}(x) | x) = 1 - \gamma$$

for some $\gamma \in [0, 1]$. The lack of dependence of γ on x follows from the law of the likelihood as, by that law, the ratio $L(\theta | x)/L(\theta_{MLE}(x) | x)$ has an interpretation independent of the model and x, as the strength of the evidence supporting θ versus the best supported value. As such, a likelihood region is specified by selecting a value $\gamma \in [0, 1]$ and putting

$$C_\gamma(x) = \{\theta : L(\theta | x)/L(\theta_{MLE}(x) | x) \geq 1 - \gamma\}. \qquad (3.1)$$

So the law of the likelihood leads to $c(x) = L(\theta_{MLE}(x) | x)(1 - \gamma)$ for some $\gamma \in [0, 1]$. The function $L(\cdot | x)/L(\theta_{MLE}(x) | x)$ is called the *relative likelihood function* and we refer to $C_\gamma(x)$ as the $(1-\gamma)$-*likelihood region* for θ. So $C_\gamma(x)$ contains those θ values for which the data provides support of at least $100(1-\gamma)\%$ of the maximum support and this provides the interpretation of γ. Royall (1997) argues, based on an urn model, for the specific choice of $1-\gamma = 1/8$ as giving a likelihood region containing all those θ for which there is "strong evidence" that they correspond to the true value. Once γ is chosen, then the size of $C_\gamma(x)$ is used to assess the accuracy of the estimate $\theta_{MLE}(x)$.

Notice that, if $\theta \in C_\gamma(x)$, then

$$L(\theta'\,|\,x)/L(\theta\,|\,x) \leq L(\theta_{MLE}(x)\,|\,x)/L(\theta\,|\,x) \leq (1-\gamma)^{-1}$$

so no element of Θ is supported over θ by a factor greater than $(1-\gamma)^{-1}$. Furthermore, if $\theta' \notin C_\gamma(x)$, then $L(\theta_{MLE}(x)\,|\,x)/L(\theta'\,|\,x) > (1-\gamma)^{-1}$ and so there is always at least one value in $C_\gamma(x)$ that is supported over θ' by a factor greater than $(1-\gamma)^{-1}$.

The hypothesis $H_0 = \{\theta_0\}$ is assessed by computing the relative likelihood at θ_0,

$$p_{\theta_0}(x) = L(\theta_0\,|\,x)/L(\theta_{MLE}(x)\,|\,x). \tag{3.2}$$

If $p_{\theta_0}(x) \in [0,1]$ is large, then the interpretation is that there is evidence in favor of H_0, as $L(\theta_0\,|\,x)$ is close to the maximum support, and if $p_{\theta_0}(x)$ is small, then the interpretation is that there is evidence against H_0. The value of $p_{\theta_0}(x)$ is taken as the likelihood-based strength of this evidence. Notice that $p_{\theta_0}(x)$ can indicate evidence either for H_0 as well as against H_0. So, using Royall's specification, if $p_{\theta_0}(x) > 1/8$, then there is strong evidence in favor of H_0.

The following result relates $p_{\theta_0}(x)$ and $C_\gamma(x)$.

Lemma 3.2.1 *The functions p_{θ_0} and C_γ are related by $p_{\theta_0}(x) = 1 - \inf\{\gamma : \theta_0 \in C_\gamma(x)\}$ and $C_\gamma(x) = \{\theta : p_\theta(x) \geq 1 - \gamma\}$.*

Proof. It is the case that $1 - \inf\{\gamma : \theta_0 \in C_\gamma(x)\} = 1 - \inf\{\gamma : L(\theta_0\,|\,x)/L(\theta_{MLE}(x) \geq 1 - \gamma\} = 1 - \inf\{\gamma : \gamma \geq 1 - L(\theta_0\,|\,x)/L(\theta_{MLE}(x))\} = p_{\theta_0}(x)$. Furthermore, $C_\gamma(x) = \{\theta : L(\theta\,|\,x)/L(\theta_{MLE}(x)\,|\,x) \geq 1 - \gamma\} = \{\theta : p_\theta(x) \geq 1 - \gamma\}$. ∎

The content of Lemma 3.2.1 is that likelihood regions C_γ can be defined via (3.1) first and then we obtain $p_{\theta_0}(x)$ given by (3.2), or start with (3.2) and derive (3.1). This equivalence between assessing the error in an estimate and hypothesis assessment will be seen to arise in other contexts as well.

The following principle of inference follows from the likelihood preference ordering and the law of the likelihood.

The Likelihood Principle: two model-data combinations with proportional likelihoods must produce identical inferences.

The likelihood principle requires identical parameter spaces but can obviously be generalized to allow for parameter spaces that are bijectively equivalent. In fact, this characterizes an important property of likelihood inferences, as they are invariant under 1 to 1 relabelings of the model parameter. This *reparameterization invariance* is a natural requirement of any valid theory of inference and so this result is stated explicitly.

Lemma 3.2.2 *Consider model $\{f_\theta : \theta \in \Theta\}$ for data $x \in \mathscr{X}$ and suppose $\Psi : \Theta \to \Psi$ is a bijection (1 to 1 and onto). Let the model $\{g_\psi : \psi \in \Psi\}$ for data $x \in \mathscr{X}$ be defined by $g_\psi = f_{\Psi^{-1}(\psi)}$. If $\theta_{MLE}(x), C_\gamma(x)$ denote the MLE and the $(1-\gamma)$-likelihood region for θ, then $\psi_{MLE}(x) = \Psi(\theta_{MLE}(x))$ and $\Psi C_\gamma(x)$ give the MLE and the $(1-\gamma)$-likelihood region for ψ.*

Proof. Suppose $L(\theta \,|\, x) = k f_\theta(x)$ is a likelihood for θ. Then

$$L'(\psi \,|\, x) = k g_\psi(x) = k f_{\Psi^{-1}(\psi)}(x) = L(\Psi^{-1}(\psi) \,|\, x)$$

is a likelihood for ψ. Therefore, $L'(\Psi(\theta) \,|\, x) \leq L(\theta_{MLE}(x) \,|\, x)$ for all θ with equality if and only if $\theta = \theta_{MLE}(x)$. This implies that $\psi_{MLE}(x) = \Psi(\theta_{MLE}(x))$. Furthermore,

$$
\begin{aligned}
C'_\gamma(x) &= \{\psi : L'(\psi \,|\, x)/L'(\psi_{MLE}(x) \,|\, x) \geq 1 - \gamma\} \\
&= \{\Psi(\theta) : L'(\Psi(\theta) \,|\, x)/L'(\Psi(\theta_{MLE}(x)) \,|\, x) \geq 1 - \gamma\} \\
&= \{\Psi(\theta) : L(\theta \,|\, x)/L(\theta_{MLE}(x) \,|\, x) \geq 1 - \gamma\} = \Psi C_\gamma(x)
\end{aligned}
$$

and this completes the proof. ∎

The law of the likelihood implies that the value $r = L(\theta_1 \,|\, x)/L(\theta_2 \,|\, x)$ has the same interpretation for every application as a measure of the strength of the evidence for θ_1 relative to θ_2. While the likelihood preference ordering seems uncontroversial, doubts arise as to whether $L(\theta \,|\, x)/L(\theta_{MLE}(x) \,|\, x)$ is indeed measuring the strength of the evidence in favor of θ. Consider the following somewhat artificial but simple example presented in Evans (1989) and based on an example in Stone (1976).

Example 3.2.1 *Do Likelihood Ratios Measure the Strength of Evidence?*
Let Θ_k be the set of all words of length M or shorter formed from k symbols $\{a_1,\dots,a_k\}$ by concatenation. So ϕ is the single word of length 0, there are k words of length 1, k^2 words of length 2, etc. Let $l(\theta)$ denote the length of the word θ. Now choose $\delta > 0$ to be very small. Let $\mathscr{X}_k = \Theta_k$ and when $l(\theta) < M$ define

$$
f_\theta(x) = \begin{cases}
1/(k+1) + \delta & x = \theta \\
1/(k+1) - \delta/k & x = \theta a_i \text{ for } i = 1,\dots,k, \\
0 & \text{otherwise,}
\end{cases}
$$

and when $l(\theta) = M$ define $f_\theta(x) = 1$ when $x = \theta$ and $f_\theta(x) = 0$ when $x \neq \theta$.

Suppose that x is observed with $1 < l(x) < M$. Then $\theta_{MLE}(x) = x$ and, letting $x'(x)$ be the word obtained from x by deleting the last symbol,

$$
L(\theta \,|\, x)/L(\theta_{MLE}(x) \,|\, x) = \begin{cases}
1 & \theta = x \\
(1 - (k+1)\delta/k)/(1 + (k+1)\delta) & \theta = x'(x) \\
0 & \text{otherwise.}
\end{cases}
$$

Note that there are only two values with nonzero likelihood and $(1 - (k+1)\delta/k)/(1 + (k+1)\delta)$ can be made arbitrarily small by choosing k large enough. Furthermore, this implies that, whenever $\gamma < 1$, then the $(1-\gamma)$-likelihood region satisfies $C_\gamma(x) = \{x\}$ for all large k.

When $l(\theta) > 0$, then $P_{\theta}(\theta = x'(x)) = k/(k+1) - \delta$ and this can be made arbitrarily close to $1 - \delta$ for k large enough. So an example can be constructed, by choice of k, δ and M, such that for many data values x, the likelihood ratio for the MLE x is huge and the accuracy of this estimate, as measured by the size of $C_{\gamma}(x)$, is high, while at the same time we are virtually certain that the true value is given by $x'(x)$, the only other possible value for θ.

This example can be made simpler by taking Θ to be the set of all finite words formed by concatenation; see Evans (1989). Attention is restricted here to finite parameter and sample spaces to show that the anomalous behavior does not depend intrinsically on infinity. ■

Example 3.2.1 suggests that $L(\theta \,|\, x)/L(\theta_{MLE}(x) \,|\, x)$ is not measuring the strength of the evidence in favor of θ and casts doubt on the law of the likelihood from which it is derived. In Example 4.5.5 we reconsider this example using the definition of statistical evidence via relative belief and show that this apparent contradiction is somewhat avoided. For example, typically there is evidence in favor of both $\theta = x$ and $\theta = x'(x)$ and, as with likelihood, categorical evidence against all other values.

It should be noted that the likelihood $L(\theta \,|\, x)$ does not give a definition of the evidence that θ is the true value. This is easily seen since $L(\cdot \,|\, x)$ is only defined up to a positive constant multiple. Rather, the values $L(\theta \,|\, x)$ give a definition of *relative evidence* for the full model parameter. This interpretation of likelihood as a measure of relative evidence is also supported by the developments in Chapter 4. It could be argued, however, that the relative likelihood $L(\theta \,|\, x)/L(\theta_{MLE}(x) \,|\, x)$ is indeed being used as if it is the statistical evidence that θ is the true value. The relative likelihood has a very unclear interpretation and, as illustrated in Example 3.2.1, there are concerns associated with its calibration.

3.2.2   Inferences for a Marginal Parameter

The lack of a likelihood-based definition of evidence causes difficulties for the theory when $\#(\Theta) > 2$. For consider the problem of deriving inferences for a parameter of interest $\psi = \Psi(\theta)$ when Ψ is not 1 to 1. For example, there may be interest in assessing the hypothesis that $\Psi(\theta) = \psi_0$ corresponding to the *composite hypothesis* $H_0 = \Psi^{-1}\{\psi_0\}$ for θ. In particular, suppose we want to assess whether or not the true value of θ is in a subset $H_0 \subset \Theta$, so we take $\Psi = I_{H_0}$ and $\psi_0 = 1$. This is the problem of nuisance parameters and for this the pure likelihood approach is effectively silent.

The most commonly recommended approach is to construct inferences for ψ, just as one does for θ, but instead use the *profile likelihood function* given by

$$L^{\Psi}(\psi \,|\, x) = \sup_{\theta \in \Psi^{-1}\{\psi\}} L(\theta \,|\, x).$$

The profile likelihood induces a preference ordering on Ψ given by $\psi_1 \preceq \psi_2$ whenever $L^{\Psi}(\psi_1 \,|\, x) \leq L^{\Psi}(\psi_2 \,|\, x)$. Notice that the profile MLE is $\psi_{MLE}(x) = \Psi(\theta_{MLE}(x))$ and $L^{\Psi}(\psi_{MLE}(x) \,|\, x) = L(\theta_{MLE}(x) \,|\, x)$. Furthermore, a γ-profile likelihood region is given by

$$C_{\Psi, \gamma}(x) = \{\psi : L^{\Psi}(\psi \,|\, x)/L^{\Psi}(\psi_{MLE}(x) \,|\, x) \geq 1 - \gamma\}$$

and we can assess the hypothesis $H_0 = \Psi^{-1}\{\psi_0\}$ by

$$p_{\psi_0}(x) = L^{\Psi}(\psi_0 \,|\, x)/L^{\Psi}(\psi_{MLE}(x) \,|\, x).$$

Profile likelihood inferences are also invariant under reparameterizations.

It turns out that, under weak conditions, $C_{\Psi,\gamma}(x) = \Psi C_\gamma(x)$.

Lemma 3.2.3 If $\theta_\psi(x) = \arg\sup_{\theta \in \Psi^{-1}\{\psi\}} L(\theta \,|\, x)$ exists for each $\psi \in \Psi$, then $C_{\Psi,\gamma}(x) = \Psi C_\gamma(x)$.

Proof. Suppose $\theta \in C_\gamma(x)$, so $L(\theta \,|\, x) \geq (1-\gamma)L(\theta_{MLE}(x))$, which implies

$$L^{\Psi}(\Psi(\theta) \,|\, x) = \sup_{\theta' \in \Psi^{-1}\Psi(\theta)} L(\theta' \,|\, x) \geq (1-\gamma)L^{\Psi}(\psi_{MLE}(x) \,|\, x).$$

Therefore, $\Psi(\theta) \in C_{\Psi,\gamma}(x)$ and so $\Psi C_\gamma(x) \subset C_{\Psi,\gamma}(x)$. Now suppose $\psi \in C_{\Psi,\gamma}(x)$, so $L^{\Psi}(\psi \,|\, x) \geq (1-\gamma)L^{\Psi}(\psi_{MLE}(x) \,|\, x)$. This implies

$$L(\theta_\psi(x) \,|\, x) \geq (1-\gamma)L(\theta_{MLE}(x) \,|\, x)$$

and so $\theta_\psi(x) \in C_\gamma(x)$. This in turn implies that $\psi = \Psi(\theta_\psi(x)) \in \Psi C_\gamma(x)$. ∎

This result says that quite often the profile likelihood region $C_{\Psi,\gamma}(x)$ is obtained by projecting the likelihood region $C_\gamma(x)$ via the mapping Ψ. When $L(\cdot \,|\, x)$ is continuous and $\Psi^{-1}\{\psi\}$ is compact for each ψ, then the condition of the lemma is immediately satisfied.

While the profile likelihood seems like an intuitively reasonable approach to the nuisance parameter problem, it has one fundamental flaw, namely, in general $L_\Psi(\cdot \,|\, x)$ is not a likelihood function and so $L^{\Psi}(\psi_1 \,|\, x)/L^{\Psi}(\psi_2 \,|\, x)$ is not a likelihood ratio. The consequence of this is that the argument for the form of the inferences for ψ is not based on the likelihood preference ordering and the law of the likelihood and so yet another principle is needed to justify this. Consider a simple example.

Example 3.2.2 *A Profile Likelihood Is Not a Likelihood.*

Suppose that the sampling model is given by Table 3.1 and our interest is in $\Psi = I_{H_0}$ where $H_0 = \{0,1\}$. Then $L^{\Psi}(0 \,|\, 1) = 1/5$, $L^{\Psi}(1 \,|\, 1) = 1/2$ and $L^{\Psi}(0 \,|\, 2) = 4/5, L^{\Psi}(1 \,|\, 2) = 2/3$. For this to define a likelihood there has to be a function $T : \{1,2\} \to \{1,2\}$ such that the likelihoods based on the model induced by T are proportional to $L^{\Psi}(\cdot \,|\, 1)$ and $L^{\Psi}(\cdot \,|\, 2)$. Since $L^{\Psi}(\cdot \,|\, 1)$ and $L^{\Psi}(\cdot \,|\, 2)$ are not proportional, T must be 1 to 1 to produce two distinct likelihood functions. Clearly there is no such

Table 3.1 *The model in Example 3.2.2.*

θ	$f_\theta(1)$	$f_\theta(2)$
0	$1/2$	$1/2$
1	$1/3$	$2/3$
2	$1/5$	$4/5$

T because the fact that it is 1 to 1 implies that its model is effectively given by Table 3.1. ∎

The following example shows that profile likelihood inferences are not always reasonable. This leads to doubts that it is appropriate to formulate axioms of inference based on the profile likelihood.

Example 3.2.3 *Profile Likelihood in Regression.*
Suppose we have k predictor variables $x_1, \ldots, x_k$ and for $i = 1, \ldots, n$, the observed responses are $y_i \mid x_1 = x_{i1}, \ldots, x_k = x_{ik} \sim N(\beta_1 x_{i1} + \cdots + \beta_k x_{ik}, \sigma^2 I)$ where $\beta \in R^k, \sigma^2 > 0$ are unknown and our interest is in $\psi = \Psi(\beta, \sigma^2) = \sigma^2$. When $X = (x_{ij}) \in R^{n \times k}$ is of rank k, then it is easy to show that $\psi_{MLE}(y) = (y - Xb)'(y - Xb)/n$ where $b = (X'X)^{-1}X'y$. Also, $(y - Xb)'(y - Xb)/\sigma^2 \sim$ chi-squared$(n - k)$ and so $E(\psi_{MLE}(y)) = (1 - k/n)\sigma^2$. Therefore, whenever k is large relative to n, the profile likelihood MLE will very likely be a serious underestimate of the true value of σ^2. Other estimates such as the standard estimate $(y - Xb)'(y - Xb)/(n - k)$ do not suffer from this problem. ∎

Sometimes other likelihoods, such as marginal and conditional likelihoods, are defined to resolve the problem with nuisance parameters and these do satisfy the property of being likelihoods. Unfortunately, these only work in limited contexts and so can't be considered as a general resolution of the problem. As such, these likelihoods are not discussed here.

3.2.3 Prediction Problems

For a prediction problem with a concealed response $y \in \mathcal{Y}$, the aim is to predict y having observed x. Suppose that y has model $\{g_\delta(\cdot \mid x) : \delta \in \Delta\}$ with $\delta = \Delta(\theta)$ and $\delta_{true} = \Delta(\theta_{true})$. After observing x, the joint likelihood can be formed for the unknowns θ and y as $L(\theta, y \mid x) = kg_{\Delta(\theta)}(y \mid x)f_\theta(x)$ for some $k > 0$. Given that our interest is in y, it is necessary to remove θ to form a *predictive likelihood* for y. As surveyed in Bjornstad (1990), various approaches have been suggested for a predictive likelihood, with perhaps the most obvious being the *profile predictive likelihood* given by

$$L^{\mathcal{Y}}(y \mid x) = \sup_\theta L(\theta, y \mid x). \tag{3.3}$$

The value $y_{MLE}(x)$ is then obtained as the predictor of y and profile likelihood regions for this quantity can be formed to assess its accuracy.

Note that, given $L(\theta, y \mid x)$, nothing prevents us from forming the profile likelihood for θ as $L^\Theta(\theta \mid x) = \sup_y L(\theta, y \mid x)$. In general, however, this will not be proportional to $L(\theta \mid x)$ and different inferences will be obtained, illustrating the inconsistency of the profiling idea. A further negative for the profile predictive likelihood is that inferences are not generally invariant, in the sense of Lemma 3.2.2, under transformations of y. Again (3.3) is not generally a likelihood and there seems to be no solution to defining a fully satisfactory predictive likelihood.

3.2.4   Summarizing the Pure Likelihood Approach

Various arguments are advanced to support pure likelihood inferences. Certainly, these inferences seem intuitively reasonable and, as opposed to many other commonly used inference methods, they satisfy reparameterization invariance. Birnbaum's Theorem is often interpreted as saying that anyone who accepts the frequentist principles of sufficiency and conditionality must accept the likelihood principle, but this conclusion is not correct; see Section 3.3.3. The lack of a general definition of a measure of statistical evidence, and the attendant problems that this produces in the nuisance parameter problem, has to be viewed as a major deficiency in pure likelihood theory. Also, it is not clear how one checks a model using the pure likelihood approach and so the principle of empirical criticism is violated. One could embed the model in a larger model and use the likelihood based on the bigger model, but this just leaves us with the problem of checking the larger model. Overall, pure likelihood theory does not lead to a fully satisfactory theory of inference because of these issues.

Adding a prior to the ingredients, however, allows us to fill in these gaps. In Royall (1997) Bayesian methods are rejected because of their inherent subjectivity. But, as discussed in Section 1.5, virtually all of statistical practice is subjective in nature, with the possible exception of the data. The subjectivity in pure likelihood theory arises through the choice of the model.

Other aspects of the pure likelihood approach are discussed in Chapter 4. A review of the use of the pure likelihood approach to evidence in statistical genetics can be found in Strug, Hodge, Chiang, Pal, Corey and Rohde (2010). Further discussion of the likelihood function and its relevance in inference can be found in Berger and Wolpert (1988), Pawitan (2013) and Reid (2013).

3.3   Sufficiency, Ancillarity and Completeness

Consider the set $\mathscr{I}$ of all *inference bases* $I = (M,x)$ where $M = (\mathscr{X}_M, \{f_{M,\theta} : \theta \in \Theta_M\})$, $\mathscr{X}_M$ is a sample space, $\{f_{M,\theta} : \theta \in \Theta_M\}$ is a collection of probability density functions on $\mathscr{X}_M$, with respect to some support measure μ_M on $\mathscr{X}_M$, indexed by $\theta \in \Theta_M$, and $x \in \mathscr{X}_M$ is the observed data. Birnbaum (1962) considered the meaning of statistical evidence in this context.

Birnbaum did not attempt to define statistical evidence, asserting only that two inference bases I_1 and I_2 are deemed to be equivalent if they contain the same amount of evidence about the unknown model parameter. Statistical principles are considered as being equivalence relations on $\mathscr{I}$ that express the fact that two inference bases contain the same amount of evidence. For example, the likelihood principle says that two inference bases with proportional likelihoods are equivalent with respect to the amount of evidence they express about the unknown distribution. In Birnbaum (1962) a theorem is proved that purportedly establishes that two commonly accepted frequentist principles, sufficiency and conditionality, are equivalent to the likelihood principle. This is controversial because frequentist inferences use aspects of I beyond the likelihood and adherence to the likelihood principle would disallow p-values, confidence regions, etc.

As established in Evans (2013), it would appear that Birnbaum's Theorem is not correctly stated. While this is a negative result, our discussion of this involves much that is relevant to the concept of statistical evidence and to other developments in this text. For the proofs of results we will restrict ourselves here to the case where $\mathscr{X}_M$ and Θ_M are finite to avoid what are essentially mathematical problems that are resolved by invoking regularity conditions for more general situations.

First recall that a *relation R* with domain D is a subset $R \subset D \times D$. Saying $(x,y) \in R$ means that the objects x and y have a property in common. A relation R is *reflexive* if $(x,x) \in R$ for all $x \in D$, *symmetric* if $(x,y) \in R$ implies $(y,x) \in R$, and *transitive* if $(x,y) \in R, (y,z) \in R$ implies that $(x,z) \in R$. If a relation R is reflexive, symmetric and transitive, then R is called an *equivalence relation*. When R is an equivalence relation and $(x,y) \in R$, then x and y are thought to possess the property to the same degree. We say that relation R on D *implies* relation R' on D whenever $R \subset R'$.

A *statistical relation* is a relation $R \subset \mathscr{I} \times \mathscr{I}$ and we will call such an R a *statistical principle* when R is an equivalence relation. For example, the *likelihood relation* L on $\mathscr{I}$ is defined by $(I_1, I_2) \in L$ whenever $\Theta_{M_1} = \Theta_{M_2}$ and there exists $k > 0$ such that $f_{M_1,\theta}(x_1) = k f_{M_2,\theta}(x_2)$ for every $\theta \in \Theta_{M_1}$. The following result is obvious.

Lemma 3.3.1 *L is a statistical principle.*

The likelihood principle as stated can be generalized in an obvious way. For suppose $I_1 = (M_1, x_1)$ and $I_2 = (M_2, x_2)$, and there exists a bijection $h : \Theta_{M_1} \to \Theta_{M_2}$ and a constant $k > 0$, such that $f_{M_1,\theta}(x_1) = k f_{M_2,h(\theta)}(x_2)$ for every $\theta \in \Theta_{M_1}$. It then seems reasonable to consider $(I_1, I_2) \in L$. This generalization is ignored here as it is not relevant to our arguments. Effectively we require that I_1 and I_2 have the same parameter space any time they are related via a statistical relation.

3.3.1   The Sufficiency Principle

A function T on $\mathscr{X}_M$ is a *sufficient statistic* for M whenever the conditional distribution of x given $T(x)$ is independent of θ. The idea is that the information in the data beyond the value of a sufficient statistic is therefore irrelevant and can be discarded. Sufficiency leads to a reduction in the data. A sufficient statistic T for a model M is said to be a *minimal sufficient statistic* for M if, whenever U is another sufficient statistic for M, there is a function h_U such that $T = h_U \circ U$. A minimal sufficient statistic makes the maximal reduction in the data via sufficiency.

Consider now the definition of the *sufficiency relation S*. First we show that a minimal sufficient statistic always exists for models with $\mathscr{X}_M$ and Θ_M finite. It is assumed that, for each $x \in \mathscr{X}_M$, there exists $\theta \in \Theta_M$ such that $f_{M,\theta}(x) > 0$, so points are not allowed in $\mathscr{X}_M$ that can't possibly be observed. Two points $x_1, x_2 \in \mathscr{X}_M$ are said to be equivalent whenever there exists constant $k > 0$ such that $f_{M,\theta}(x_1) = k f_{M,\theta}(x_2)$ for every $\theta \in \Theta$ and the equivalence class containing x is denoted by $[x]$. The following result provides a minimal sufficient statistic.

Lemma 3.3.2 $T(x) = [x]$ *is a minimal sufficient statistic for M.*

Proof. If $z \in [x]$, then there exists $k(z) > 0$ such that $f_{M,\theta}(z) = k(z)f_{M,\theta}(x)$ for every θ. Suppose that θ is true. If $f_{M,\theta}(x) > 0$, then $f_{M,\theta}(z) > 0$ for every $z \in [x]$ and the conditional probability of x given $T(x) = [x]$ based on $f_{M,\theta}$ equals

$$f_{M,\theta}(x) / \sum_{z \in [x]} f_{M,\theta}(z) = f_{M,\theta}(x) / \sum_{z \in [x]} k(z)f_{M,\theta}(x) = 1 / \sum_{z \in [x]} k(z). \qquad (3.4)$$

If $f_{M,\theta}(x) = 0$, then $f_{M,\theta}(z) = 0$ for every $z \in [x]$ and so the probability of $[x]$ based on $f_{M,\theta}$ is 0. Therefore, the conditional probability of x given $T(x) = [x]$ based on $f_{M,\theta}$ can be defined arbitrarily. There is a θ' such that $f_{M,\theta'}(x) > 0$ and the conditional probability of x given $T(x) = [x]$ based on $f_{M,\theta'}$ equals (3.4). So if the conditional probability of x given $T(x)$ based on $f_{M,\theta}$ is defined by (3.4), this conditional probability is independent of θ and T is sufficient. Now suppose that U is a sufficient statistic for M. Then

$$f_{M,\theta}(x) = f_M(x \,|\, U(x))f_{M,\theta,U}(U(x)) \qquad (3.5)$$

where $f_M(\cdot \,|\, U(x))$ is the conditional probability function given $U(x)$ and $f_{M,\theta,U}$ is the marginal for U. Since $f_{M,\theta}(x) > 0$ for at least one θ, it must be that $f_M(x \,|\, U(x)) > 0$. If $U(x_1) = U(x_2)$, then $f_{M,\theta,U}(U(x_1)) = f_{M,\theta,U}(U(x_2))$ and from (3.5)

$$f_{M,\theta}(x_1) = f_M(x_1 \,|\, U(x_1))f_{M,\theta,U}(U(x_2)) = \frac{f_M(x_1 \,|\, U(x_1))}{f_M(x_2 \,|\, U(x_2))} f_{M,\theta}(x_2),$$

which implies that $T(x_1) = T(x_2)$. Therefore, T is a minimal sufficient statistic. ∎

With additional conditions Lemma 3.3.2 can be extended to quite general models.

Any 1 to 1 function of a minimal sufficient statistic is also minimal sufficient. So the minimal sufficient statistic here can always be taken to be real valued since $[x]$ takes only finitely many values. Also, if T and T' are both minimal sufficient for model M, then $T = h \circ T'$ for some function h, since T' is sufficient and T is minimal sufficient. If $T'(x_1) \neq T'(x_2)$ but $h(T'(x_1)) = h(T'(x_2))$, then T' would not be minimal sufficient and so h must be 1 to 1.

Let T_i denote a minimal sufficient statistic for model M_i with marginal model M_{i,T_i}. Define the sufficiency relation by $(I_1, I_2) \in S$ whenever there is a 1 to 1 map h between the sample spaces of M_{1,T_1} and M_{2,T_2} such that $M_{1,T_1} = M_{2,h \circ T_2}$ and $T_1(x_1) = h(T_2(x_2))$. The following result proves that S is an equivalence relation.

Lemma 3.3.3 S *is a statistical principle and* $S \subset L$.

Proof. Consider inference base $I = (M, x)$ and suppose T, T' are minimal sufficient statistics for M. Then, as discussed, there is a 1 to 1 function h such that $T = h \circ T'$, which implies $M_T = M_{h \circ T'}$ and $T(x) = h(T'(x))$ so $(I, I) \in S$ and S is reflexive. If $(I_1, I_2) \in S$ via h, then $M_{1,T_1} = M_{2,h \circ T_2}$ and so T_1 and $h \circ T_2$ have the same distribution for each θ. This implies that $h^{-1} \circ T_1$ and T_2 have the same distribution for each θ and so $M_{1,h^{-1} \circ T_1} = M_{2,T_2}$. Also, $T_1(x_1) = h(T_2(x_2))$ implies $T_2(x_2) = h^{-1}(T_1(x_1))$ so $(I_2, I_1) \in S$, which proves S is symmetric. If $(I_1, I_2) \in S$ via h_1 and $(I_2, I_3) \in S$ via h_2, then $(I_1, I_3) \in S$ via $h = h_1 \circ h_2$, and so S is transitive. Now (3.5) implies that a likelihood function obtained from (M, x) is proportional to a likelihood function obtained from $(M_T, T(x))$ when T is a minimal sufficient statistic for M. When $(I_1, I_2) \in S$, then $M_{1,T_1} = M_{2,h \circ T_2}$ and $T_1(x_1) = h(T_2(x_2))$, which implies that a likelihood function obtained from $(M_{1,T_1}, T_1(x_1))$ is proportional to a likelihood function obtained from $(M_{2,T_2}, T_2(x_2))$. Therefore, a likelihood function obtained from I_1 is proportional to a likelihood function obtained from I_2 so $(I_1, I_2) \in L$. We conclude that $S \subset L$. ∎

3.3.2 The Conditionality Principle

A function A on $\mathscr{X}_M$ is an *ancillary statistic* for model M if the marginal model induced by A is given by one probability distribution, namely, the distribution of A is independent of $\theta \in \Theta_M$. So the value of $A(x)$ can tell us nothing about the true value of θ. For $x \in \mathscr{X}_M$ the *conditional model* given $A(x)$ is $\{f_{M,\theta}(\cdot \mid A(x)) : \theta \in \Theta_M\}$ where $f_{M,\theta}(\cdot \mid A(x))$ is the density for the data given $A(x)$. The *conditionality relation* C is defined by $(I_1, I_2) \in C$ whenever $\Theta_{M_1} = \Theta_{M_2}$, $x_1 = x_2$ and there exists ancillary statistic A for M_1 such that the conditional model given $A(x_1)$ is M_2, or with the roles of I_1 and I_2 reversed. The motivation for this relation comes from a simple example.

Example 3.3.1 *Two Measuring Instruments.*
Suppose that a measurement $x \in R^1$ is to be made using one of two instruments where $M_i = (R^1, \{f_{M_i, \theta} : \theta \in R^1\})$, with $f_{M_i, \theta}$ a $N(\theta, \sigma_i^2)$ density, is the appropriate model when using instrument i. Furthermore, suppose that the σ_i^2 are known with $\sigma_1^2 \gg \sigma_2^2$ and instrument i is used with known probability p_i with $p_1 + p_2 = 1$. Before the instrument is selected, the appropriate model for the response (i, x) is $M = (\{1, 2\} \times R^1, \{p_i f_{M_i, \theta} : \theta \in R^1, i \in \{1, 2\}\})$. Therefore, $A(i, x) = i$ is ancillary for M. It seems obvious that once (i, x) is observed, so that it is known instrument i was used to obtain x, then any inferences about θ should be based on the conditional model M_i and data x, as the precisions of the instruments are so different. This is characteristic of applications of conditionality, as it is felt that a more appropriate assessment of the accuracy of the inference is made when using the conditional model. ∎

For additional examples where it is readily apparent that C is appropriate, see, for example, Cox and Hinkley (1974) and Fraser (1979).

Unfortunately, C has a significant defect, as it is not an equivalence relation.

Lemma 3.3.4 *C is reflexive and symmetric but is not transitive and $C \subset L$.*

Proof. The reflexivity, symmetry and $C \subset L$ are obvious. The lack of transitivity follows via a simple example. Consider the model M with $\mathscr{X}_M = \{1, 2\}^2$, $\Theta_M = \{1, 2\}$ and with $f_{M,\theta}$ given by Table 3.2. Now note that $U(x_1, x_2) = x_1$ and $V(x_1, x_2) = x_2$ are both ancillary, and the conditional models, when $(x_1, x_2) = (1, 1)$ is observed, are given by Tables 3.3 and 3.4. The only ancillary for both these conditional models

Table 3.2 *Unconditional distributions.*

(x_1, x_2)	$(1,1)$	$(1,2)$	$(2,1)$	$(2,2)$
$f_{M,1}(x_1, x_2)$	$1/6$	$1/6$	$2/6$	$2/6$
$f_{M,2}(x_1, x_2)$	$1/12$	$3/12$	$5/12$	$3/12$

Table 3.3 *Conditional distributions given U = 1.*

(x_1, x_2)	$(1,1)$	$(1,2)$	$(2,1)$	$(2,2)$
$f_{M,1}(x_1, x_2 \mid U = 1)$	$1/2$	$1/2$	0	0
$f_{M,2}(x_1, x_2 \mid U = 1)$	$1/4$	$3/4$	0	0

Table 3.4 *Conditional distributions given V = 1.*

(x_1, x_2)	$(1,1)$	$(1,2)$	$(2,1)$	$(2,2)$
$f_{M,1}(x_1, x_2 \mid V = 1)$	$1/3$	0	$2/3$	0
$f_{M,2}(x_1, x_2 \mid V = 1)$	$1/6$	0	$5/6$	0

is the trivial ancillary (the constant map). Therefore, there are no applications of C that lead to the inference base I_2, given by Table 3.3 with data $(1,1)$, being related to the inference base I_3, given by Table 3.4 with data $(1,1)$. But both of I_2 and I_3 are related under C to the inference base I_1 given by Table 3.2 with data $(1,1)$. ∎

Note that even under relabelings, the inference bases I_2 and I_3 in Lemma 3.3.4 are not equivalent.

If R is a relation on D, then the equivalence relation $\bar{R}$ generated by R is the smallest equivalence relation containing R. So $\bar{R}$ is the intersection of all equivalence relations on D containing R. For an arbitrary reflexive relation R there is the following characterization of $\bar{R}$.

Lemma 3.3.5 *If R is a reflexive relation on D, then*

$$\bar{R} = \left\{ \begin{array}{c} (x,y) : \exists n, x_1, \ldots, x_n \in D \text{ with } x = x_1, y = x_n \text{ and for each } i = 1, \ldots, n \\ \text{either } (x_i, x_{i+1}) \in R \text{ or } (x_{i+1}, x_i) \in R \end{array} \right\}.$$

Proof. Since $R \subset \bar{R}$, then $\bar{R}$ is reflexive. If $(x,y) \in \bar{R}$, then there exists $n, x_1, \ldots, x_n \in D$ with $x = x_1, y = x_n$ and $(x_i, x_{i+1}) \in R$ or $(x_{i+1}, x_i) \in R$ and so $(y,x) \in \bar{R}$ using $x_i' = x_{n-i+1}$ instead of the x_i. If $(x,y), (y,z) \in \bar{R}$, then we have $(x,z) \in \bar{R}$ simply by concatenating the chains that put $(x,y) \in \bar{R}$ and $(y,z) \in \bar{R}$. ∎

Note that for statistical relations R, as characterizations of statistical evidence, it makes sense to assume that R is reflexive since inference base I must contain the same amount of information as itself.

It may be, however, that $\bar{R}$ does not have a meaningful interpretation, at least as it relates to the property being expressed by R. In fact, it is entirely possible that $\bar{R} = D \times D$ and so says nothing. As an example, suppose that $D = \{2,3,4,\ldots\}$ and $(x,y) \in R$ when x and y have a common factor bigger than 1. Then R is reflexive and symmetric but not transitive. If $x, y \in D$, then $(x, xy) \in R, (xy, y) \in R$ so $\bar{R} = D \times D$ and $\bar{R}$ is saying nothing. It seems that each situation, where a relation R is completed by extending it to an equivalence relation, must be examined to see whether or not this extension has any meaningful content for the application.

Accordingly, if $(I_1, I_2) \in C$ is to mean that I_1 and I_2 contain an equivalent amount of information under C, then it is necessary to expand C to $\bar{C}$ so that it is an equivalence relation. But this implies that the two inference bases I_2 and I_3 presented in the proof of Lemma 3.3.4 contain an equivalent amount of information and yet they are not directly related via C. Rather, they are related only because they are conditional models obtained from a supermodel that has two essentially different maximal ancillaries. An ancillary A is a *maximal ancillary* if, whenever $A = g \circ A'$ and A' is ancillary, then g is a bijection.

Saying that the inference bases I_2 and I_3 in Lemma 3.3.4 contain an equivalent amount of statistical information is clearly a substantial generalization of C. Note that, when $(1,1)$ is observed, the MLE is $\hat{\theta}(1,1) = 1$. To measure the accuracy of this estimate, consider the conditional probabilities based on the two inference bases, namely,

$$P_1(\hat{\theta}(x_1,x_2) = 1 \,|\, U = 1) = 1/2, P_2(\hat{\theta}(x_1,x_2) = 2 \,|\, U = 1) = 3/4,$$
$$P_1(\hat{\theta}(x_1,x_2) = 1 \,|\, V = 1) = 1/3, P_2(\hat{\theta}(x_1,x_2) = 2 \,|\, V = 1) = 5/6.$$

Therefore, the accuracy of $\hat{\theta}$ is quite different depending on whether we use I_2 or I_3. It seems unlikely that one would want to interpret these inference bases as containing an equivalent amount of evidence in a frequentist formulation of statistics without further justification. Of course, there is no reason to necessarily accept the equivalences given by a generated equivalence relation, unless the equivalence relation is seen to express the essence of the basic relation.

Lemma 3.3.5 is used to prove the following result found in Evans, Fraser and Monette (1986).

Theorem 3.3.1 $C \subset \bar{C} = L$ *where the first containment is proper.*

Proof. Clearly $C \subset \bar{C}$ and this containment is proper by Lemma 3.3.4 since $\bar{C}$ is an equivalence relation and C is not. It remains to show that $\bar{C} = L$.

If $(I_1, I_2) \in \bar{C}$, then Lemma 3.3.5 implies $(I_1, I_2) \in L$ and so $\bar{C} \subset L$ as L is an equivalence relation with $C \subset L$.

Now suppose that $(I_1, I_2) \in L$. Then $f_{M_1,\theta}(x_1) = k f_{M_2,\theta}(x_2)$ for every θ for some $k > 0$. Assume first that $k > 1$. Now construct a new inference base $I_1^* = (M_1^*, (1, x_1))$ where $\mathscr{X}_{M_1^*} = \{0,1\} \times \mathscr{X}_{M_1}$, and $\{f_{M_1^*,\theta} : \theta \in \Theta_{M_1}\}$ is given by Table 3.5 where $x_{10}, x_{100}, \ldots$ are the elements of $\mathscr{X}_{M_1}$ not equal to x_1 and $p \in [0,1)$ satisfies $p/(1 - p) = 1/k$. Then $U(i,x) = i$ is ancillary as is V given by $V(i,x) = 1$ when $x = x_1$ and $V(i,x) = 0$ otherwise. Conditioning on $U(i,x) = 1$ implies that $(I_1^*, I_1) \in C$ while conditioning on $V(i,x) = 1$ implies that $(I_1^*, I) \in C$ where $I = ((\{0,1\}, \{p_\theta : \theta \in \Theta_{M_1}\}), 1)$ and p_θ is the Bernoulli$(f_{M_1,\theta}(x_1)/c)$ probability function. Now, using I_2, construct I_2^* by replacing p by $1/2$ and $f_{M_1,\theta}(x_1)$ by $f_{M_2,\theta}(x_2)$ in Table 3.5 and obtain that $(I_2^*, I) \in C$ since $f_{M_1,\theta}(x_1)/c = f_{M_2,\theta}(x_2)$. Lemma 3.3.5 gives that $(I_1, I_2) \in \bar{C}$. If $k \le 1$, then start the construction process with I_2 instead. This proves that $\bar{C} = L$. ∎

Table 3.5 *The model E_1^*.*

	x_1	x_{10}	x_{100}	$\cdots$
$i = 1$	$pfM(x_1)$	$pf_{M_1,\theta}(x_{10})$	$pf_{M_1,\theta}(x_{100})$	$\cdots$
$i = 0$	$1 - p - pf_{M_1,\theta}(x_1)$	$pf_{M_1,\theta}(x_1)$	0	$\cdots$

So Theorem 3.3.1 shows that the proof that $C = L$ provided in Evans, Fraser and Monette (1986) is really a proof that $\bar{C} = L$. Therefore, accepting the relation C is not really equivalent to accepting L unless it is agreed that the additional elements of $\bar{C}$ make sense. This is essentially equivalent to saying that it doesn't matter which maximal ancillary is conditioned on and it is unlikely that this is acceptable to most frequentist statisticians, as it implies acceptance of L.

There is therefore a defect in C, namely, it is not an equivalence relation due to the general nonexistence of unique maximal ancillaries. There is one positive outcome from Theorem 3.3.1, however, as it is clear that any resolution of the problems associated with C as an expression of statistical evidence must take the form of an equivalence relation. Theorem 3.3.1 shows that there is no equivalence relation between C and L that might do the trick unless L is accepted and all frequentist properties of inferences are deemed irrelevant.

As such, it is necessary to look inside C for a resolution. The natural candidate for this is discussed in Basu (1959). An ancillary statistic A_{lam} for the model is a *laminal ancillary* if A_{lam} is a function of every maximal ancillary for the model and any other ancillary with this property is a function of A_{lam}. So, if A is a maximal ancillary, there is a function h such that $A_{lam} = h \circ A$ and, if A'_{lam} is ancillary and also a function of every maximal ancillary, then there is a function h' such that $A'_{lam} = h' \circ A_{lam}$. The naturalness of A_{lam} arises from the fact that the equivalences that arise due to the laminal ancillary correspond to what is in a sense the best resolution among the conflicts presented when there are multiple nonequivalent maximal ancillaries.

Fraser (2004) and Ghosh, Reid and Fraser (2010) provide thorough reviews of the use of ancillarity in frequentist statistics. In Chapter 5 we will see applications of ancillary statistics in model checking and checking for prior-data conflict.

3.3.3   Birnbaum's Theorem

It is now shown that the proof in Birnbaum (1962) actually establishes the following result.

Theorem 3.3.2 *(Birnbaum's Theorem)* $S \cup C \subset L \subset \overline{S \cup C}$.

Proof. The first containment is obvious. For the second, suppose that $(I_1, I_2) \in L$. Construct a new inference base $I = (M, y)$ from I_1 and I_2 as follows. Let M be given by $\mathscr{X}_M = (\{1\} \times \mathscr{X}_{M_1}) \cup (\{2\} \times \mathscr{X}_{M_2})$,

$$f_{M,\theta}(1,x) = \begin{cases} (1/2)f_{M_1,\theta}(x) & \text{when } x \in \mathscr{X}_{M_1} \\ 0 & \text{otherwise,} \end{cases}$$

$$f_{M,\theta}(2,x) = \begin{cases} (1/2)f_{M_2,\theta}(x) & \text{when } x \in \mathscr{X}_{M_2} \\ 0 & \text{otherwise.} \end{cases}$$

Then

$$T(i,x) = \begin{cases} (i,x) & \text{when } x \notin \{x_1, x_2\} \\ \{x_1, x_2\} & \text{otherwise} \end{cases}$$

is sufficient for M and so $((M,(1,x_1)),(M,(2,x_2))) \in S$ by Lemma 3.3.3. Also, $h(i,x) = i$ is ancillary for M and thus

$$((M,(1,x_1)),(M_1,x_1)) \in C, ((M,(2,x_2)),(M_2,x_2)) \in C.$$

By Lemma 3.3.5 $((M_1,x_1),(M_2,x_2)) \in \overline{S \cup C}$ and this completes the proof. ∎

Note that Birnbaum's proof only establishes the containments with no equalities but the following result does this.

Theorem 3.3.3 $S \cup C$ *is properly contained in L while* $L = \overline{S \cup C}$.

Proof. First it is shown that $S \cup C \subset L$ is proper. Suppose that M_1 has $\mathscr{X}_{M_1} = \{0,1\}, \Theta_{M_1} = \{1/5, 1/3\}$ with $f_{M_1,\theta}(x) = \theta^x (1-\theta)^{1-x}$ and M_2 has $\mathscr{X}_{M_1} = \{0,1,2\}, \Theta_{M_2} = \{1/5, 1/3\}$ with $f_{M_2,\theta}(0) = \theta, f_{M_2,\theta}(1) = \theta(1-\theta)$ and $f_{M_2,\theta}(2) = (1-\theta)^2$. Suppose further that $x_1 = 1$ and $x_2 = 0$ are observed so $f_{M_1,\theta}(1) = \theta = f_{M_2,\theta}(0)$. Note that the full data is minimal sufficient for both M_1 and M_2 and that both of these models have only trivial ancillaries. Therefore, if $I_1 = (M_1,1)$ and $I_2 = (M_2,0)$, then $(I_1,I_2) \notin S, (I_1,I_2) \notin C$ but $(I_1,I_2) \in L$, which proves that $S \cup C$ is properly contained in L. To prove that the second containment is exact, Lemma 3.3.5 implies that $(I_1,I_2) \in \overline{S \cup C}$, which implies that I_1 and I_2 give rise to proportional likelihoods, as this is true for each element of $S \cup C$, and so $\overline{S \cup C} \subset L$. ∎

So it is not true, as usually stated for Birnbaum's Theorem, that S and C are together equivalent to L but rather $\overline{S \cup C}$ is equal to L. Acceptance of $\overline{S \cup C}$ is not entailed, however, by acceptance of both S and C, as it is necessary to examine the additional relationships added to $S \cup C$ to see if these make sense. If one wishes to say that acceptance of S and C implies the acceptance of $\overline{S \cup C}$, then a compelling argument is required for these additions.

Birnbaum's Theorem and Theorem 3.3.1 imply the following result.

Corollary 3.3.1 $S \cup C \subset \bar{C} = L$ *where the first containment is proper. Furthermore,* $S \subset \bar{C}$ *and this containment is proper.*

It is interesting to note that including S in the modified statement of Birnbaum's Theorem is not required to reason to L. This is a reassuring result, as S seems to be a correct characterization of statistical evidence in the sense of Birnbaum but C is not, at least as C is currently stated.

So, although it is commonly claimed that accepting S and C is equivalent to accepting L, this is not really the case. As such, Birnbaum's Theorem does not provide support for the likelihood principle.

Again, there is a positive outcome from this discussion. For Theorem 3.3.2 shows, as discussed in Evans, Fraser and Monette (1986), that S and C contradict one another in the proof in the following sense. The evidence that is discarded as irrelevant by S is exactly the information that is used by C to form the conditional model. As such, as recommended by Durbin (1970), it seems necessary to restrict attention to ancillaries that are functions of a minimal sufficient statistic, as this avoids such contradictions. This restriction is applied in Chapter 5.

There is a considerable literature concerned with Birnbaum's Theorem and the reader is referred to Kalbfleisch (1975), Holm (1985), Berger and Wolpert (1988),

Barndorff-Nielsen (1995), Helland (1995), Robins and Wasserman (2000) and Mayo (2014).

3.3.4 Completeness

The concept of completeness establishes a connection between sufficiency and ancillarity. A statistic T is said to be *complete* for model M if, whenever a function h satisfies $E_{\theta,T}(h(T)) = 0$ for all $\theta \in \Theta$, then $P_{\theta,T}(\{t : h(t) = 0\}) = 1$ for every $\theta \in \Theta$. Completeness has an interesting relationship with sufficiency and ancillarity.

Theorem 3.3.4 *(Basu's Theorem) If T is a complete, minimal sufficient statistic, then T is statistically independent of every ancillary statistic A.*

Proof. Suppose that A is ancillary and g is such that $E(g(A))$ exists and is finite. Then

$$0 = E(g(A) - E(g(A))) = E_{\theta}(g(A) - E(g(A))) = E_{\theta,T}(E(g(A) - E(g(A)))|T))$$

and so $h(T) = E(g(A) - E(g(A))|T)$ satisfies $E_{\theta,T}(h(T)) = 0$ for all $\theta \in \Theta$. By the completeness of T we have that $E(g(A)|T) = E(g(A))$ with $P_{\theta,T}$ probability 1. Since this holds for every bounded g, this implies that A and T are independent. ∎

Basu's Theorem says that when there is a complete, minimal statistic, all ancillaries can be ignored for inference about θ, because conditioning on such an ancillary doesn't change anything. Note that when a statistical model is an exponential model, then under mild conditions a minimal sufficient statistic is complete; see Lehmann and Romano (2005).

3.4 p-Values and Confidence

This section is concerned with frequentist concepts associated with characterizing statistical evidence. For example, the *p*-value is probably the most commonly used statistical methodology and is routinely interpreted as a measure of evidence.

3.4.1 p-Values and Tests of Significance

Following Cox and Hinkley (1974), a *p*-value is defined as follows. Suppose there is a model $\{f_\theta : \theta \in \Theta\}$ for observed data x and a hypothesis $H_0 \subset \Theta$. Let $T : \mathscr{X} \to R^1$ be a *test statistic* such that larger values of $T(x)$ indicate a bigger discrepancy from H_0 and where the marginal distribution $P_{H_0,T}$ of T is fixed for all $\theta \in H_0$. A *p-value* based on T is then defined by

$$P_{H_0,T}(T \geq T(x)) = P_{H_0,T}([T(x),\infty)) \tag{3.6}$$

where $T(x)$ is considered as fixed. When (3.6) is small, then there is supposedly evidence against H_0. Furthermore, it is asserted that the smaller (3.6) is, then the more evidence there is against H_0. Sometimes the computation of a *p*-value is referred to as a *test of significance*.

A simple, well-known example illustrates the *p*-value concept.

Example 3.4.1 *The Z-Test.*

Suppose $x = (x_1, \ldots, x_n)$ is a sample from an $N(\mu, 1)$ distribution where $\mu \in R^1$ is unknown and interest is in assessing the hypothesis $\mu \in H_0 = \{0\}$. Then $Z(x) = \sqrt{n}\bar{x} \sim N(0, 1)$ under H_0. If $T(x) = |Z(x)|$, then clearly the bigger $T(x)$ is the more discrepant are the data with respect to H_0. Accordingly, a relevant *p*-value is $P_{H_0,T}(T \geq T(x)) = 2(1 - \Phi(\sqrt{n}|\bar{x}|))$. ∎

Undoubtedly this is a very artificial example, as it is rare we would know the variance and not know the mean, but it turns out to be rich enough to produce questions about *p*-values for which there do not appear to be good answers.

For example, given a model and a hypothesis H_0, how are we to determine T? The following example illustrates the necessity of obtaining the right T.

Example 3.4.2 *The Z-Test Modified.*

Suppose $x = (x_1, \ldots, x_n)$ is as in Example 3.4.1 but now what is reported to us is only whether or not $\sqrt{n}\bar{x}$ is greater than or equal to 1.64485, which is the 0.95 quantile of the standard normal. So we are told the value of $T'(x) = I_{\{x:\sqrt{n}\bar{x} \geq 1.64485\}}(\sqrt{n}\bar{x})$ and that is all. Clearly "large" values of T' correspond to evidence against $H_0 = \{0\}$ and (3.6) equals 0.05 when $T(x) = 1$. If, as is commonly done, a significant result is noted when the recorded *p*-value is less than or equal to 5%, then whenever $T'(x) = 1$ there is evidence against H_0. But notice that, if $1.64485 \leq \sqrt{n}|\bar{x}| < 1.96$, then the inference based on T will not record a significant result while that based on T' will. So when using T', the *p*-value methodology allows us to conclude evidence against H_0, while that based on the value of $\sqrt{n}|\bar{x}|$ does not. ∎

The concept of a *p*-value is incomplete without a theory that tells us how to choose T and presumably this should be based on the meaning of evidence. The likelihood ratio statistic $T(x) = \sup_{\theta \in H_0^c} L(\theta \,|\, x) / \sup_{\theta \in H_0} L(\theta \,|\, x)$ is often recommended, but we know from the discussion of profile likelihood that this is generally not a true likelihood ratio, and so can't be justified by the law of the likelihood.

It is often asserted that people misinterpret large values of (3.6) as evidence in favor of H_0. For consider Example 3.4.1 and put $U = 2(1 - \Phi(\sqrt{n}|\bar{x}|))$. A simple calculation reveals that U is uniformly distributed when H_0 is true. So when H_0 is true, no particular value is expected for the *p*-value and $U = 0.96$ cannot be interpreted as more evidence in favor of H_0 than $U = 0.16$. The uniform distribution of the *p*-value under H_0 is commonly cited as a necessary characteristic of a *p*-value. It is undoubtedly quite natural, however, to expect that any reasonable measure of evidence must be able to provide evidence in favor of H_0, as well as evidence against H_0. For when drawing inferences about θ it is assumed that the model is correct so that one of the values in Θ *is* true. For example, if $\Theta = \{\theta_1, \theta_2\}$, then evidence for θ_1 must be evidence against θ_2 and conversely. These same considerations apply to any statistical model irrespective of the cardinality of Θ. This failure to provide evidence in favor of a hypothesis is a substantial weakness of the *p*-value concept. It is undoubtedly connected to the fact that the *p*-value is not derived based on a more fundamental concept of evidence.

Consider the interpretation of (3.6) as a probability. If an independent replication of the data collection were carried out, then (3.6) is the conditional probability that the independent value of T is at least as large as the observed value $T(x)$. So (3.6) is a conditional probability for this joint process given having observed the first outcome. There does not appear to be a good justification for considering this enlarged model, which seems artificial.

From a practical point of view it is also reasonable to ask if it is really the case that (3.6) being small is evidence against H_0. In Example 3.4.1 suppose $T(x) = 5$ so that (3.6) equals 6.4×10^{-5}, surely a small p-value. But suppose that $n = 10^4$. This implies that $|\bar{x}| = 5/10^2 = 0.05$. Now consider our discussion in Section 1.4 concerning the use of continuity. It is natural to ask the question: is a deviation of 0.05 from the null meaningful? Of course, this is application dependent but, if our measurements are in centimeters, then a deviation of only 0.05 centimeters would actually appear to be evidence in favor of H_0 and the p-value is misleading. This tells us that a p-value is not the whole story when assessing a hypothesis. It is also necessary to ask if we have indeed detected a real difference from the null.

For us p-values do not serve as a suitable characterization of statistical evidence. Concerns about p-values have been raised in, for example, Berger and Sellke (1987), Berger and Delampady (1987), Ioannidis (2005), Royall (1997), Schervish (1996), Thompson (2007) and Ziliak and McCloskey (2009). Still there is some relevance for the concept. In Chapter 4 a posterior probability like a p-value has a use in hypothesis assessment, not as a measure of the evidence but as a measure of the strength of the evidence and this leads to quite different conclusions. In Chapter 5 something very like a p-value plays a role in model checking and checking for prior-data conflict. Checking these ingredients are both situations where there are no alternatives. The role of the p-value in these contexts is simply to measure whether or not an observed value is in a tail or some other region of relatively low probability and, as such, it is a measure of *surprise*.

3.4.2 *Neyman–Pearson Tests*

The *Neyman–Pearson theory of hypothesis testing* is an attempt to resolve the issue of choosing T. For this, consider a decision procedure $\delta(x) = I_R(x)$ where $R \subset \mathscr{X}$ is a rejection region. The interpretation is that when $x \in R$, then H_0 is rejected and otherwise accept H_0. Note that, as opposed to significance testing, in this approach evidence can be found for H_0, by accepting, as well as evidence against H_0, by rejecting. The presentation is, however, categorical as always H_0 is either accepted or rejected and this seems contrary to a meaningful presentation of evidence which should be more graded.

There are a variety of difficulties associated with the Neyman–Pearson approach but perhaps the greatest is that determining δ is dependent on the choice of a loss function L where $L(\theta, \delta(x))$ is the loss incurred if decision $\delta(x)$ is made when θ is true. There are a number of possible choices for L and inevitably this is a subjective choice which, as discussed in Section 1.6, can't be checked for its suitability against the data. As such, this violates the principle of empirical criticism. Even with some of the standard choices of L, the theory fails in that it does not always provide answers to reasonable problems; see Lehmann and Romano (2005). Furthermore, optimal testing procedures are often seen to require the use of arbitrary randomization devices that have no connection with the problem but are necessary for optimality on average in repeated performances. For example, this arises in testing the value of a proportion in a binomial model. In fact, nobody uses randomized tests even though they are optimal. Royall (1997) contains considerable additional discussion of the negative attributes of these procedures.

Given that Neyman–Pearson theory is not directly concerned with characterizing evidence, but rather making decisions, it is not considered further here. An elementary development can be found in Evans and Rosenthal (2010). It turns out, however, that the key result in that approach, namely, the *Neyman–Pearson theorem*, plays an important role in establishing optimality results for the inferences derived in Chapter 4. So this famous result also finds its place in an evidence-based approach to inference. Interestingly, however, in that context it does not suffer from the limitation encountered in frequency theory that sometimes it works and sometimes it does not.

3.4.3   Rejection Trials and Confidence Regions

A compromise between significance testing and Neyman–Pearson testing is referred to by Royall (1997) as *rejection trials*. For this, choose a cut-off α such that the *p*-value less than or equal to α leads to saying that there is evidence against H_0 and otherwise neither accept H_0 nor say that there is evidence in favor of H_0. The smaller the cut-off α is, supposedly the stronger the evidence is that is being demanded for the rejection of H_0. The value α is called the *size of the test*. So one chooses α and then only claims evidence against H_0 when $P_{H_0,T}(T \geq T(x)) \leq \alpha$. While $\alpha = 0.05$ is common, more rigorous assessments are sometimes required, as when asserting the existence of the Higgs particle the value $\alpha = 3 \times 10^{-7}$ was used; see Draper (2014).

Suppose there is a parameter of interest $\psi = \Psi(\theta)$, and for each possible value $\psi \in \Psi$, there is a statistic T_ψ such that P_{θ,T_ψ} is the same for every $\theta \in \Psi^{-1}\{\psi\}$. So putting $T_{\psi,\alpha} = \inf\{t : P_{\theta,T_\psi}([t,\infty)) < \alpha\}$ and rejecting whenever $T_\psi(x) > T_{\psi,\alpha}$ gives a size α test for $H_0 = \Psi^{-1}\{\psi\}$. Now, for given observed data x, let

$$C(x) = \{\psi : T_\psi(x) \leq T_{\psi,\alpha}\} \qquad (3.7)$$

and note that $C(x)$ is a random subset of Ψ consisting of those ψ values that will not be rejected by the corresponding size α test. The following result establishes a confidence property for C.

Lemma 3.4.1 *For C defined as in (3.7)*

$$P_\theta(\Psi(\theta) \in C(x)) \geq 1 - \alpha \qquad (3.8)$$

for all $\theta \in \Theta$.

Proof. Now supposing $x \sim P_\theta$, then

$$P_\theta(\Psi(\theta) \in C(x)) = P_\theta(\{x : T_{\Psi(\theta)}(x) \leq T_{\Psi(\theta),\alpha}\})$$
$$= 1 - P_\theta(\{x : T_{\Psi(\theta)}(x) > T_{\Psi(\theta),\alpha}\}) \geq 1 - \alpha$$

and this holds for every $\theta \in \Theta$. ∎

The probability $P_\theta(\Psi(\theta) \in C(x))$ is called the *coverage probability* of the random set $C(x) \subset \Psi$ when θ is true. If the coverage probability of C satisfies (3.8) for all $\theta \in \Theta$, then C is a $(1 - \alpha)$-*confidence region* for $\Psi(\theta)$. So Lemma 3.4.1 shows that a $(1 - \alpha)$-confidence region arises naturally from a set of size α tests for the parameter of interest.

Conversely, a size α test can be obtained from a $(1 - \alpha)$-confidence region.

Lemma 3.4.2 *If $C(x)$ is a $(1 - \alpha)$-confidence region for $\psi = \Psi(\theta)$, then rejecting $H_0 = \Psi^{-1}\{\psi\}$ whenever $\psi \notin C(x)$ is a size α test for H_0.*

Proof. We have that $P_\theta(\Psi(\theta) \notin C(x)) = 1 - P_\theta(\Psi(\theta) \in C(x)) \leq 1 - (1 - \alpha) = \alpha$ for every $\theta \in \Theta$. ∎

An odd feature of confidence theory is its connection with hypothesis testing rather than estimation. For surely the purpose of quoting a confidence region $C(x)$ for $\psi = \Psi(\theta)$ is to assess the accuracy of an estimate contained in it by looking at the size of $C(x)$. Yet, after determining $C(x)$, there is still the problem of choosing an estimate. Contrast this with pure likelihood theory where the approach determines an estimate, a region like $C(x)$ that contains the estimate and which also connects naturally with hypothesis assessment. It seems more natural for a theory of statistical inference to treat estimation and hypothesis assessment in a integrated way as opposed to being completely separate problems.

Once again there is the problem that it may not be possible to find a suitable confidence region for a parameter of interest $\psi = \Psi(\theta)$ and so the theory is incomplete. Also, as the following example shows, even when a confidence region exists, it may be absurd.

Example 3.4.3 *An Absurd Confidence Interval.*

Suppose $x \in R^1$ has density $f_\theta(x) = (1 - \theta)f(x) + \theta f(x - 1)$ where f is the $N(0, 1)$ density function and $\Theta = [0, 1]$. Then, as discussed in Plante (1991),

$$C(x) = \begin{cases} [0, 1] & -1.68148 \leq x \leq 2.68148 \\ \phi & \text{otherwise} \end{cases}$$

is a 0.95 confidence interval for θ. Notice that the confidence interval is always absurd in that it is either the full parameter space or the null set. In fact, this is also an *unbiased confidence region* in the sense that the probability of covering a false value is always bounded above by the probability of covering the true value, namely, $P_\theta(\theta' \in C(x)) \leq P_\theta(\theta \in C(x))$ for all $\theta, \theta' \in \Theta$. Furthermore, it is *uniformly most accurate* in that it minimizes the probability of covering a false value among all unbiased 0.95 confidence regions for θ. ∎

Example 3.4.3 is very simple and it displays the not uncommon phenomenon of confidence regions giving absurd answers. Further examples of this behavior can be found in Fraser (1979), including an early example due to Welch (1939). The papers Plante (1991, 1994) provide discussion of attempts to build a theory of inference based on the concept of confidence.

Perhaps the most telling arguments against rejection trials are associated with the somewhat perverse logic underlying the sequential collection of data and the assessment of evidence. Cornfield (1966) discusses this point and argues that the assessment of evidence should be independent of the *stopping rule* used when collecting data sequentially. The following example illustrates the motivation for this.

Example 3.4.4 *P-values and Collecting Additional Data.*
Consider the situation described in Example 3.4.1 where the investigator takes a sample of m and computes the p-value to be 0.06. This seems pretty close to the magical 0.05 level in common use in many disciplines and so he decides to collect a further n values in hopes of rejecting H_0 at the 0.05 level. But now the appropriate probability of rejection under H_0 is 0.05, which is the probability of rejecting after the first m observations, plus the probability of rejection after the additional sample of n, given that H_0 was not rejected after m observations. This probability is always greater than 0.05. So the investigator can never reject H_0 at the 0.05 level no matter how many additional observations are taken. It seems perfectly reasonable for a scientist to look at the evidence as it is being acquired and draw inferences based on this. The rejection region criterion seems to disallow a very natural approach to the acquisition of evidence. ■

The pure likelihood approach leads to inferences that are independent of the stopping rule as do Bayesian approaches to inference. Stopping rules do play a role, however, in the methods discussed in Chapter 5 for model checking and checking for prior-data conflict. So, when adopting a Bayesian approach to inference, the method of data collection does have an influence on a statistical analysis but not on inferences about the true value of θ.

3.4.4 Summarizing the Frequentist Approach

A theory of inference based on frequentist ideas suffers from two main defects. First, there does not seem to be a good answer to the question of why it is necessary to consider the frequency properties of statistical procedures. Without this justification, basing statistical practice on the principle of frequentism seems like a weak foundation. Even with a good argument for evaluating statistical procedures based on their frequentist properties, there is the problem of which aspects of the data should be conditioned on for the evaluation. Many examples exist that point to the necessity of conditioning in the frequentist approach. The unresolved problem posed by multiple maximal ancillaries, however, leaves us with additional doubts concerning the soundness of determining inferences this way. Second, it seems almost misleading to refer to *the* frequentist theory of inference because it does not exist in the sense that such a theory can be applied to statistical problems with a guaranteed sensible answer or, for that matter, even an answer.

3.5 Bayesian Inferences

Bayesian inference is often characterized as being subjective and therefore unscientific. As discussed in Section 1.5, however, this criticism is basically applicable to all of statistics. The choice of a model is subjective as is the choice of the prior. We do not argue as some do, however, that the choice of these ingredients is based on some inherently coherent subjective thought process and so is justified. Rather, the model and the prior are seen simply as choices, made with the best of intentions and judgment, but still possibly fundamentally in error because of their subjective nature. The developments in Section 4.6 and Chapter 5 constitute our attempt to deal with this apparent weakenss of statistical analyses.

One might argue, however, if there is a concern that our subjective choices may lead us to error, why compound the problem by adding the additional ingredient — the prior? Our answer is that, with this addition, a sound theory of statistical evidence can be constructed and, without the prior, this does not seem to be available. So the developments in Chapter 4 do include a prior, and so could be considered to be Bayesian in nature, but they are somewhat different from common Bayesian approaches, which are now described.

3.5.1 Basic Concepts

So now another ingredient is added to the model $\{f_\theta : \theta \in \Theta\}$ and data x for the statistical analysis, namely, the *prior distribution* on Θ. While the model expresses beliefs about the possible values of the true distribution in the form of a set, the prior expresses a priori beliefs about what the true distribution is given the model or, equivalently, what the true value of θ is within Θ. Suppose that the *prior probability measure* Π has *prior density* π with respect to volume measure ν on Θ. Therefore, the a priori (before observing the data) probability that the true value of θ is in $A \subset \Theta$ is given by

$$\Pi(A) = \int_A \pi(\theta)\,\nu(d\theta). \tag{3.9}$$

For $\Psi : \Theta \to \Psi$ the a priori probability that the true value of $\psi = \Psi(\theta)$ is in $A' \subset \Psi$ is given by $\Pi_\Psi(A') = \Pi(\Psi^{-1}A')$. For a smooth Ψ we will write

$$\Pi_\Psi(A') = \int_{A'} \pi_\Psi(\psi)\,\nu_\Psi(d\psi) \tag{3.10}$$

where

$$\pi_\Psi(\psi) = \int_{\Psi^{-1}\{\psi\}} \pi(\theta)J_\Psi(\theta)\,\nu_{\Psi^{-1}\{\psi\}}(d\theta)$$

is the *marginal prior density* of $\psi = \Psi(\theta)$ with respect to volume measure ν_Ψ on Ψ, $J_\Psi(\theta) = (\det((d\Psi(\theta))^t d\Psi(\theta)))^{-1/2}$, $d\Psi(\theta)$ is the differential of Ψ evaluated at θ, and $\nu_{\Psi^{-1}\{\psi\}}$ is volume measure on $\Psi^{-1}\{\psi\}$. Appendix A provides more details on this formula for π_Ψ.

Choosing a model $\{f_\theta : \theta \in \Theta\}$ for the data x and a prior π for θ is in effect prescribing a joint probability measure for (θ, x). For, if P_θ is the probability measure associated with f_θ, then for $B \subset \mathcal{X}$,

$$P_\theta(B) = \int_B f_\theta(x)\,\mu(dx)$$

and P_θ can be thought of as the conditional probability measure of x given θ. Therefore, by the multiplication rule, the joint probability measure of (θ, x) is given by

$\Pi \times P_\theta$ where

$$\Pi \times P_\theta(A \times B) = \int_A \left(\int_B f_\theta(x) \mu(dx) \right) \pi(\theta) \nu(d\theta)$$

$$= \int_{A \times B} \pi(\theta) f_\theta(x) \nu \times \mu(d\theta, dx). \tag{3.11}$$

Now (3.11) expresses the joint probability measure of (θ, x) as the product of the marginal for θ times the conditional for x given θ. This can be reversed to express the joint as the product of the marginal for x times the conditional for θ given x. The marginal probability measure M for x is called the *prior predictive* for x and is given by

$$M(B) = \Pi \times P_\theta(\Theta \times B) = \int_\Theta \pi(\theta) \left(\int_B f_\theta(x) \mu(dx) \right) \nu(d\theta)$$

$$= \int_B \left(\int_\Theta \pi(\theta) f_\theta(x) \nu(d\theta) \right) \mu(dx) = \int_B m(x) \mu(dx)$$

where

$$m(x) = \int_\Theta \pi(\theta) f_\theta(x) \nu(d\theta) \tag{3.12}$$

is the average of the conditional densities of x with respect to the prior. So (3.12) is the *prior predictive density* of x, with respect to volume measure μ on $\mathcal{X}$.

The conditional probability measure of θ given x, referred to as the *posterior probability measure* of θ, is given by

$$\Pi(A|x) = \frac{\int_A \pi(\theta) f_\theta(x) \nu(d\theta)}{m(x)} = \int_A \pi(\theta|x) \nu(d\theta) \tag{3.13}$$

where

$$\pi(\theta|x) = \frac{\pi(\theta) f_\theta(x)}{m(x)} \tag{3.14}$$

is the *posterior probability density* of θ with respect to volume measure ν. So, again by the multiplication rule, we can write the joint probability measure of (θ, x) as

$$\Pi \times P_\theta = \Pi(\cdot|x) \times M \tag{3.15}$$

or, in terms of the joint density with respect to volume measure $\nu \times \mu$, as $\pi(\theta) f_\theta(x) = \pi(\theta|x) m(x)$.

The principle of conditional probability, as discussed in Section 2.1.2, implies that the a posteriori (after observing the data) probability that the true value of θ is in $A \subset \Theta$, as given by (3.13), *must* replace (3.9) as the measure of our belief that the true value of θ is in A. Notice that this only refers to probability statements about θ and does not say that other ingredients, such as the prior, can't be used in determining the inferences made. There seems to be a common misconception that Bayesian inferences are only based on the posterior but, with the exception of probability statements as determined by the principle of conditional probability, there is nothing to support this point of view.

Before observing the data x, the prior predictive distribution M is available for probability statements about this unknown x just as the prior Π is available for probability statements about the unknown θ. After observing x, there is no need for prediction but there are still uses for M with respect to model checking and checking for prior-data conflict, as discussed in Chapter 5. So the factorization (3.15) corresponds to a logical split of the joint probability measure into a part relevant for inferences about θ and a part relevant for checking the ingredients put into the analysis.

For a smooth $\Psi : \Theta \to \Psi$ the *marginal posterior probability measure* of $A' \subset \Psi$ equals

$$\Pi_\Psi(A' \mid x) = \Pi(\Psi^{-1}A' \mid x) = \int_{A'} \pi_\Psi(\psi \mid x)\, \nu_\Psi(d\psi)$$

where the *marginal posterior density* of $\psi = \Psi(\theta)$, with respect to volume measure ν_Ψ, is given by

$$\pi_\Psi(\psi \mid x) = \int_{\Psi^{-1}\{\psi\}} \pi(\theta \mid x) J_\Psi(\theta)\, \nu_{\Psi^{-1}\{\psi\}}(d\theta). \tag{3.16}$$

The posterior $\Pi_\Psi(\cdot \mid x)$ expresses beliefs about what the true value of ψ is after observing the data x.

The *conditional prior* distribution of θ given $\psi = \Psi(\theta)$ is concentrated on $\Psi^{-1}\{\psi\}$. For $A \subset \Psi^{-1}\{\psi\}$ this is given by

$$\Pi(A \mid \psi) = \int_A \pi(\theta \mid \psi)\, \nu_{\Psi^{-1}\{\psi\}}(d\theta)$$

where

$$\pi(\theta \mid \psi) = \frac{\pi(\theta) J_\Psi(\theta)}{\pi_\Psi(\psi)}$$

is the *conditional prior density* of θ given that $\psi = \Psi(\theta)$, with respect to volume measure $\nu_{\Psi^{-1}\{\psi\}}$ on $\Psi^{-1}\{\psi\}$. The *conditional posterior* of θ, given that $\psi = \Psi(\theta)$, is

$$\begin{aligned}
\Pi(A \mid \psi, x) &= \frac{\int_A \pi(\theta \mid \psi) f_\theta(x)\, \nu_{\Psi^{-1}\{\psi\}}(d\theta)}{\int_{\Psi^{-1}\{\psi\}} \pi(\theta \mid \psi) f_\theta(x)\, \nu_{\Psi^{-1}\{\psi\}}(d\theta)} \\
&= \int_A \frac{\pi(\theta \mid \psi) f_\theta(x)}{m(x \mid \psi)}\, \nu_{\Psi^{-1}\{\psi\}}(d\theta) = \int_A \pi(\theta \mid \psi, x)\, \nu(d\theta).
\end{aligned}$$

So we see that the *conditional posterior density* of θ, given that $\psi = \Psi(\theta)$, with respect to volume measure $\nu_{\Psi^{-1}\{\psi\}}$, equals

$$\pi(\theta \mid \psi, x) = \frac{\pi(\theta \mid \psi) f_\theta(x)}{m(x \mid \psi)}$$

where

$$m(x \mid \psi) = \int_{\Psi^{-1}\{\psi\}} \pi(\theta \mid \psi) f_\theta(x)\, \nu_{\Psi^{-1}\{\psi\}}(d\theta)$$

is the *conditional prior predictive density* of x, given $\psi = \Psi(\theta)$, with respect to μ.

The Bayesian formulation has a useful consistency property.

Lemma 3.5.1 *If the model for the data x is given by $\{m(\cdot\mid\psi):\psi\in\Psi\}$ and the prior is π_Ψ, then the posterior density of $\psi=\Psi(\theta)$ obtained from these ingredients equals $\pi_\Psi(\cdot\mid x)$, as specified in (3.16).*

Proof. Given the model and prior, the posterior density of ψ equals

$$\frac{\pi_\Psi(\psi)m(x\mid\psi)}{\int_\Psi \pi_\Psi(\psi)m(x\mid\psi)\,\nu_\Psi(d\psi)}=\frac{\pi_\Psi(\psi)\int_{\Psi^{-1}\{\psi\}}\pi(\theta\mid\psi)f_\theta(x)\,\nu_{\Psi^{-1}\{\psi\}}(d\theta)}{m(x)}$$

$$=\int_{\Psi^{-1}\{\psi\}}\frac{\pi(\theta)J_\Psi(\theta)f_\theta(x)}{m(x)}\,\nu_{\Psi^{-1}\{\psi\}}(d\theta)=\int_{\Psi^{-1}\{\psi\}}\pi(\theta\mid x)J_\Psi(\theta)\,\nu_{\Psi^{-1}\{\psi\}}(d\theta),$$

which agrees with (3.16). ∎

So for inferences about ψ, the original model and parameter can be forgotten and we can instead work with the model $\{m(\cdot\mid\psi):\psi\in\Psi\}$ and the prior π_Ψ.

Note that if θ is considered as being "random" with distribution given ψ specified by $\pi(\cdot\mid\psi)$, then indeed $m(\cdot\mid\psi)$ is the sampling density of x. Actually, there seems to be no merit in distinguishing between the situations where ψ can be considered as random or not. Rather a joint probability model for (θ,x) is being specified that describes beliefs about these quantities.

Observe that, for any $k>0$, the function $L_\Psi(\cdot\mid x)=km(x\mid\cdot)$ is a likelihood, commonly referred to as an *integrated likelihood* since

$$L_\Psi(\psi\mid x)=km(x\mid\psi)=k\int_{\Psi^{-1}\{\psi\}}\pi(\theta\mid\psi)f_\theta(x)\,\nu_{\Psi^{-1}\{\psi\}}(d\theta)$$

$$=\int_{\Psi^{-1}\{\psi\}}\pi(\theta\mid\psi)L(\theta\mid x)\,\nu_{\Psi^{-1}\{\psi\}}(d\theta).$$

So an integrated likelihood arises via averaging the likelihood for θ with respect to the conditional distribution of the full parameter given the parameter of interest. As opposed to the profile likelihood, an integrated likelihood is indeed a likelihood.

The following result is a consequence of Lemma 3.5.1.

Lemma 3.5.2 *If $\Pi_\Psi(A')=0$, then $\Pi_\Psi(A'\mid x)=0$. So the posterior distribution of ψ is absolutely continuous with respect to its prior distribution and the density of $\Pi_\Psi(\cdot\mid x)$ with respect to Π_Ψ equals*

$$\pi_\Psi(\psi\mid x)/\pi_\Psi(\psi). \qquad (3.17)$$

Proof. From Lemma 3.5.1,

$$\Pi_\Psi(A'\mid x)=\frac{\int_{A'}\pi_\Psi(\psi)m(x\mid\psi)\,\nu_\Psi(d\psi)}{\int_\Psi \pi_\Psi(\psi)m(x\mid\psi)\,\nu_\Psi(d\psi)}.$$

Therefore, $\Pi_\Psi(A')=0$ implies that $\int_{A'}\pi_\Psi(\psi)m(x\mid\psi)\,\nu_\Psi(d\psi)=0$ and so $\Pi_\Psi(A'\mid x)=0$. Also,

$$\Pi_\Psi(A'\mid x)=\int_{A'}\frac{\pi_\Psi(\psi\mid x)}{\pi_\Psi(\psi)}\pi_\Psi(\psi)\nu_\Psi(d\psi)=\int_{A'}\frac{\pi_\Psi(\psi\mid x)}{\pi_\Psi(\psi)}\Pi_\Psi(d\psi)$$

and so (3.17) is the density of $\Pi_\Psi(\cdot\mid x)$ with respect to Π_Ψ. ∎

Lemma 3.5.2 has an important practical consequence. For if a prior is chosen that places 0 probability in some region, then a positive posterior probability will never be assigned to this region, namely, we are categorically ruling out the possibility that such values can be true. So one has to be careful to not exclude possible values of parameters in a model unless absolutely certain that these values are impossible. The density (3.17) plays a key role in our measure of evidence in Chapter 4.

Another important consistency property is known as *Bayesian updating.* For this, suppose that x is observed with $\pi(\cdot\,|x)$ as the posterior density for θ. Now suppose additional data y is observed with conditional model $\{g_\theta(\cdot\,|x) : \theta \in \Theta\}$ and with the same true value of θ. The following result says that it doesn't matter if we calculate a new posterior by treating $\pi(\cdot\,|x)$ as the prior and y as the data with model $\{g_\theta(\cdot\,|x) : \theta \in \Theta\}$, or calculate a new posterior by using the original prior π and (x,y) as the data with the model given by the joint density $f_\theta(x)g_\theta(y|x)$.

Lemma 3.5.3 *(Bayesian Updating) Suppose that the model $\{f_\theta : \theta \in \Theta\}$ for data x and the conditional model $\{g_\theta(\cdot\,|x) : \theta \in \Theta\}$ for data y have the same true value of θ. Suppose either (i) data x is observed and prior π is used to obtain posterior $\pi(\cdot\,|x)$, then y is observed and $\pi(\cdot\,|x)$ used as the prior to obtain the posterior for θ or (ii) the model with joint density $f_\theta(x)g_\theta(y|x)$ for (x,y) together with the prior π is used to obtain the posterior for θ. The same posterior for θ is obtained from (i) and (ii).*

Proof. Under (i) the final posterior is given by

$$\frac{\pi(\theta\,|x)g_\theta(y|x)}{\int_\Theta \pi(\theta\,|x)g_\theta(y|x)\,v(d\theta)} = \frac{\pi(\theta)f_\theta(x)g_\theta(y|x)/m(x)}{\int_\Theta (\pi(\theta)f_\theta(x)g_\theta(y|x)/m(x))\,v(d\theta)}$$
$$= \frac{\pi(\theta)f_\theta(x)g_\theta(y)}{m(x,y)}$$

and the final expression is the posterior density of θ under (ii). ∎

Consider now probability statements for a response value $y \in \mathscr{Y}$ where, after observing x, the value of y is somehow concealed from us. For example, it may be that y will be observed in the future or is *missing data.* Suppose that y has the conditional model $\{g_\delta(\cdot\,|x) : \delta \in \Delta\}$ with $\delta = \Delta(\theta)$ where $g_\delta(\cdot\,|x)$ is a density with respect to volume measure $\mu_{\mathscr{Y}}$ on $\mathscr{Y}$. The connection between the models arises via assuming that $\delta_{true} = \Delta(\theta_{true})$. The joint density of (θ,y,x) is given by $\pi(\theta)f_\theta(x)g_{\Delta(\theta)}(y|x)$. After observing x, the joint posterior of (θ,y) is given by

$$\frac{\pi(\theta)f_\theta(x)g_{\Delta(\theta)}(y|x)}{\int_\Theta \int_{\mathscr{Y}} \pi(\theta)f_\theta(x)g_{\Delta(\theta)}(y|x)\,\mu_{\mathscr{Y}}(dy)\,v(d\theta)} = \frac{\pi(\theta)f_\theta(x)g_{\Delta(\theta)}(y|x)}{m_{\mathscr{X}}(x)}$$
$$= \pi(\theta\,|x)g_{\Delta(\theta)}(y|x)$$

and so the posterior density of y equals

$$m_{\mathscr{Y}}(y|x) = \int_\Theta \pi(\theta\,|x)g_{\Delta(\theta)}(y|x)\,v(d\theta)$$

and is called the *posterior predictive density* of y given x. Note that the prior predictive density of y, with respect to $\mu_{\mathscr{Y}}$, equals

$$m_{\mathscr{Y}}(y) = \int_{\Theta} \pi(\theta) \int_{\mathscr{X}} f_{\theta}(x) g_{\Delta(\theta)}(y\,|\,x) \,\mu_{\mathscr{X}}(dx) \,\nu(d\theta)$$

$$= \int_{\mathscr{X}} m_{\mathscr{X}}(x) \int_{\Theta} \pi(\theta\,|\,x) g_{\Delta(\theta)}(y\,|\,x) \,\nu(d\theta) \,\mu_{\mathscr{X}}(dx)$$

$$= \int_{\mathscr{X}} m_{\mathscr{Y}}(y\,|\,x) m_{\mathscr{X}}(x) \,\mu_{\mathscr{X}}(dx)$$

and this is used to make probability statements about y before observing x.

One of the virtues of the Bayesian approach is that prediction problems are treated exactly the same as inference problems for parameters. There is no need to develop different methods.

3.5.2 Likelihood, Sufficiency and Conditionality

It is worth noting that the posterior distribution only depends on the data through the value of a minimal sufficient statistic.

Lemma 3.5.4 *Suppose T is a minimal sufficient statistic for the model $\{f_{\theta} : \theta \in \Theta\}$. Then the posterior density of θ satisfies $\pi(\theta\,|\,x) = \pi(\theta\,|\,T(x))$.*

Proof. By the factorization theorem $f_{\theta}(x) = g(x) h_{\theta}(T(x))$ for some nonnegative functions g and h. Then from (3.14) and (3.12),

$$\pi(\theta\,|\,x) = \frac{\pi(\theta) g(x) h_{\theta}(T(x))}{\int_{\Theta} \pi(\theta) g(x) h_{\theta}(T(x)) \,\nu(d\theta)} = \frac{\pi(\theta) h_{\theta}(T(x))}{\int_{\Theta} \pi(\theta) h_{\theta}(T(x)) \,\nu(d\theta)} = \pi(\theta\,|\,T(x)).$$

■

So there is an automatic reduction of the data to the minimal sufficient statistic for the probability assignments in a Bayesian analysis after the data is observed.

If two model-data combinations have proportional likelihoods, and the same prior is used with both models, then the posterior distributions are the same. This is often interpreted as implying the likelihood principle but in reality this only says that the probability assignments will be the same, not necessarily the inferences. Provided the priors are the same, the approach to inference presented in Chapter 4 does lead to identical inferences. If the priors are not the same, however, inferences may well be different for marginal parameters and so the likelihood principle is violated. It is also commonly claimed that the likelihood principle prevents activities like model checking or checking the prior, as discussed in Chapter 5. The likelihood principle, however, is a statement about θ and does not proscribe such activities. Overall, the point of view taken here does not recognize the likelihood principle as a necessary or even valid principle of inference.

Also, it is clear that, because the posterior is based on conditioning on the full data, ancillary statistics play no role in determining the relevant probability assignments. For example, it is simple to prove that, if the original model is replaced by the conditional model given the value of an ancillary, then the posterior formed based on this model and the prior is the same posterior based on the original model and the prior. So, at least for inference about θ, the lack of a unique maximal ancillary is irrelevant for any Bayesian formulation.

3.5.3   MAP-Based Inferences

One commonly used approach to dealing with the inference problems discussed in Section 1.8 is so-called *MAP-based inferences*. Here *MAP* stands for *maximum a posteriori*.

The *maximum a posteriori preference ordering* says that $\psi_1 \preccurlyeq \psi_2$, namely, ψ_2 is preferred at least as much as ψ_1, whenever $\pi_\Psi(\psi_1 \mid x) \leq \pi_\Psi(\psi_2 \mid x)$. So, at least when there is a discrete prior, ψ_2 is preferred at least as much as ψ_1, whenever the posterior probability of ψ_2 is at least as large as the posterior probability of ψ_1. Based on the discussion in Section 1.4, when densities are defined appropriately as limits, this interpretation holds for the continuous case too.

It is an immediate consequence of the MAP preference ordering that the estimate of ψ is given by the posterior mode

$$\psi_{MAP}(x) = \arg\sup \pi_\Psi(\psi \mid x), \tag{3.18}$$

as it is the most preferred value. It is not necessary that $\psi_{MAP}(x)$ be unique, but under reasonable conditions it is and this is assumed throughout the remainder of this section.

The accuracy of $\psi_{MAP}(x)$ is assessed by quoting a γ-*highest posterior density region*) (*hpd region*

$$C_{\Psi,\gamma}(x) = \{\psi : G_\Psi(\pi_\Psi(\psi \mid x) \mid x) \geq 1 - \gamma\}, \tag{3.19}$$

where $G_\Psi(\cdot \mid x)$ is the posterior cumulative distribution function (cdf) of $\pi_\Psi(\cdot \mid x)$ and the "size" of this region is taken as the measure of accuracy. The measure of size used is that relevant to the application and could be Euclidean volume but doesn't have to be.

To assess the hypothesis $H_0 = \{\psi_0\}$ it would seem to be implicit in this approach to use the posterior probability $\pi_\Psi(\psi_0 \mid x)$. Of course this will not work in continuous contexts without discretization. In general, however, the value $\pi_\Psi(\psi_0 \mid x)$ may be small simply because $\pi_\Psi(\psi_0)$ is small. Therefore, in such contexts it seems more natural to use the *Bayesian p-value*

$$G_\Psi(\pi_\Psi(\psi_0 \mid x) \mid x) = \Pi_\Psi(\pi_\Psi(\psi \mid x) \leq \pi_\Psi(\psi_0 \mid x) \mid x), \tag{3.20}$$

namely, the posterior probability that the true value of ψ has posterior probability no greater than the posterior probability of the hypothesized value. If (3.20) is small, then ψ_0 is in a region of low posterior probability for ψ and so it would seem that there is evidence against H_0.

To discuss the properties of (3.19) and (3.20), first consider a general problem where there is a probability measure P on a set Ω and a function $h : \Omega \to [0, \infty)$. Let G denote the cdf of h, namely, $G(k) = P(h(x) \leq k)$ with quantile function

$G^{-1}(\gamma) = \inf\{k : G(k) \geq \gamma\}$ and note that, because of the right continuity of G, then $G(G^{-1}(\gamma)) \geq \gamma$. Now for $\gamma \in [0, 1]$ define

$$C_\gamma = \{\omega : G(h(\omega)) \geq 1 - \gamma\} = \{\omega : h(\omega) \geq G^{-1}(1 - \gamma)\}$$

and $C_\gamma^{strict} = \{\omega : G(h(\omega)) > 1 - \gamma\}$. The following result covers the construction of several types of credible region relevant to developments in this text.

Proposition 3.5.1 *The following properties hold for the sets C_γ:*
(i) $C_{\gamma_1} \subset C_{\gamma_2}$ when $\gamma_1 \leq \gamma_2, C_0 = \{\omega : h(\omega) = \sup h\}$ and $C_1 = \Omega$,
(ii) $P(C_\gamma) \geq \gamma$,
(iii) $G(h(\omega)) = 1 - \inf\{\gamma : \omega \in C_\gamma\}$,
(iv) $C_{\gamma+\delta} \downarrow C_\gamma$ as $\delta \downarrow 0$,
(v) $C_\gamma^{strict} \subset C_\gamma, C_{\gamma-\delta} \uparrow C_\gamma^{strict}$ as $\delta \downarrow 0$ and $C_\gamma^{strict} = C_\gamma$ whenever $P(\{\omega : G(h(\omega)) = 1 - \gamma\}) = 0$.

Proof. (i) These properties are obvious. (ii) We have $P(C_\gamma) = P(G(h(\omega)) \geq 1 - \gamma) = P(h(\omega) \geq G^{-1}(1 - \gamma)) = 1 - P(h(\omega) < G^{-1}(1 - \gamma))$. If $P(h(\omega) < G^{-1}(1 - \gamma)) > 1 - \gamma$, then there exists $\varepsilon > 0$ such that $P(h(\omega) \leq G^{-1}(1 - \gamma) - \varepsilon) > 1 - \gamma$ since $(-\infty, G^{-1}(1 - \gamma) - \varepsilon] \uparrow (-\infty, G^{-1}(1 - \gamma))$ as $\varepsilon \downarrow 0$ and so $G(G^{-1}(1 - \gamma) - \varepsilon) \uparrow P(h(\omega) < G^{-1}(1 - \gamma))$. But this contradicts the definition of $G^{-1}(1 - \gamma)$ and so $P(h(\omega) < G^{-1}(1 - \gamma)) \leq 1 - \gamma$, which implies $P(C_\gamma) \geq \gamma$. (iii) We have that $1 - \inf\{\gamma : \omega \in C_\gamma\} = 1 - \inf\{\gamma : G(h(\omega)) \geq 1 - \gamma\} = G(h(\omega))$. (iv) Suppose that $\omega \in \lim_{\delta\downarrow 0} C_{\gamma+\delta}$. Then $G(h(\omega)) \geq 1 - \gamma - \delta$ for all $\delta > 0$ and so $G(h(\omega)) \geq 1 - \gamma$, which implies $\omega \in C_\gamma$ and $C_\gamma \subset C_{\gamma+\delta}$ for all $\delta > 0$ implies that $C_\gamma \subset \lim_{\delta\downarrow 0} C_{\gamma+\delta}$. (v) If $\omega \in \lim_{\delta\downarrow 0} C_{\gamma-\delta}$, then $G(h(\omega)) \geq 1 - \gamma + \delta$ for all δ small enough, which implies $\omega \in C_\gamma^{strict}$. If $\omega \in C_\gamma^{strict}$, then $G(h(\omega)) > 1 - \gamma$ implies that $G(h(\omega)) \geq 1 - \gamma + \delta$ for all δ small enough and so $\omega \in \lim_{\delta\downarrow 0} C_{\gamma-\delta}$. ∎

Taking $\omega = \psi$ and $h = \pi_\Psi(\cdot \mid x)$, then

$$C_{\Psi,\gamma}(x) = \{\psi : \pi_\Psi(\psi \mid x) \geq G_\Psi^{-1}(1 - \gamma \mid x)\},$$

and applying Proposition 3.5.1 gives properties of hpd regions and the Bayesian p-value.

Properties of hpd Regions and the Bayesian p-value:
(i) says that hdp regions are nested and always contain $\psi_{MAP}(x)$,
(ii) says that a γ-hpd region is indeed a γ-*credible region* as it contains at least γ of the posterior probability for ψ,
(iii) together with (3.19) shows an equivalence between the definition of the Bayesian p-value and hpd credible regions, namely,

$$\Pi_\Psi(\pi_\Psi(\psi \mid x) \leq \pi_\Psi(\psi_0 \mid x) \mid x) = 1 - \inf\{\gamma : \psi_0 \in C_{\Psi,\gamma}(x)\},$$

(iv) says that hpd regions are always continuous from above and
(v) says that hpd regions are also continuous from below at any γ such that $G_\Psi(\cdot \mid x)$ assumes the value $1 - \gamma$ with probability 0.

So combining (iv) and (v) implies that γ-hpd regions are continuous in γ whenever the posterior distribution of $\pi_\Psi(\cdot\,|\,x)$ is continuous.

A negative characteristic of MAP-based inferences is that, with a continuous parameter space, the inferences are not invariant under reparameterizations. For if $\lambda = \Lambda(\psi)$ is 1 to 1 and smooth, then the density of λ is given by

$$\pi_\Lambda(\lambda\,|\,x) = \pi_\Psi(\Lambda^{-1}(\lambda)\,|\,x)J_\Lambda(\Lambda^{-1}(\lambda)) \tag{3.21}$$

with respect to volume measure on the range space Λ. Therefore, whenever $J_\Lambda(\Lambda^{-1}(\lambda))$ is not a constant function of λ, the possibility exists that $\lambda_{MAP}(x) \neq \Lambda(\psi_{MAP}(x))$. In general, if $C_{\Psi,\gamma}(x), C_{\Lambda,\gamma}(x)$ denote the γ-hpd regions for ψ and λ, respectively, it is not the case that $C_{\Lambda,\gamma}(x) = \Lambda C_{\Psi,\gamma}(x)$. Furthermore, by an appropriate choice of Λ, it is often possible to make $\Pi_\Lambda(\pi_\Lambda(\lambda\,|\,x) \leq \pi_\Lambda(\lambda_0\,|\,x)\,|\,x)$ any particular value in $(0,1)$ by choice of Λ.

The lack of invariance under reparameterizations in the continuous parameter case can be viewed as a weakness for MAP-based inferences. When ψ is a discrete-valued parameter, however, invariance immediately holds. If the argument of Section 1.4 is invoked, namely, viewing the continuous context as an approximation to something that is essentially finite, then the lack of invariance can be avoided by an appropriate discretization. This requires a choice of a parameterization and then a choice of a discretization based on what are believed to be differences of practical importance between parameter values. Beliefs as expressed by a prior, however, are in a practical sense highly dependent on the parameterization chosen, as this is determined by our understanding of what are realistic values for the parameter in question, and similarly, what are meaningful differences. A parameter shouldn't be treated simply as a mathematical variable in a statistical problem with a somewhat arbitrary choice of prior, else the subsequent analysis has little meaning! The approach to inference discussed in Chapter 4 does lead to invariant inferences in the continuous context and this is very convenient. It is still necessary, however, to take into account the ultimate discrete nature of applications if inferences are to be sensible. So we don't take the lack of invariance for MAP-based inferences as being a reason to necessarily rule them out. Druilhet and Marin (2007) contains proposals for modifying the MAP-based approach so that invariance is achieved based on connections with the reference prior approach. Optimality results for the credible regions produced then follow from results in Evans, Guttman and Swartz (2006).

Perhaps the biggest criticism that can be levelled at MAP-based inferences, however, lies with the implicit definition of the statistical evidence that ψ is the true value as given by the posterior probability $\pi_\Psi(\psi\,|\,x)$. This seems almost certainly wrong, as it confounds measuring belief with measuring evidence. Note that, if the prior probability $\pi_\Psi(\psi)$ is big, then $\pi_\Psi(\psi\,|\,x)$ will also tend to be big even when ψ is not the true value, unless there is a sufficiently large amount of data. A similar comment applies when $\pi_\Psi(\psi)$ is small and ψ is true. The inference methods of Chapter 4 do not suffer from this defect, as they are based on how probabilities change rather than their magnitudes. Furthermore, property (iii) of Proposition 3.5.1 has a much more natural interpretation in that context.

Another criticism lies with the relatively few positive results that have been obtained in support of MAP-based inferences. For example, it is straightforward to see that among all γ-credible regions for ψ, a γ-hpd region has the smallest ν_Ψ volume. But this is one of the few optimality properties that seems obtainable and this contrasts unfavorably with the inferences of Chapter 4.

3.5.4   Quantile-Based Inferences

First define the credible regions for this approach. Suppose that ψ is real valued and, for $\alpha \in [0,1]$, let $\psi_\alpha(x) = \inf \{\psi_0 : \Pi_\Psi((-\infty, \psi_0] \,|\, x) \geq \alpha\}$ be the α-th quantile of the posterior of ψ. Then $\Pi_\Psi((-\infty, \psi_\alpha(x)] \,|\, x) \geq \alpha$ while $\Pi_\Psi((-\infty, \psi_0] \,|\, x) < \alpha$ for any $\psi_0 < \psi_\alpha(x)$.

A γ-credible interval for ψ is then given by $C_{\Psi,\gamma}(x) = [\psi_{\alpha(1-\gamma)}(x), \psi_{\gamma+\alpha(1-\gamma)}(x)]$ since

$$\Pi_\Psi([\psi_{\alpha(1-\gamma)}(x), \psi_{\gamma+\alpha(1-\gamma)}(x)] \,|\, x) \geq \gamma + \alpha(1-\gamma) - \alpha(1-\gamma) = \gamma.$$

So this interval is formed by discarding at most $\alpha(1-\gamma)$ of the posterior probability in the left tail and at most $(1-\alpha)(1-\gamma)$ of the probability in the right tail. It is typical to take $\alpha = 1/2$, so that an equal amount is discarded on each tail, but there is no reason to do this unless the posterior density is symmetric about its center. Note that

$$\psi_{\alpha(1-\gamma)}(x) \leq \psi_\alpha(x) \leq \psi_{\alpha+\gamma(1-\alpha)}(x) = \psi_{\gamma+\alpha(1-\gamma)}(x)$$

so $\psi_\alpha(x)$ is always in the interval. Also, letting $\gamma \to 0$, this credible interval implicitly defines the estimate $\psi_\alpha(x)$, although the limit could be an interval. When $\alpha = 1/2$, we get a posterior median.

Assessing the hypothesis $H_0 = \{\psi_0\}$, when ψ has a continuous posterior distribution, leads to the p-value

$$1 - \inf\{\gamma : \psi_0 \in C_{\Psi,\gamma}(x)\} = 1 - \inf\{\gamma : \psi_0 \in [\psi_{\alpha(1-\gamma)}(x), \psi_{\gamma+\alpha(1-\gamma)}(x)]\}$$
$$= \begin{cases} (1-\alpha)^{-1}(1 - \Pi_\Psi((-\infty, \psi_0] \,|\, x)) & \text{if} \quad \psi_0 \geq \psi_\alpha(x), \\ \alpha^{-1}\Pi_\Psi((-\infty, \psi_0] \,|\, x) & \text{if} \quad \psi_0 \leq \psi_\alpha(x). \end{cases}$$

When $\alpha = 1/2$, the p-value equals two times the right-tail probability, when ψ_0 is to the right of the posterior median, and two times the left-tail probability, when ψ_0 is to the left of the posterior median. When ψ has a discrete posterior distribution, the p-value is even more involved.

While these inferences would seem to have some intuitive support, and they are invariant under reparameterizations in the continuous case, there are numerous concerns that can be raised. The unnaturalness of the p-value is certainly an issue, but it is also worth noting that it is not at all clear how to define these inferences for parameters ψ of dimension greater than one. Even in one dimension it is difficult to derive interesting properties that would lend support for this approach. Most significantly, however, these inferences are not based on a measure of statistical evidence. In fact, it is not clear that there is even an implicit definition of evidence. As such, quantile-based inferences are not viewed as being appropriate.

3.5.5   Loss-Based Inferences

Consider adding a loss function $L : \Theta \times \Psi \to [0, \infty)$ to the ingredients of the problem of making an inference about ψ. The Bayesian formulation of the decision problem then leads immediately to an estimate of ψ. This is given by a *Bayes rule*

$$\psi_{Bayes}(x) = \arg\inf_{\psi} r(\psi \,|\, x) = \arg\inf_{\psi} \int_{\Theta} L(\theta, \psi)\pi(\theta \,|\, x)\, v(d\theta)$$

where $r(\cdot \,|\, x)$ denotes the *posterior risk*, namely, $r(\psi_0 \,|\, x)$ is the average posterior loss incurred when estimating the true value of ψ by ψ_0. It is clear that ψ_{Bayes} minimizes the *prior risk* $r(\delta) = \int_{\Theta} \int_{\mathscr{X}} L(\theta, \delta(x)) f_{\theta}(x)\pi(\theta)\, v(d\theta)$ among all estimators $\delta : \mathscr{X} \to \Psi$.

Inferences can be derived from this as follows. A γ-*lowest posterior loss region* for ψ is defined by

$$C_{\Psi, \gamma}(x) = \{\psi_0 : r(\psi_0 \,|\, x) \le c_{\gamma}(x)\}$$

where

$$c_{\gamma}(x) = \inf\{k : \Pi_{\Psi}(r(\psi \,|\, x) \le k \,|\, x) \ge \gamma\}$$

so $c_{\gamma}(x)$ is the γ-quantile of the posterior distribution of $r(\cdot \,|\, x)$. By the right continuity of $\Pi_{\Psi}(r(\psi_0 \,|\, x) \le k \,|\, x)$ as a function of k, then $\Pi_{\Psi}(C_{\Psi, \gamma}(x) \,|\, x) \ge \gamma$ and so a γ-lowest posterior loss region is a γ-credible region. Clearly, $C_{\Psi, \gamma_1}(x) \,|\, x) \subset C_{\Psi, \gamma_2}(x) \,|\, x)$ whenever $\gamma_1 \le \gamma_2$. Furthermore, $r(\psi_{Bayes}(x) \,|\, x)$ is the minimum value for $r(\cdot \,|\, x)$ and thus $\psi_{Bayes}(x) \in C_{\Psi, \gamma}(x)$ for every γ. Presuming that the Bayes rule is unique, then $C_{\Psi, 0}(x) = \{\psi_{Bayes}(x)\}$. For assessing the hypothesis $H_0 = \{\psi_0\}$, there is the p-value $1 - \inf\{\gamma : r(\psi_0 \,|\, x) \le c_{\gamma}(x)\} = \Pi_{\Psi}(r(\psi \,|\, x) > r(\psi_0 \,|\, x) \,|\, x)$.

Consider a commonly used loss function.

Example 3.5.1 *Quadratic Loss.*

Suppose Ψ is a convex subset of a Euclidean space and $L(\theta, \psi_0) = \|\Psi(\theta) - \psi_0\|^2$. Then, assuming the expectation $E_{\Pi_{\Psi}(\cdot \,|\, x)}(\psi)$ is finite,

$$r(\psi_0 \,|\, x) = \int_{\Psi} \|\psi - \psi_0\|^2\, \Pi_{\Psi}(d\psi \,|\, x)$$

$$= \int_{\Psi} \left\|\psi - E_{\Pi_{\Psi}(\cdot \,|\, x)}(\psi)\right\|^2 \Pi_{\Psi}(d\psi \,|\, x) + \left\|E_{\Pi_{\Psi}(\cdot \,|\, x)}(\psi) - \psi_0\right\|^2.$$

Therefore, $C_{\Psi, \gamma}(x)$ is a sphere centered at the Bayes rule $E_{\Pi_{\Psi}(\cdot \,|\, x)}(\psi)$ containing at least γ of the posterior probability for ψ. ∎

As discussed in Section 1.6, the loss function is another ingredient to the problem and, by the principle of empirical criticism, a method is needed for checking this against the data to see if it is reasonable in the application. It does not seem clear how to do this for L. For example, why is quadratic loss used in so many statistical problems? Certainly, it has some intuitive support, but in general it seems to be used more for mathematical convenience than anything else. Again loss-based inferences are not based on a characterization of statistical evidence and so we do not view this approach as appropriate.

Bernardo (2005) contains much discussion about loss-based inferences. This paper also introduces the notion of a loss function induced intrinsically by a model that leads to inferences that are invariant under reparameterizations. This has connections with reference priors discussed in Section 5.3.2.

3.5.6 Bayes Factors

Suppose that $A \subset \Theta$ and $0 < \Pi(A) < 1$. The *Bayes factor* in favor of A is then defined by

$$BF(A \mid x) = \frac{Odds(A \mid x)}{Odds(A)} \qquad (3.22)$$

where

$$Odds(A) = \frac{\Pi(A)}{\Pi(A^c)} \quad \text{and} \quad Odds(A \mid x) = \frac{\Pi(A \mid x)}{\Pi(A^c \mid x)}$$

give the prior and posterior *odds in favor* of A, respectively. So $BF(A \mid x)$ is measuring the change in the odds in favor of the true value of θ being in A from a priori to a posteriori and as such is measuring change in belief when belief is expressed in terms of odds. A value $BF(A \mid x) < 1$ is interpreted as evidence that A is not true, a value $BF(A \mid x) > 1$ is interpreted as evidence that A is true and $BF(A \mid x) = 1$ is not evidence for or against A being true. Furthermore, it is assumed that the larger or smaller $BF(A \mid x)$ is, the stronger the evidence is. Note that $\Pi(A \mid x)$ can be obtained as

$$\Pi(A \mid x) = \frac{\Pi(A)BF(A \mid x)}{1 - \Pi(A) + \Pi(A)BF(A \mid x)}$$

and so, considering $\Pi(A)$ constant, large values of $BF(A \mid x)$ correspond to large values of $\Pi(A \mid x)$ and conversely. Also, it is possible that $\Pi(A \mid x) = 1$ even when $\Pi(A) < 1$ and then the Bayes factor is infinite. When there is a hypothesis $H_0 \subset \Theta$, then $BF(H_0 \mid x)$ is taken as the measure of the evidence that H_0 is true.

When $\Pi(A) > 0$, a slightly simpler measure of change of belief is given by the relative belief ratio $RB(A \mid x) = \Pi(A \mid x)/\Pi(A)$. A relative belief ratio is measuring the change in the probability of A from a priori to a posteriori and as such is measuring change in belief when belief is expressed in terms of probability. Relative belief ratios, and their relationship with Bayes factors, are discussed in Chapter 4, where it is seen that relative belief ratios present, in many ways, a more natural, simpler measure of change in belief.

It is common in the statistics literature to define a Bayes factor as a ratio of two densities. The following result shows that this is implied by (3.22). The definition (3.22) is preferred here, as it is then clear that the Bayes factor is a measure of evidence through change in belief, as discussed in Chapter 4.

Lemma 3.5.5 *Suppose that $A \subset \Theta$ satisfies $0 < \Pi(A) < 1$ and T is a minimal sufficient statistic for the model $\{f_\theta : \theta \in \Theta\}$. Then*

$$BF(A \mid x) = \frac{m(x \mid A)}{m(x \mid A^c)} = \frac{m_T(T(x) \mid A)}{m_T(T(x) \mid A^c)}$$

where $m(\cdot|A)$ is the prior predictive of x given $\theta \in A$ and $m_T(\cdot|A)$ is the prior predictive of T given $\theta \in A$.

Proof. We can write

$$\Pi(A\,|\,x) = \frac{\int_A f_\theta(x)\,\Pi(d\theta)}{\int_\Theta f_\theta(x)\,\Pi(d\theta)} = \frac{\Pi(A)\int_A f_\theta(x)\,\Pi(d\theta\,|\,A)}{\int_\Theta f_\theta(x)\,\Pi(d\theta)}$$

and so

$$Odds(A\,|\,x) = \frac{\Pi(A\,|\,x)}{\Pi(A^c\,|\,x)} = \frac{\Pi(A)}{\Pi(A^c)}\frac{\int_A f_\theta(x)\,\Pi(d\theta\,|\,A)}{\int_{A^c} f_\theta(x)\,\Pi(d\theta\,|\,A^c)},$$

which completes the proof of the first equality. The result concerning T follows as in Lemma 3.5.4. ∎

So the Bayes factor is also given by the ratio of the prior predictive densities obtained by conditioning on the event and its complement. As such, in the context of hypothesis assessment, $BF(H_0\,|\,x)$ can be considered as a generalized likelihood ratio.

Various concerns can be raised about the Bayes factor. Perhaps the most obvious is that the Bayes factor in favor of a hypothesis H_0 is not defined when $\Pi(H_0) = 0$ and this could occur simply because H_0 is a lower dimensional subset of Θ. Certainly situations like this arise in applications. A solution to this, as in Jeffreys (1961), is to require the specification of three ingredients for such a problem, namely, a prior probability $p > 0$ for H_0, and two conditional priors $\Pi(\cdot\,|\,H_0)$ and $\Pi(\cdot\,|\,H_0^c)$. Then the overall prior is taken to be the mixture $\Pi = p\Pi(\cdot\,|\,H_0) + (1-p)\Pi(\cdot\,|\,H_0^c)$. Therefore, H_0 has prior probability $\Pi(H_0) = p$ and $BF(H_0\,|\,x) = m(x\,|\,H_0)/m(x\,|\,H_0^c)$ as in Lemma 3.5.5. This requires a different specification of the prior depending on what H_0 is and so seems quite unnatural and contrary to the basic intuition of what a prior represents. An early reference to Bayes factors is Jeffreys (1935).

In many cases $H_0 = \Psi^{-1}\{\psi_0\}$ for some smooth Ψ and, in that case, when $\Pi(H_0) = 0$, it seems much more natural to define $BF(H_0\,|\,x)$ as a limit, recalling that continuous models serve as approximations to contexts that are finite. This approach to defining the Bayes factor is discussed in Chapter 4.

There are situations, however, where H_0 is not generated by the specification of such a Ψ and then it is not clear how to define a Bayes factor when $\Pi(H_0) = 0$. This is similar to the Borel paradox discussed in Example 2.1.2. So the problem is not fully specified and the resolution entails making a somewhat arbitrary choice of a Ψ generating H_0 via $H_0 = \Psi^{-1}\{\psi_0\}$.

A much more serious issue associated with Bayes factors is their calibration as a measure of evidence. This is illustrated by the following well-known example.

Example 3.5.2 *The Jeffreys–Lindley Paradox.*

Suppose $x = (x_1, \ldots, x_n)$ is a sample from an $N(\mu, 1)$ distribution, where $\mu \in R^1$ is unknown, so $T(x) = \bar{x} \sim N(\mu, 1/n)$ is a minimal sufficient statistic. For the prior Π, use the mixture $p\delta_0 + (1-p)N(0, \tau_0^2)$ where δ_0 is the probability measure degenerate at 0 for some $p > 0$. Suppose $\Psi(\mu) = \mu$, and the goal is to assess $H_0 = \{0\}$.

Now $m_T(\cdot\,|\,H_0)$ is the $N(0,1)$ density. Under H_0^c, then $\sqrt{n}\mu \sim N(0, n\tau_0^2)$ independent of $z = \sqrt{n}\bar{x} - \sqrt{n}\mu \sim N(0,1)$ and so $\sqrt{n}\bar{x} = \sqrt{n}\mu + z \sim N(0, 1 + n\tau_0^2)$, implying that

$$m_T(\sqrt{n}\bar{x} \,|\, H_0^c) = \frac{1}{\sqrt{2\pi}\sqrt{1+n\tau_0^2}} \exp\left\{ -\frac{n\bar{x}^2}{2(1+n\tau_0^2)} \right\}.$$

Therefore, by Lemma 3.5.5,

$$BF(H_0 \,|\, x) = \frac{m_T(T(x)\,|\,H_0)}{m_T(T(x)\,|\,H_0^c)} = \sqrt{1+n\tau_0^2}\exp\left\{ -\frac{n\bar{x}^2}{2}\frac{n\tau_0^2}{1+n\tau_0^2} \right\}. \tag{3.23}$$

From (3.23), for fixed $\sqrt{n}\bar{x}$, it is seen that $BF(H_0\,|\,x) \to \infty$ as $n\tau_0^2 \to \infty$. So, for example, as the prior becomes increasingly diffuse by taking τ_0^2 bigger and bigger, the evidence in favor of H_0 being true would appear to get stronger and stronger, irrespective of what the value of $\sqrt{n}\bar{x}$ is. This seems extremely anomalous as, when $\sqrt{n}\bar{x} = 5$, the standard frequentist p-value is

$$2(1 - \Phi(|\sqrt{n}\bar{x}|)) = 2(1 - 0.9999997) = 6.0 \times 10^{-7}$$

and surely this is overwhelming evidence against H_0. The paradox arises as we might expect Bayesian methodology and standard frequentist methodology to be similar when using an extremely diffuse prior and here the two approaches contradict one another.

Consider the behavior of the Bayes factor. As $\tau_0^2 \to \infty$, the bulk of the prior probability is moving further and further away from H_0. As such, whatever its value, $\sqrt{n}\bar{x}$ looks more and more reasonable as a value from the $N(0, 1)$ distribution, at least when considered as a possible value from an $N(\mu, 1)$ distribution for the μ's where the bulk of the prior probability lies. So actually the Bayes factor seems to be doing the right thing as a measure of evidence.

As discussed in Section 3.4.1, it is not at all clear that a p-value is an appropriate measure of evidence. In Example 4.4.2, however, it will be shown that the p-value $2(1 - \Phi(|\sqrt{n}\bar{x}|))$ is the limiting value of a measure of the strength of the evidence presented by the Bayes factor. This provides a partial resolution of the paradox because there can be evidence in favor of a hypothesis, as expressed by the Bayes factor, but this evidence can be very weak. This is the case when $|\sqrt{n}\bar{x}| = 5$ and τ_0^2 is large. The size of the Bayes factor does not measure the strength of the evidence and it needs to be calibrated.

This is only a partial resolution of the paradox, however, because it does not tell us how to choose the hyperparameter τ_0^2 in the prior. It seems clear that choosing τ_0^2 very large is introducing bias in favor of H_0 into the problem by inducing a large Bayes factor in favor of H_0. This is a common phenomenon with diffuse priors and must be guarded against. Certainly by choosing τ_0^2 large, prior-data conflict, as discussed in Chapter 5, is being avoided but at the same time this has to be balanced against the potential for bias. Section 4.6 introduces measures of bias in favor of and against a hypothesis based on a definition of evidence. The bias induced by a prior can be measured and controlled. As such, this presents what we consider a full resolution of the Jeffreys–Lindley paradox.

Sometimes this paradox is also discussed in terms of what happens as $n \to \infty$ as, for fixed $\sqrt{n}\bar{x}$ and τ_0^2, (3.23) converges to ∞ while $2(1 - \Phi(|\sqrt{n}\bar{x}|))$ is fixed. But this formulation of the problem seems somewhat artificial as, when $\sqrt{n}\bar{x}$ is not fixed and $n \to \infty$, then (3.23) converges to 0 and $2(1 - \Phi(|\sqrt{n}\bar{x}|)) \to 0$ when H_0 is false, and (3.23) converges to ∞ and $2(1 - \Phi(|\sqrt{n}\bar{x}|))$ is uniformly distributed when H_0 is true. Therefore, the paradox is not so apparent in this case.

More discussion on the Jeffreys–Lindley paradox and its impact can be found in Robert (1993). ∎

The following result has some relevance when considering the relationship between Bayes factors and relative belief ratios.

Lemma 3.5.6 *Suppose that* $H_0 = \{\theta_0\}, \Pi_0(H_0) = 0$ *and* $\Pi = p\delta_0 + (1-p)\Pi_0$, *where* δ_0 *is the probability measure on* Θ *degenerate at* θ_0. *Then*

$$BF(H_0 \,|\, x) = \pi_0(\theta_0 \,|\, x)/\pi_0(\theta_0),$$

the ratio of the posterior and prior densities at θ_0 *under the prior* Π_0.

Proof. Let $m_0(x) = \int_\Theta f_{\theta_0}(x)\pi_0(\theta)\,v(d\theta)$. Therefore, using Lemma 3.5.5,

$$BF(H_0 \,|\, x) = \frac{f_{\theta_0}(x)}{m_0(x)} = \frac{f_{\theta_0}(x)\pi_0(\theta_0)}{m_0(x)} \frac{1}{\pi_0(\theta_0)} = \frac{\pi_0(\theta_0 \,|\, x)}{\pi_0(\theta_0)},$$

as stated. ∎

Lemma 3.5.6 shows that the Bayes factor, as defined by Jeffreys, equals the relative belief ratio defined under the prior Π_0 whenever H_0 corresponds to a point null hypothesis about the full model parameter. This result does not hold more generally, namely, for Bayes factors for composite hypotheses.

The following example demonstrates another problem associated with Bayes factors as defined by Jeffreys.

Example 3.5.3 *Information Inconsistency for Bayes Factors.*
Suppose $x = (x_1,\ldots,x_n)$ is a sample from an $N(\mu,\sigma^2)$ distribution, where $\mu \in R^1, \sigma^2 > 0$ are unknown, so $T(x) = (\bar{x},s^2)$, where $\sqrt{n}\bar{x} \sim N(\mu,1)$ independently of $s^2 = (x - \bar{x}1)'(x - \bar{x}1) \sim \sigma^2\text{chi-squared}(n-1)$ is a minimal sufficient statistic. The prior Π is given by

$$\mu \,|\, \sigma^2 \sim p\delta_{\mu_0} + (1-p)N(\mu_0, \tau_0^2\sigma^2)$$
$$1/\sigma^2 \sim \text{gamma}_{\text{rate}}(\alpha_0, \beta_0)$$

where δ_{μ_0} is the probability measure degenerate at μ_0 and $p, \tau_0^2, \alpha_0, \beta_0$ are hyperparameters. Suppose $\Psi(\mu) = \mu$, and we want to assess $H_0 = \{\mu_0\}$. Note that this is a composite hypothesis.

Therefore,

$$m_T(\bar{x},s^2 \,|\, H_0) = c(s^2)^{\frac{n-1}{2}-1}[n(\bar{x} - \mu_0)^2 + s^2 + 2\beta_0]^{-(\frac{n}{2}+\alpha_0)},$$
$$m_T(\bar{x},s^2 \,|\, H_0^c) = \frac{c(s^2)^{\frac{n-1}{2}-1}}{(1+n\tau_0^2)^{1/2}}[n(1+n\tau_0^2)^{-1}(\bar{x} - \mu_0)^2 + s^2 + 2\beta_0]^{-(\frac{n}{2}+\alpha_0)},$$

where $c = (n/2\pi)^{1/2}(1/2)^{n+\alpha_0-1/2}\beta_0^{\alpha_0}\Gamma((n+2\alpha_0)/2/\Gamma((n-1)/2)\Gamma(\alpha_0)$, and so

$$BF(H_0\,|\,x) = (1+n\tau_0^2)^{1/2}\left[\frac{(1+n\tau_0^2)^{-1}n(\bar{x}-\mu_0)^2+s^2+2\beta_0}{n(\bar{x}-\mu_0)^2+s^2+2\beta_0}\right]^{n/2+\alpha_0}.$$

Now notice that as $\sqrt{n}\,|\bar{x}-\mu_0|\to\infty$ with s^2 fixed, then

$$BF(H_0\,|\,x)\to(1+n\tau_0^2)^{-n/2-\alpha_0+1/2}.$$

While this limiting value is typically small, it is not 0. One might expect, however, that, as the data become more and more divergent from the hypothesized value, the Bayes factor would converge to 0 and give definitive evidence against H_0. There doesn't seem to be a good explanation for this nonzero limit and it is referred to as information inconsistency.

There are several comments that can be made about this phenomenon. For any fixed τ_0^2, a value of $\sqrt{n}\,|\bar{x}-\mu_0|$ suitably large will lead to prior-data conflict that will be detected using the methods of Chapter 5. This in turn leads to a modification of the prior, typically by increasing the value of τ_0^2, which makes the limit smaller. Also, in Example 4.5.6 it is shown that defining the Bayes factor as a limit avoids the inconsistency in this problem. Further discussion of the information consistency associated with the Bayes factor, particularly in the context of improper reference priors, can be found in Berger, Bayarri and Mulder (2014). ∎

Another issue concerning Bayes factors is raised in Lavine and Schervish (1999). In essence, we can have $A\subset B$ and yet $BF(A\,|\,x)>BF(B\,|\,x)$. In fact, it is possible to have $BF(B\,|\,x)<1$, so that there is evidence against B, and yet $BF(A\,|\,x)>1$, so that there is evidence in favor of A. This seems anomalous, as it would seem that evidence that A is true is also evidence that B is true. The same issue arises with relative belief ratios. As it is much easier to understand why this occurs in that context, and why this *is* appropriate behavior, a thorough discussion on this issue is deferred to Chapter 4.

Another point concerns the appropriateness of comparing $BF(A\,|\,x)$ and $BF(B\,|\,x)$. Common usage seems to be based on the idea that there is a universal scale on which a Bayes factor is measuring evidence. But the support for this is weak at best and examples such as Example 3.5.2 should serve to caution those who use Bayes factors in this way. When comparing the values of Bayes factors, it is necessary to be sure that an apple is being compared with an apple, otherwise the comparison is not meaningful. This is the purpose of measuring the strength of the evidence, as introduced in Chapter 4. Certainly it is meaningful, however, to examine the relationship between the evidence for A and the evidence for B. When this is done in Chapter 4, it is seen that the fact that the evidence for A can be greater than the evidence for B when $A\subset B$ is not really anomalous. As such, we disagree with the assertion in Lavine and Schervish (1999) that the Bayes factor is incoherent. The Bayes factor does not behave appropriately as a measure of belief, but it does behave appropriately as a measure of evidence.

A good review of Bayes factors can be found in Kass and Raftery (1995). In particular, a commonly used scale for the calibration of Bayes factors is discussed there.

3.5.7 Hierarchical Bayes

So far the situation has been considered where the prior π for the model parameter θ is specified to obtain the joint distribution for (θ, x) with density $\pi(\theta)f_\theta(x)$. In many situations, however, rather than eliciting the prior on θ directly, it is supposed that it is a member of a family $\{\pi_1(\cdot \,|\, \lambda) : \lambda \in \Lambda\}$ and λ is elicited. This is typically a simpler approach to eliciting a prior; see Section 5.3.1 for further discussion on elicitation. The variable λ is referred to as a *hyperparameter* in this situation.

Alternatively, a prior π_2 may be placed on λ but then it is necessary to elicit π_2. Of course, this process may be continued by supposing that π_2 is a member of a family and then elicit the hyperparameter associated with this family or put yet another prior on this family of priors. This approach is referred to as *hierarchical Bayes*. It is clear, however, that this process must stop at some point and the problem addressed of selecting the prior at the final stage of the hierarchy.

3.5.8 Empirical Bayes

Suppose, in the hierarchical Bayes context, the hierarchy stops at the second stage and, rather than selecting a π_2 on λ, it is decided instead to estimate λ based on the data x, obtaining $\lambda(x)$. Recall that the posterior density of θ given λ equals

$$\pi_1(\theta \,|\, \lambda, x) = \pi_1(\theta \,|\, \lambda)f_\theta(x)/m_1(x \,|\, \lambda)$$

where $m_1(x \,|\, \lambda) = \int_\Theta \pi_1(\theta \,|\, \lambda)f_\theta(x)\nu(d\theta)$ is the prior predictive of x given λ. In *empirical Bayes* the function $\pi_1(\cdot \,|\, \lambda(x), x)$ is used as the posterior density for θ.

To implement empirical Bayes, the estimate $\lambda(x)$ needs to be specified. If θ were irrelevant for the problem, then it would be natural to integrate out θ and treat $m_1(x \,|\, \lambda)$ as the likelihood for λ. So $\lambda(x)$ is often taken to be the MLE based on $m_1(x \,|\, \lambda)$.

Empirical Bayes procedures are commonly recommended. This seems to be partly based on the fact that it is an intuitively reasonable way to avoid the requirements of a full Bayesian analysis and, in a number of problems, the inferences are seen to have good frequentist properties. A strong theoretical basis for empirical Bayes does not appear to exist, however, and from a foundational point of view one can voice objections, as it violates what we see as one of the most fundamental principles of inference. For notice that

$$\pi_1(\theta \,|\, \lambda(x), x) = \pi_1(\theta \,|\, \lambda(x))f_\theta(x)/m_1(x \,|\, \lambda(x)) \qquad (3.24)$$

but $\pi_1(\theta \,|\, \lambda(x))f_\theta(x)$ is not the joint conditional distribution of (θ, x) given $\lambda(x)$ and thus $\pi_1(\cdot \,|\, \lambda(x), x)$ is not a true posterior. So (3.24) does not arise via a valid application of the principle of conditional probability.

Our goal here is a theory of statistics free of such inconsistencies and so we do not consider empirical Bayes methods as a sound foundation for a theory of statistics. In Section 5.7, however, methods are considered for modifying a given prior, when a prior-data conflict is found to exist, and these can be considered as being in the same spirit as empirical Bayes. In practice, such methods may be necessary to produce an inference. Thus empirical Bayes ideas can be viewed as a compromise on our gold standard for inference. The text Carlin and Louis (1996) is an excellent reference for further discussion on this topic.

3.5.9   Bayesian Frequentism

The frequentist approach to statistics, as discussed in Section 1.7, considers a sequence of independent data sets $x_1, x_2, \ldots$ generated from f_θ and compares statistical procedures with respect to their properties in such sequences, for each $\theta \in \Theta$. Typically, there is some measure of merit and one looks for statistical procedures that perform well, if not optimally, with respect to that measure on average in a sequence, for each $\theta \in \Theta$. For example, a γ-confidence region $C_{\Psi,\gamma}(x)$ for $\Psi(\theta)$ is required to satisfy

$$P_\theta(\Psi(\theta) \in C_{\Psi,\gamma}(x)) = \lim_{n \to \infty} \frac{1}{n} \sum_{i=1}^{n} I_{C_{\Psi,\gamma}(x_i)}(\Psi(\theta)) \geq \gamma$$

whenever $x_1, x_2, \ldots$ is generated from f_θ, for each $\theta \in \Theta$. A γ-confidence region $C_\gamma(x)$ that minimizes the frequency of covering a false value, namely, minimizes

$$P_\theta(\psi' \in C_{\Psi,\gamma}(x)) = \lim_{n \to \infty} \frac{1}{n} \sum_{i=1}^{n} I_{C_{\Psi,\gamma}(x_i)}(\psi')$$

for each $\psi' \neq \Psi(\theta)$ and all θ, might then be taken as an appropriate choice of confidence region.

Bayesian inferences can also be considered in a frequentist framework. For example, consider the sequence of independent pairs $(\theta_1, x_1), (\theta_2, x_2), \ldots$ where $\theta_i \sim \Pi$ and $x_i \mid \theta_i \sim P_{\theta_i}$. Then, if $C_\gamma(x)$ is a γ-credible region for $\Psi(\theta)$, we have that

$$\Pi \times P_\theta(\Psi(\theta) \in C_{\Psi,\gamma}(x)) = \lim_{n \to \infty} \frac{1}{n} \sum_{i=1}^{n} I_{C_{\Psi,\gamma}(x_i)}(\Psi(\theta_i))$$

$$= \int_{\mathscr{X}} \Pi(\Psi(\theta) \in C_{\Psi,\gamma}(x) \mid x) M(dx) \geq \gamma$$

and so C_γ has a confidence property. Furthermore, if for each x the γ-credible region $C_\gamma(x)$ is chosen to possess some optimal property, such as minimizing $\Lambda(C_\gamma(x))$ for some measure Λ, then

$$E_{\Pi \times P_\theta}(\Lambda(C_{\Psi,\gamma}(x))) = \lim_{n \to \infty} \frac{1}{n} \sum_{i=1}^{n} \Lambda(C_{\Psi,\gamma}(x_i))$$

is minimized among all γ-credible regions for $\Psi(\theta)$.

It is not our intention to advocate that Bayesian procedures be considered with respect to their Bayesian frequentist properties. In fact, as noted here, these seem redundant once these properties are determined a posteriori. It does lead us to wonder, however, why much of statistical theory has focused on the frequentist properties of inferences with θ fixed. It seems completely reasonable, and perhaps even more realistic, to imagine a sequence where the θ values change according to a somewhat weakly informative prior over the effective range of values one might expect in an application.

It will be seen in Chapter 4, however, that relative belief inferences do have various optimality properties a priori and these can be thought of as (Bayesian) frequentist in nature. Furthermore, the assessment of the bias induced by a prior can be considered in this way. There is no contradiction or conflict in this, however, as the inferences are determined by principles, such as the principle of conditional probability, that supersede consideration of frequentist behavior. Still, frequentist behavior is important and cannot be ignored. Accordingly, there is a satisfying union of Bayesian and frequentist ideas, albeit somewhat outside traditional ways of thinking about these approaches.

3.5.10   Summarizing the Bayesian Approach

Bayesian inferences require the specification of both a model and a prior. These ingredients have been treated as being given in this chapter. In Chapter 5 discussion is provided on how to go about choosing the model and prior in an application and how to check for the reasonableness of the choices made.

Given the ingredients, it is necessary to specify how to produce inferences for the problems discussed in Section 1.8. Various approaches have been adopted for this but, as expressed here, none of these provides an entirely satisfactory treatment based clearly on statistical evidence. Bayes factors can be considered as a measure of statistical evidence, but concerns can be raised about the way these are used and defined. These issues are addressed in Chapter 4. Further discussion of Bayesian methodology can be found in DeGroot (1970), Zellner (1971), Box and Tiao (1973), Bernardo and Smith (1993), Leonard and Hsu (1999), Robert (2001), Press (2003), Gelman, Carlin, Stern and Rubin (2004), Berger (2006), Lindley (2006) and Kadane (2011).

3.6   Fiducial Inference

The Bayesian approach is at least partially characterized by probability distributions on the model parameter θ. For some this is anomalous because parameters are not "random." For our developments there is no anomaly because, as discussed in Chapter 2, probabilities measure beliefs and have nothing to do with randomness. Still, some have pursued the goal of obtaining distributions on parameters without the need to place a prior on θ that represents beliefs. There are several approaches taken to this, with perhaps the most prominent currently being the *reference prior* approach. Reference priors are discussed in Section 5.3.

Another approach originated with Fisher (1930) and is known as *fiducial inference*. Basic to fiducial inference is the existence of a *pivotal quantity* $T(\theta, x)$, namely, a function of the parameter and data which has a distribution independent of θ, say with probability measure P_T. The fiducial idea is then to fix the data and write for

$A \subset \Theta$,

$$P_{fiducial,x}(A) = P_T\left(\{t : t = T(\theta,x) \text{ for some } \theta \in A\}\right). \qquad (3.25)$$

So $P_{fiducial,x}(A)$ would appear to be a probability statement about θ. A basic example is given by the following.

Example 3.6.1 *Fiducial Distribution.*

Suppose that $S(x)$ is a real-valued statistic and $\Theta \subset R^1$. Let $T(\theta,x)$ denote the distribution function of S evaluated at $S(x)$ when θ is true. When S has a continuous distribution for each θ, then $T(\theta,x)$ has a fixed $U(0,1)$ distribution and so is pivotal. If $T(\theta,x)$ is continuous and increasing in θ for each x, then

$$\begin{aligned} P_{fiducial,x}(\theta \le \theta_0) &= P_T\left(\{t : t = T(\theta,x) \text{ for some } \theta \le \theta_0\}\right) \\ &= T(\theta_0,x) - \inf_{\theta \le \theta_0} T(\theta,x). \end{aligned} \qquad (3.26)$$

A similar expression is obtained when $T(\theta,x)$ is continuous and decreasing in θ for each x. ∎

Various issues have been raised about fiducial probabilities. For example, it is clear from (3.25) and (3.26) that, while $P_{fiducial,x}$ is additive, it is not necessarily the case that $P_{fiducial,x}(\Theta) = 1$ unless $T(\cdot,x)$ is surjective for each x. Even when $P_{fiducial,x}$ is a probability measure, it is not at all clear what principle is being applied to obtain this distribution, as it does not arise from an application of the principle of conditional probability. In fact, it is manifestly true that $P_{fiducial,x}$ does not have the properties of a conditional probability such as satisfying the multiplication rule.

So a new principle is necessary to justify fiducial and it is unresolved what this would be to provide a general theory of inference. For example, clearly it is necessary to restrict attention to pivotals such that $P_{fiducial,x}$ is a probability distribution and these do not exist for general problems. Furthermore, there seems to be an inherent dependence on continuous models and, for the reasons cited in Chapter 1, this seems inappropriate to us. For many problems, there are several possible pivotals and choosing among them is not data based, which violates the principle of empirical criticism.

It is true, however, that sometimes the fiducial approach produces inferences that conform with what are generally believed to be appropriate. Perhaps the *structural theory of inference* as developed by Fraser (1961, 1968, 1979) does the best job of explaining why this is the case. A simple example illustrates this.

Example 3.6.2 *Structural Inference.*

Suppose $x = (x_1,\ldots,x_n)$ is a sample from $f_\theta(w) = f(w - \theta)$ where f is some fixed density on R^n and $\Theta = R^1$. Now consider $(\Theta, +)$ as a group of transformations acting on R^n via $\theta \cdot x = x + \theta 1_n$ where $1_n \in R^n$ is a vector of ones. The basic model can now be written as a *structural model*, namely,

$$x = \theta 1_n + z \qquad (3.27)$$

where $z = (z_1,\ldots,z_n) \sim f$. Of course, z is not observed but it is known that $z \in \Theta \cdot x =$

$\{\theta 1_n + x : \theta \in \Theta\}$, the *orbit* of x induced by the group acting on R^n, which is the line through x parallel to $\mathscr{L}\{1_n\}$. The function $D(x) = x - \bar{x}1_n$ indexes these orbits and, since $x = \bar{x}1_n + D(x)$, the position of x on the orbit specified by $D(x)$, is given by $\bar{x}$. From (3.27) then $D(x) = D(z)$ so that observing $D(x)$ indeed tells us that z is in this orbit.

The density of D is given by

$$f_{\theta D}(d) = \int_{D^{-1}\{d\}} f(x - \theta 1_n) J_D(x) \, \mu_{D^{-1}\{d\}}(dw)$$

$$= \int_{D^{-1}\{d\}} f(z) J_D(z) \, \mu_{D^{-1}\{d\}}(dz) = f_D(d)$$

which is independent of θ so D is ancillary. Therefore, by the principle of conditional probability, we have the conditional distribution of $\bar{z}$ given $D(z)$ for the unobserved value of z.

These conditional probabilities transfer to probabilities for θ via the *structural equation* obtained from (3.27), namely,

$$\bar{x} = \theta + \bar{z}, \tag{3.28}$$

where $\bar{x}$ is fixed as it is observed. It is clear from (3.28) that these same probabilities can be obtained using the pivotal $T(\theta, x) = \bar{x} - \theta$, although the additional aspect of using the conditional distribution of the pivotal as opposed to its marginal distribution is now part of the argument.

As a specific example, suppose that f is the $N(0,1)$ distribution. Then $\bar{z}$ and $D(z)$ are statistically independent and a 0.95 structural interval for θ is $\bar{x} \pm z_{0.975}/\sqrt{n}$, which is also the usual 0.95 confidence interval for θ. In general, structural intervals will also have the confidence property.

The structural approach can be extensively generalized beyond this simple context and in fact it can treat many of the most important problems in statistics. Difficulties with this approach lie with its reliance on continuity, as the same structure rarely applies in discrete situations. While structural models can be thought of as approximations to a finite reality, as a general theory, the fact that it doesn't cover the more basic contexts can be seen as a weakness.

Structural theory is an important contribution, however, as it explains why Fisher's fiducial approach at times produces reasonable inferences. In our view, this is because the inferences can be considered as arising as limiting Bayesian inferences based on what is known as the *right Haar prior*. When the parameter space is not compact, then the right Haar prior is typically improper. This leads to the need to consider limiting inferences based on a sequence of priors obtained by conditioning the right Haar prior to an increasing sequence of compact sets converging to the whole parameter space. As discussed in Section 5.3, a variety of issues arise with improper priors. ∎

Fiducial inference remains an active area of research in statistics as researchers seek the Holy Grail of making probability statements for parameters without the need to specify a proper prior probability distribution. For example, Hannig (2009) and

Martin and Liu (2009) reflect current developments in this area. It is fair to say, however, that the basic problems noted for fiducial inference remain unresolved.

3.7 Concluding Comments

A variety of approaches have been developed to deal with the concept of statistical evidence and many of these have been discussed in this chapter. This has not been an exhaustive presentation, rather only giving the main features of an approach and then the reasons it is felt to be unsatisfactory as a basis for a theory of inference.

While our conclusion is that none of the approaches discussed is fully adequate, they all contain aspects of what one would want a theory of statistical inference based on evidence to possess. Pure likelihood and Bayes factors seem the most direct in that they speak somewhat more explicitly about how to measure statistical evidence. The approach to measuring statistical evidence developed in Chapter 4 can be seen as building on likelihood and Bayes factors to form a more complete theory of statistical inference based on measuring evidence.

Chapter 4

Measuring Statistical Evidence Using Relative Belief

4.1 Introduction

In this chapter, a method for measuring the statistical evidence that a given statement is true is provided based on a combination of data, model and prior. This statement could be that a certain quantity takes a specified value or is in some specified range of values. All inferences are derived from the measure of evidence.

A key aspect of dealing with statistical evidence is its calibration. For example, if there is evidence in favor of a value being true or being false, it needs to be specified if this is strong or weak evidence. The assessment of the strength of the evidence is separate from the actual measurement of the evidence. The calibration of evidence takes place within the context of the ingredients specified for an inference problem. We do not look for a universal measurement scale on which evidence is to be calibrated. In fact, such a quest seems fruitless but, given that the evidence can be calibrated meaningfully in a given application, the lack of such a scale doesn't matter.

Throughout this chapter a number of results are derived that suggest the method for measuring statistical evidence has many appropriate properties and is related to some standard concepts like likelihood and the Bayes factor. Even something much like a p-value is relevant to our assessment of evidence. In the end, any such method of measuring statistical evidence must be shown to have suitable properties, otherwise it should be discarded. For example, it is reasonable to have some skepticism toward the principle of conditional probability for updating beliefs, but the many good properties of conditional probability suggest it is a natural candidate for such a rule.

The method for measuring evidence is dependent on the ingredients specified, namely, a model and a prior, in combination with the data. If these ingredients are in error, or the data is not collected correctly, then the analysis based on the evidence is suspect. Of course, one can never say that the ingredients specified are correct, but we can check these ingredients against the data. This checking is based on measuring to what extent aspects of the data are surprising in light of the ingredients specified. This is the topic of Chapter 5, but for this chapter all ingredients are assumed to be correct.

In Section 4.2 the relative belief ratio as the measure of statistical evidence is introduced and some basic properties derived. In Section 4.3 the relative belief ratio is compared to other possible choices and is seen to arise from some very simple requirements for a measure of evidence. Section 4.4 is concerned with calibrating the relative belief ratio. Section 4.5 discusses the solutions to the basic inference problems implied by measuring statistical evidence using the relative belief ratio. Section 4.6 discusses the important issue of assessing the bias introduced into a statistical analysis by a prior. Section 4.7 presents numerous optimality properties possessed by the inferences derived via relative belief. It is these properties that lead to a conviction concerning the usefulness of the approach.

4.2   Relative Belief Ratios and Evidence

A first priority is to distinguish between the concepts of belief and statistical evidence. Belief is measured by probability. Evidence is what leads to change in belief. So we start with certain beliefs, acquire evidence and accordingly modify our beliefs. Given that beliefs are measured by probability, the principle of conditional probability prescribes how to modify beliefs given the evidence. Of course, it is assumed that the evidence is generated by a valid information generator, as discussed in Section 2.1.2, so that an unambiguous application of conditional probability can be made. It is then clear that the evidence that a certain statement is true is to be measured by the change in probability. There are many ways to measure this change and we will employ one that seems the most natural, as it leads to a simpler theory than others. Statistical inferences are based on change in belief rather than belief because this change is induced by the data which, in ideal circumstances, *is* objective. Accordingly, our inferences are less influenced by the subjectively chosen model and prior when compared to other approaches.

Throughout this section let us suppose that there is a probability measure P, which could be the joint measure induced from the model and prior, for a concealed response $\omega \in \Omega$ and the information is obtained that $\omega \in C \subset \Omega$ where $P(C) > 0$. The principle of conditional probability then implies that the probability assigned to the event $A \subset \Omega$ changes from $P(A)$ to $P(A|C) = P(A \cap C)/P(C)$. As such, the evidence in favor of A is measured by the change from $P(A)$ to $P(A|C)$.

The following is considered here as a fundamental principle of inference.

Principle of Evidence: If $P(A|C) > P(A)$, then there is evidence in favor of A being true because the belief in A has increased. If $P(A|C) < P(A)$, then there is evidence against A being true because the belief in A has decreased. If $P(A|C) = P(A)$, then there is neither evidence in favor of A nor against A because the belief in A has not changed.

For example, if $C \subset A$, so there is evidence corroborating A, then $P(A|C) = 1 > P(A)$ whenever $P(A) \neq 1$. When $A \cap C = \phi$, so the evidence completely contradicts A, then $P(A|C) = 0 < P(A)$ whenever $P(A) \neq 0$. As simple as this seems, the principle of evidence has significant consequences for a theory of statistical inference.

It seems natural then to measure the amount of evidence concerning the truth of A by some real-valued function of $(P(A), P(A|C))$ where larger values mean more evidence in favor and smaller values mean more evidence against. For any selected evidence function, any strictly increasing function of this is to be considered as an equivalent measure of evidence. In Section 4.3 several possible measures of evidence are discussed but we focus first on the ratio of $P(A|C)$ to $P(A)$, as it has many nice properties and it is our ultimate choice for measuring evidence.

The discussion is separated into the discrete and continuous cases. This is to emphasize that it is the properties in the discrete case that are essential in determining whether or not the approach is sound. For the continuous case, it must be kept in mind that this is concerned with approximations.

4.2.1 Basic Definition of a Relative Belief Ratio

The relative belief ratio is defined as follows.

Definition 4.2.1 *The relative belief ratio of the event $A \subset \Omega$, given that $C \subset \Omega$ has occurred, where $P(A) > 0$ and $P(C) > 0$, is*

$$RB(A|C) = \frac{P(A|C)}{P(A)} = \frac{P(A \cap C)}{P(A)P(C)}.$$

If $RB(A|C) > 1$, then the occurrence of C is evidence in favor of $\omega \in A$, as this information has increased belief in $\omega \in A$. If $RB(A|C) < 1$, then the occurrence of C is evidence against $\omega \in A$, as this information has decreased belief in $\omega \in A$. If $RB(A|C) = 1$, then the occurrence of C has no effect on the belief that $\omega \in A$ and so provides no evidence.

As a simple example consider the Bayesian context.

Example 4.2.1 *Bayesian Relative Belief Ratios (discrete case).*

Suppose that $\Omega = \Theta \times \mathscr{X}$ with $P = \Pi \times P_\theta$ and $C = \Theta \times \{x\}$, namely, the data has been observed. Further suppose that $P_\theta(\{x\}) = f_\theta(x)$ and $\Pi(\{\theta\}) = \pi(\theta)$ so all the probability measures involved are discrete. Therefore, $P(C) = \sum_{\theta \in \Theta} f_\theta(x)\pi(\theta) = m(x) > 0$ and for $B \subset \Theta$, so $A = B \times \mathscr{X}$ with $P(A) = \Pi(B) > 0$, then

$$RB(A|C) = \frac{P(A|C)}{P(A)} = \frac{\Pi(B|x)}{\Pi(B)},$$

is the ratio of the posterior probability of B to the prior probability of B. When our interest is in parameter $\psi = \Psi(\theta)$, so $A = \Psi^{-1}\{\psi\} \times \mathscr{X}$, then with π_Ψ and $\pi_\Psi(\cdot|x)$ denoting the prior and posterior probability functions of Ψ, the relative belief ratio equals

$$RB(A|C) = \frac{\pi_\Psi(\psi|x)}{\pi_\Psi(\psi)}.$$

Subsequently, the notation $RB_\Psi(\psi|x) = \pi_\Psi(\psi|x)/\pi_\Psi(\psi)$ will be used.

So in the fully discrete case, an appropriate definition of the evidence that $\psi = \Psi(\theta)$ is true is given by the ratio of the posterior to the prior density. This result will be shown to apply in great generality and furthermore, the relative belief ratio serves as the basis for statistical inference about ψ. ∎

The value $RB(A|C)$ indicates by what factor beliefs in A have increased or decreased. For example, $RB(A|C) = 2$ means that belief, as measured by probability, has doubled. Clearly, the larger $P(A)$ is, the harder it is for $RB(A|C)$ to be large while when $P(A)$ is small $RB(A|C)$ can be very large. The fact that $RB(A|C) = 2$ when $P(A) = 1/4$ seems much more significant than $RB(A|C) = 2$ when $P(A) = 1/100$ indicates that there is not a universal scale on which $RB(A|C)$ is measuring evidence.

The following summarizes some basic properties of $RB(A|C)$ that relate directly to its interpretation.

Proposition 4.2.1 *Suppose $A, C \subset \Omega$, $P(A) > 0$ and $P(C) > 0$. The following properties then hold,*
(i) $0 \leq RB(A|C)$ with $RB(A|C) = 0$ iff $P(A|C) = 0$,
(ii) $RB(A|C) = 1$ iff A and C are independent,
(iii)

$$P(A|C) \leq RB(A|C) \leq 1/P(A) \qquad (4.1)$$

with $P(A|C) = RB(A|C)$ iff $P(A) = 1$ (and so $RB(A|C) = 1$) and $RB(A|C) = 1/P(A)$ iff $P(A|C) = 1$.

While Proposition 4.2.1 is obvious, the results have implications for the interpretation of $RB(A|C)$ as a measure of evidence. Property (i) says that $RB(A|C) = 0$ is equivalent to having categorical evidence that A is false. Property (ii) says that having no evidence one way or the other is equivalent to the evidence being independent of the event in question. Property (iii) shows that there is not a universal scale on which evidence is being measured by $RB(\cdot|C)$, as the upper bound is dependent on the prior probability of the event in question. So, for example, it is not necessarily appropriate to compare $RB(A|C)$ with $RB(B|C)$ for different events. If $P(A)$ is small, then it is possible to get large values of $RB(A|C)$ but not otherwise. To an extent this explains a commonly noted asymmetry between finding evidence against and evidence for the truth of A. The lower bound is universal while the upper bound is attained only when we have categorical evidence of the truth of A and it depends on the initial probability of the event. Also the measure of belief that A is true is always strictly smaller than the measure of the evidence that A is true, unless we know categorically that A is true before receiving any evidence.

Notice that $P(A|C)$ may be small while $RB(A|C)$ is large. It is not correct to interpret small values of $P(A|C)$ as evidence against A, as this may arise simply because $P(A)$ is very small. This inequality indicates that, if $RB(A|C)$ is small, then so is $P(A|C)$ and if $P(A|C)$ is large, then $RB(A|C)$ cannot be small, although it can still be less than 1. This property has relevance to our discussion of the strength of the evidence given by $RB(A|C)$ in Section 4.4.

The following immediate result has some significance for inference.

Proposition 4.2.2 *(Savage–Dickey Ratio) Suppose $A, C \subset \Omega$, $P(A) > 0$ and $P(C) > 0$, then $RB(A|C) = RB(C|A)$.*

So there is evidence in favor of A, having observed C, if and only if there is evidence in favor of C, having observed A. While the measure of the evidence is the same for these two situations, it turns out that the common value can represent different strengths of evidence.

It is useful to consider how the evidence for A given C relates to the evidence for A^c given C, the evidence for A given C^c, etc. The following formulas all follow via simple calculations.

Proposition 4.2.3 *Suppose* $A, C \subset \Omega$, $P(A) > 0$ *and* $P(C) > 0$, *then*
(i) when $P(A^c) > 0$

$$RB(A^c|C) = \frac{1 - P(A)RB(A|C)}{1 - P(A)},$$

so $RB(A|C) < 1$ *iff* $RB(A^c|C) > 1$ *and* $RB(A|C) = 1$ *iff* $RB(A^c|C) = 1$,
(ii) when $P(C^c) > 0$

$$RB(A|C^c) = \frac{1 - P(C)RB(A|C)}{1 - P(C)},$$

so $RB(A|C) < 1$ *iff* $RB(A|C^c) > 1$ *and* $RB(A|C) = 1$ *iff* $RB(A|C^c) = 1$,
(iii) when $P(A^c) > 0$ *and* $P(C^c) > 0$

$$RB(A^c|C^c) = \frac{1 - P(A) - P(C) + P(A)P(C)RB(A|C)}{1 - P(A) - P(C) + P(A)P(C)},$$

so $RB(A|C) < 1$ *iff* $RB(A^c|C^c) < 1$ *and* $RB(A|C) = 1$ *iff* $RB(A^c|C^c) = 1$.

Proposition 4.2.3 shows that relative belief ratios behave appropriately as measures of evidence under complementation. For example, evidence for A is indeed evidence against A^c, etc.

The behavior of relative belief ratios with respect to combinations of events is also of interest. For example, we would like to know the relationship between the evidence for $A \cap B$ and the evidence for A and B separately.

Proposition 4.2.4 *Suppose* $A, C \subset \Omega$, $P(A) > 0$ *and* $P(C) > 0$. *The following properties hold.*
(i) If $P(A \cap B) > 0, P(B \cap C) > 0$, *then*

$$RB(A \cap B|C) = \frac{RB(A|B \cap C)RB(B|C)}{RB(A|B)},$$

and if A *and* B *are independent given* C, *then*

$$RB(A \cap B|C) = \frac{RB(A|C)RB(B|C)}{RB(A|B)},$$

and when A *and* B *are also independent, then*

$$RB(A \cap B|C) = RB(A|C)RB(B|C).$$

(ii) If $P(A \cap C) > 0, P(C \cap D) > 0, P(D) > 0$, then

$$RB(A \mid C \cap D) = RB(A \mid C) \frac{RB(A \cap C \mid D)}{RB(C \mid D)}$$

where

$$\frac{RB(A \cap C \mid D)}{RB(C \mid D)} = \frac{P(A \mid C \cap D)}{P(A \mid C)}$$

is the conditional relative belief ratio of A, given that we observe D, having already observed C.

(iii) (additivity) if $A \cap B = \phi$ and $P(B) > 0$, then

$$RB(A \cup B \mid C) = RB(A \mid C)P(A \mid A \cup B) + RB(B \mid C)P(B \mid A \cup B),$$

which is the conditional average given $A \cup B$, of $RB(A \mid C)$ and $RB(B \mid C)$,

(iv) (general additivity) if $P(A \cap B) > 0$, then

$$RB(A \cup B \mid C) = RB(A \mid C)P(A \mid A \cup B) + RB(B \mid C)P(B \mid A \cup B) -$$
$$RB(A \cap B \mid C)P(A \cap B \mid A \cup B).$$

Proof. The first formula in (i) follows from

$$RB(A \cap B \mid C) = \frac{P(A \cap B \cap C)}{P(A \cap B)P(C)} = \frac{P(A \cap B \cap C)}{P(B \cap C)P(A)} \frac{P(A)P(B)P(B \cap C)}{P(A \cap B)P(B)P(C)}$$
$$= RB(A \mid B \cap C)RB(B \mid C)/RB(A \mid B),$$

the second formula follows from $P(A \cap B \cap C) = P(A \mid C)P(B \mid C)P(C)$, and the final formula from Proposition 4.2.1 (ii). Clearly (ii) follows from (i) and Proposition 4.2.2 and

$$\frac{RB(A \cap C \mid D)}{RB(C \mid D)} = \frac{P(A \cap C \cap D)}{P(A \cap C)} \frac{P(C)}{P(C \cap D)} = \frac{P(A \mid C \cap D)}{P(A \mid C)}$$

is the relative belief in A having observed D, given that we have already observed C. For (iii),

$$RB(A \cup B \mid C) = \frac{P(A \cup B \mid C)}{P(A \cup B)} = \frac{P(A \mid C) + P(B \mid C)}{P(A \cup B)}$$

and the result follows. For (iv), apply (iii) several times to obtain

$$RB(A \cup B \mid C) = RB((A \cap B^c) \cup (A \cap B) \cup (A^c \cap B) \mid C)$$
$$= RB(A \cap B^c \mid C) \frac{P(A \cap B^c)}{P(A \cup B)} + RB(A \cap B \mid C) \frac{P(A \cap B)}{P(A \cup B)}$$
$$+ RB(A^c \cap B \mid C) \frac{P(A^c \cap B)}{P(A \cup B)}$$
$$= RB(A \mid C) \frac{P(A)}{P(A \cup B)} + RB(A^c \cap B \mid C) \frac{P(A^c \cap B)}{P(A \cup B)}$$
$$= RB(A \mid C) \frac{P(A)}{P(A \cup B)} + RB(B \mid C) \frac{P(B)}{P(A \cup B)} - RB(A \cap B \mid C) \frac{P(A \cap B)}{P(A \cup B)}.$$

∎

Property (i) of Proposition 4.2.4 relates the evidence for both A and B being true to the evidence for each of A and B and the evidence for A when B instead of C is observed (or the evidence for B when A instead of C is observed since $RB(A|B) = RB(B|A)$). When A and B are independent, as well as conditionally independent given C, then the joint evidence is multiplicative, as one might expect.

Property (ii) shows how evidence is modified as it is obtained sequentially. So suppose C is observed first, providing evidence $RB(A|C)$. After this we observe D and the formula says that the overall evidence from observing both is the product of the evidence from each stage. Note that it is possible that $RB(A|C) > 1$ but $RB(A \cap C|D)/RB(C|D) < 1$ so that observing D after C doesn't necessarily increase the evidence for A. Notice too that the overall evidence is independent of the order in which C and D are observed. Also, due to the symmetry property $RB(A|B) = RB(B|A)$, properties (i) and (ii) are equivalent.

Property (iii) is the most interesting and instructive of the properties. It is easy to see that it follows directly from Proposition 4.2.3 (i) when $B = A^c$. This gives a form of additivity, as the relative belief ratio of the union of disjoint events is the weighted sum of their relative belief ratios, with weights equal to their conditional probabilities given that one of the events has occurred. Note that, when $P(A) > 0$ and $P(B) > 0$, then, unless $RB(A|C) = RB(B|C)$, it is always true that one of $RB(A|C) > RB(A \cup B|C)$ or $RB(B|C) > RB(A \cup B|C)$ with the other strictly less than $RB(A \cup B|C)$. In fact, it can happen that $RB(A|C) > 1$ but $RB(A \cup B|C) < 1$. The following characterizes when this occurs.

Corollary 4.2.1 *If* $A \cap B = \phi, P(A) > 0, P(B) > 0,$ *and* $P(C) > 0$ *with* $RB(A|C) > 1,$ *then* $RB(A \cup B|C) < 1$ *iff* $RB(B|C) < 1$ *and*

$$P(A|A \cup B) < \frac{1 - RB(B|C)}{RB(A|C) - RB(B|C)}.$$

The conditions of the corollary will apply when $P(A|A \cup B)$ is very small or $RB(A|C) \approx 1$. The conditions of the corollary will not apply when $RB(A|C)$ is large or $RB(B|C) \approx 1$.

It might seem that a sensible measure of evidence would not allow $RB(A|C) > RB(B|C)$ when $A \subset B$, as evidence that A is true should also be evidence that B is true. Proposition 4.2.4 (iii) shows, however, that the evidence that A is true is indeed contributing to the evidence that B is true, but this contribution is weighted by the probability of A given that B is true. When the conditional probability $P(A|B)$ is very small, then the contribution of the evidence about A to the evidence about B is very small and this seems quite reasonable. Consider the following example, which illustrates that this behavior of the relative belief ratio as a measure of evidence is appropriate.

Example 4.2.2 *Evidence of a Crime.*

Suppose that a murder is committed and it is known that an adult member of a town with m adult citizens committed the crime. Suppose that categorical evidence is obtained indicating that a person of ethnic origin a committed this crime and there are $m_1 < m$ adult members of this ethnic group residing in the town. Suppose further that the town contains a university with n adult students of which n_1 are of ethnic origin a. Assuming uniform beliefs before the evidence is obtained, the probability that a university student committed the crime is n/m and the probability that a university student of ethnic origin a committed the crime is n_1/m.

After the evidence is obtained, the probability that a university student committed the crime is n_1/m_1, which is the same as the probability that a university student of ethnic origin a committed the crime. So the relative belief ratio that a university student committed the crime is $n_1 m/m_1 n$ and the relative belief ratio that a university student of ethnic origin a committed the crime is $n_1 m/m_1 n_1 = m/m_1$. Now $m/m_1 > 1$ but $n_1 m/m_1 n < 1$ when n_1/n is small enough, namely, whenever the students of ethnic origin a comprise a small enough fraction of the student population.

So indeed there can be evidence in favor of the statement that a student of ethnic origin a committed the crime, while at the same time, there is evidence against the statement that a student committed the crime. It is certainly appropriate for the measure to indicate evidence favoring the statement that a student of ethnic origin a committed the crime. It would be inappropriate (unfair), however, to indicate evidence in favor of a student having committed the crime unless students of ethnic origin a comprised a substantial fraction of the student population. When students of ethnic origin a are a small fraction of the student population, it is completely reasonable that the measure should indicate evidence against the statement that a university student did it. Notice that the relative belief ratio that a person of ethnic origin a committed the crime is also given by m/m_1. ∎

Actually, the occurrence of $RB(A|C) > RB(B|C)$ when $A \subset B$ is not unusual. For $RB(\Omega|C) = 1$ for every C with $P(C) > 0$. So evidence that Ω is true is never acquired from learning that C is true and this is correct because Ω and C are statistically independent. It is known categorically that Ω is true without knowing that C is true, so this evidence can contribute nothing to beliefs about Ω. Furthermore, there will typically be many $A \subset \Omega$ with $RB(A|C) > 1$. This emphasizes that $RB(A|C)$ is measuring the contribution to the belief that A is true from the evidence provided by knowing that C is true. The relative belief ratio $RB(A|C)$ is *not* a measure of belief.

It is interesting to note that $RB(\Omega|C) = 1$ and Proposition 4.2.4 (iii) suggests that $RB(\cdot|C)$ behaves somewhat like a probability measure. This is expressed by the following result.

Corollary 4.2.2 *If* $\{A_1,\ldots,A_m\}$ *is a partition of* Ω *with* $P(A_i) > 0$, *then*
$$\sum_{i=1}^{m} RB(A_i|C)P(A_i) = 1.$$

A substantial generalization of this is obtained in Proposition 4.2.6 (iii).

4.2.2 General Definition of a Relative Belief Ratio

So far attention has been restricted to defining the relative belief ratio when all the relevant events have positive probabilities. Under weak conditions this is easily generalized to determine relative belief ratios based on sets having probability zero simply because they are of lower dimensionality. For this context the mathematical structure discussed in Appendix A for the determination of all densities is assumed. In particular, all densities are taken with respect to volume measure on their respective spaces.

So suppose there is a probability measure P on the space Ω with density f with respect to volume measure ν on Ω. Suppose that $\Xi = (\Xi_1, \Xi_2) : \Omega \to \Xi$ is a smooth function with density f_Ξ with respect to the volume measure ν_Ξ on Ξ. Consider now the determination of the evidence concerning the truth of the statement $\Xi_1(\omega) = \xi_1$ given that we know $\Xi_2(\omega) = \xi_2$. Let $N_{\Xi_1, \delta}(\xi_1)$ and $N_{\Xi_2, \varepsilon}(\xi_2)$ be neighborhoods converging nicely (see Appendix A) to ξ_1 and ξ_2, respectively, as $\delta, \varepsilon \to 0$. Then the following is the appropriate definition of the relative belief ratio.

Definition 4.2.2 *For $(\xi_1, \xi_2) = (\Xi_1(\omega), \Xi_2(\omega))$, the relative belief ratio at $\Xi_1(\omega) = \xi_1$, given that $\Xi_2(\omega) = \xi_2$, is defined by*

$$RB_{\Xi_1}(\xi_1 \mid \xi_2) = \lim_{\delta \to 0, \varepsilon \to 0} RB(N_{\Xi_1, \delta}(\xi_1) \mid N_{\Xi_2, \varepsilon}(\xi_2)),$$

whenever this limit exists for any nice neighborhoods $N_{\Xi_1, \delta}(\xi_1)$ and $N_{\Xi_2, \varepsilon}(\xi_2)$ converging to ξ_1 and ξ_2, respectively, as $\delta, \varepsilon \to 0$.

This definition is based on the idea that, in the general case, continuous probability is being used to approximate a finite reality and so the relative belief ratio is defined as a limit. Note that Definition 4.2.2 implies Definition 4.2.1. In fact, we only need Definition 4.2.2 for the continuous case, as all discrete situations are covered by Definition 4.2.1. It is also worth noting that the definition implies that $RB_{\Xi_1}(\xi_1 \mid \xi_2)$ is the density of $P_\Xi(\cdot \mid \xi_2)$ with respect to the support measure P_{Ξ_1}.

Under suitable regularity conditions (see Appendix A),

$$RB_\Xi(\xi_1 \mid \xi_2) = \frac{f_\Xi(\xi_1 \mid \xi_2)}{f_{\Xi_1}(\xi_1)} \tag{4.2}$$

where f_{Ξ_1} is the marginal density of ξ_1 and $f_\Xi(\cdot \mid \xi_2)$ is the conditional density of ξ_1 given that $\Xi_2(\omega) = \xi_2$. For (4.2) to hold, it is necessary that $f_{\Xi_1}(\xi_1) > 0$ but in the continuous case the relative belief ratio can be defined by Definition 4.2.2 even when $f_{\Xi_1}(\xi_1) = 0$. This is due to a certain arbitrariness in the choice of the support measure, as the following simple example illustrates.

Example 4.2.3 *Relative Belief Ratio When Marginal Density Vanishes.*
Suppose ω has a $U(-1, 1)$ distribution and

$$(\xi_1, \xi_2) = (\Xi_1(\omega), \Xi_2(\omega)) = (\omega^{1/3}, I_{(-1/2.1/2)}(\omega)).$$

Then $f_{\Xi_1}(\xi_1) = 3\xi_1^2/2$ for $\xi_1 \in (-1, 1)$. The value $\xi_1 = 0$ is meaningful and yet $f_{\Xi_1}(0) = 0$. Given $\xi_2 = 1$, we have that $f_\Xi(\xi_1 \mid 1) = 3\xi_1^2$ for $\xi_1 \in (-1/\sqrt[3]{2}, 1/\sqrt[3]{2})$. Using Definition 4.2.2, $RB_{\Xi_1}(0 \mid 1) = 2$ but (4.2) does not hold. If the factor ξ_1^2 is placed in the support measure, however, then (4.2) applies. ∎

In general, (4.2) will apply but, in certain contexts, it is necessary to remember that ultimately a relative belief ratio is defined as a limit.

As (4.2) shows, $RB_{\Xi_1}(\xi_1 \mid \xi_2)$ is independent of the support measure used on the space Ξ_1. More generally, the relative belief ratio has an important invariance property.

Proposition 4.2.5 *(Invariance of Relative Belief Ratios) Suppose that Λ is a smooth, 1 to 1 transformation of ξ_1, Υ is a smooth, 1 to 1 transformation of ξ_2 and $\lambda = \Lambda(\xi_1)$, $\tau = \Upsilon(\xi_2)$, then $RB_\Lambda(\lambda \mid \tau) = RB_{\Xi_1}(\xi_1 \mid \xi_2)$, whenever this is defined.*

Proof. This follows because $\Lambda N_{\Xi_1,\delta}(\xi_1)$ and $\Upsilon N_{\Xi_2,\varepsilon}(\xi_2)$ form nice neighborhoods of λ and ξ and their probability contents are the same as the probability contents of $N_{\Xi_1,\delta}(\xi_1)$ and $N_{\Xi_2,\varepsilon}(\xi_2)$, respectively. ∎

Notice that when (4.2) applies, then the invariance with respect to Λ follows from

$$RB_\Lambda(\lambda \mid \tau) = \frac{f_{\Lambda,\Upsilon}(\lambda \mid \tau)}{f_\Lambda(\lambda)} = \frac{f_\Xi(\Lambda^{-1}(\lambda) \mid \xi_2) J_\Lambda(\Lambda^{-1}(\lambda))}{f_{\Xi_1}(\Lambda^{-1}(\lambda)) J_\Lambda(\Lambda^{-1}(\lambda))} = \frac{f_\Xi(\xi_1 \mid \xi_2)}{f_{\Xi_1}(\xi_1)}$$

due to the cancellation of the Jacobian factors in the numerator and denominator.

The following derives the general relative belief ratio in Bayesian contexts.

Example 4.2.4 *Bayesian Relative Belief Ratios (general).*

Here $\omega = (\theta, x) \in \Theta \times \mathscr{X} = \Xi$ and $f(\theta, x) = \pi(\theta) f_\theta(x)$. Let us suppose that $(\psi, x) = (\Psi(\theta), x) = (\Xi_1(\theta, x), \Xi_2(\theta, x))$ so Ξ_1 doesn't depend on x and Ξ_2 is just projection on x and doesn't depend on θ. Following Example 4.2.1, and applying Definition 4.2.2, leads to

$$RB_\Psi(\psi \mid x) = \frac{\pi_\Psi(\psi \mid x)}{\pi_\Psi(\psi)} \tag{4.3}$$

when $\pi_\Psi(\psi) > 0$, where π_Ψ is the prior density and $\pi_\Psi(\cdot \mid x)$ is the posterior density of $\psi = \Psi(\theta)$. Therefore, $RB_\Psi(\psi \mid x) \approx RB(N_\varepsilon(\psi) \mid x)$ when $N_\varepsilon(\psi)$ is a small neighborhood of ψ. So in general $RB_\Psi(\psi \mid x)$ is approximating the change in belief in the truth of ψ from a priori to a posteriori and this is the measure of the evidence that ψ is the true value.

There is no reason to restrict to computing relative belief ratios of parameters, although this seems to cover most statistical applications. The formulation, however, allows the consideration of relative belief ratios of functions of both θ and x. ∎

There is an important consequence of Proposition 4.2.5.

Corollary 4.2.3 *For the Bayesian context $RB_\Psi(\psi \mid x) = RB_\Lambda(\lambda \mid x)$ for any smooth 1 to 1 transformation $\Lambda(\psi) = \lambda$.*

This has the satisfying implication that for inference it is irrelevant how we parameterize a problem. Of course, the parameterization chosen is of considerable importance when choosing a prior, and this needs to be remembered when continuous approximations are employed. This is exemplified by Example 1.4.2.

While it is possible to develop properties of the relative belief ratio in the general context introduced in this section, it is the statistical situation where there is a model, prior and data that is of most interest and so the focus hereafter is primarily on this situation. For example, we have the following analog of Proposition 4.2.4 for the Bayesian context.

Proposition 4.2.6 *If (4.3) holds everywhere, the following properties hold.*

(i) If $\Psi = (\Psi_1, \Psi_2)$, then

$$RB_\Psi(\psi_1, \psi_2 \mid x) = \frac{RB_{\Psi_1}(\psi_1 \mid \psi_2, x) RB_{\Psi_2}(\psi_2 \mid x)}{RB_{\Psi_1}(\psi_1 \mid \psi_2)},$$

when ψ_1 and ψ_2 are a posteriori independent given x,

$$RB_\Psi(\psi_1, \psi_2 \mid x) = \frac{RB_{\Psi_1}(\psi_1 \mid x) RB_{\Psi_2}(\psi_2 \mid x)}{RB_{\Psi_1}(\psi_1 \mid \psi_2)},$$

and, when ψ_1 and ψ_2 are also a priori independent, then

$$RB_\Psi(\psi_1, \psi_2 \mid x) = RB_{\Psi_1}(\psi_1 \mid x) RB_{\Psi_2}(\psi_2 \mid x).$$

(ii) If $x = (x_1, x_2)$ and $X_i(x_1, x_2) = x_i$, then

$$RB_\Psi(\psi \mid x_1, x_2) = RB_\Psi(\psi \mid x_1) \frac{RB_{(\Psi, X_1)}(\psi, x_1 \mid x_2)}{RB_{X_1}(x_1 \mid x_2)},$$

where

$$\frac{RB_{(\Psi, X_1)}(\psi, x_1 \mid x_2)}{RB_{X_1}(x_1 \mid x_2)} = \frac{\pi_\Psi(\psi \mid x_1, x_2)}{\pi_\Psi(\psi \mid x_1)}$$

is the conditional relative belief ratio for ψ given we observe x_2, having already observed x_1.

(iii) (additivity) If $\psi = \Psi(\theta)$, then $RB_\Psi(\psi \mid x) = E_{\Pi(\cdot \mid \psi)}(RB(\theta \mid x))$, where $\Pi(\cdot \mid \psi)$ is the conditional prior of θ given $\Psi(\theta) = \psi$ and $E_{\Pi_\Psi}(RB_\Psi(\psi \mid x)) = E_{\Pi_\Psi(\cdot \mid x)}(1/RB_\Psi(\psi \mid x)) = 1$.

Property (i) in Proposition 4.2.6 shows that, if ψ_1 and ψ_2 are both a posteriori and a priori independent, then the measure of evidence for their joint value is the product of the measures of evidence for their individual values. The general property is easily expressed in terms of densities as

$$\frac{\pi_\Psi(\psi_1, \psi_2 \mid x)}{\pi_\Psi(\psi_1, \psi_2)} = \frac{\pi_{\Psi_1}(\psi_1 \mid \psi_2, x)}{\pi_{\Psi_1}(\psi_1)} \frac{\pi_{\Psi_2}(\psi_2 \mid x)}{\pi_{\Psi_2}(\psi_2)} \Big/ \frac{\pi_{\Psi_1}(\psi_1 \mid \psi_2)}{\pi_{\Psi_1}(\psi_1)}.$$

Also note that $RB_{\Psi_1}(\psi_1 \mid \psi_2)$ is computed using only the prior.

For property (ii) it is supposed that the data is observed sequentially, first observing x_1 and then observing x_2. It is seen that the evidence accumulates multiplicatively and this is how the evidence from observing x_2 either increases or decreases the evidence based on x_1 alone. This property can be expressed directly in terms of densities as

$$\frac{\pi_\Psi(\psi \mid x_1, x_2)}{\pi_\Psi(\psi)} = \frac{\pi_\Psi(\psi \mid x_1)}{\pi_\Psi(\psi)} \frac{\pi_\Psi(\psi \mid x_1, x_2)}{\pi_\Psi(\psi \mid x_1)}.$$

Property (iii) says that the evidence for a value of the parameter of interest is the average of the evidence for the full parameter where the averaging is with respect to the conditional prior of θ given that value for the parameter of interest. Also, the average evidence is neutral when averaging $RB_\Psi(\psi\,|\,x)$ using the prior and the average reciprocal of the evidence is 1 when averaging using the posterior. An interpretation of $1/RB_\Psi(\psi\,|\,x)$ is provided after Proposition 4.3.2.

The following states the general Savage–Dickey ratio result; see Dickey (1971) and Dickey and Lientz (1970). This is based on the conditional prior predictive of x given $\Psi(\theta) = \psi$, namely, $m(x\,|\,\psi) = \int_{\Psi^{-1}\{\psi\}} f_\theta(x)\pi(\theta\,|\,\psi)\,v_{\Psi^{-1}\{\psi\}}(d\theta)$ where $\pi(\theta\,|\,\psi) = \pi(\theta)J_\Psi(\theta)/\pi_\Psi(\psi)$ for $\theta \in \Psi^{-1}\{\psi\}$ is the conditional prior density of θ given that $\Psi(\theta) = \psi$.

Proposition 4.2.7 *When $\pi_\Psi(\psi) > 0$, then*

$$RB_\Psi(\psi\,|\,x) = \frac{m(x\,|\,\psi)}{m(x)} = \frac{m_T(T(x)\,|\,\psi)}{m_T(T(x))},$$

where T is a minimal sufficient statistic for the model.

Proof. By definition

$$RB_\Psi(\psi\,|\,x) = \frac{\pi_\Psi(\psi\,|\,x)}{\pi_\Psi(\psi)} = \int_{\Psi^{-1}\{\psi\}} \frac{\pi(\theta\,|\,x)}{\pi_\Psi(\psi)} J_\Psi(\theta)\,v_{\Psi^{-1}\{\psi\}}(d\theta)$$

$$= \int_{\Psi^{-1}\{\psi\}} \frac{f_\theta(x)}{m(x)} \frac{\pi(\theta)J_\Psi(\theta)}{\pi_\Psi(\psi)}\,v_{\Psi^{-1}\{\psi\}}(d\theta)$$

$$= \frac{\int_{\Psi^{-1}\{\psi\}} f_\theta(x)\pi(\theta\,|\,\psi)\,v_{\Psi^{-1}\{\psi\}}(d\theta)}{m(x)} = \frac{m(x\,|\,\psi)}{m(x)}.$$

∎

An interesting implication of the proposition is that the relative belief ratio is not essentially an a posteriori concept. We could even imagine using the relative belief ratio in a frequentist analysis. A basic concept, however, prevents this at least for inference about θ, namely, the principle of conditional probability. It does make sense, however, to consider the a priori properties of relative belief ratios and this is discussed in Section 4.6.

4.3   Other Proposed Measures of Evidence

There are a number of different functions of $(P(A), P(A\,|\,C))$ that could possibly serve as measures of the evidence for the truth of A given having observed that C is true. Note that a function of $(P(A), P(A\,|\,C))$ can also be considered as a function of $(P(A), RB(A\,|\,C))$, at least when $P(A), P(C) > 0$.

Crupi, Tentori and Gonzalez (2007) discuss many such measures and also measures that are functions of $(P(A), P(C), P(A\,|\,C))$. Table 4.1 presents these measures, considered as functions of $(P(A), P(C), RB(A\,|\,C))$, together with the values they take when A and C converge nicely to $\Xi_1^{-1}\{\xi_1\}$ and $\Xi_2^{-1}\{\xi_2\}$, respectively, both having probability 0, as well as providing the initial reference. Note that $D(C,A) = P(A\,|\,C) - P(A)$ is perhaps the simplest comparison one can think of between $P(A)$ and $P(A\,|\,C)$.

Table 4.1 *Measures of evidence.*

Measure	Limit	Reference
$D(C,A)=P(A)(RB(A\mid C)-1)$	0	Carnap (1950)
$S(C,A)=\frac{P(A)}{1-P(C)}(RB(A\mid C)-1)$	0	Christensen (1999)
$M(C,A)=P(C)(RB(A\mid C)-1)$	0	Mortimer (1988)
$N(C,A)=\frac{P(C)}{1-P(C)}(RB(A\mid C)-1)$	0	Nozick (1983)
$C(C,A)=P(A)P(C)(RB(A\mid C)-1)$	0	Carnap (1950)
$G(C,A)=\frac{P(A)}{1-P(A)}(RB(A\mid C)-1)$	0	Rips (2001)
$R(C,A)=RB(A\mid C)-1$	$RB_{\Xi_1}(\xi_1\mid\xi_2)-1$	Finch (1960)
$L(C,A)=\frac{RB(A\mid C)-1}{RB(A\mid C)+1-2P(A)RB(A\mid C)}$	$\frac{RB_{\Xi_1}(\xi_1\mid\xi_2)-1}{RB_{\Xi_1}(\xi_1\mid\xi_2)+1}$	Kemeny and Openheim (1952)

Observe that only measures R and L deal adequately with the limiting situation and these are both equivalent to the relative belief ratio in the sense that they are 1 to 1 increasing functions of this quantity, in general for R, and in the limit for L. Furthermore, if the other measures are modified by dividing or multiplying by $P(A)$ or $P(C)$ so that they deal adequately with the limiting situation, then all are 1 to 1 increasing functions of $RB(A\mid C)$ and, in the limit, they are also 1 to 1 increasing functions of $RB_{\Xi_1}(\xi_1\mid\xi_2)$.

Crupi, Tentori and Gonzalez (2007) also propose a new measure given by

$$Z(C,A) = \begin{cases} (RB(A\mid C)-1)/(P(A)(1-P(A))) & \text{if } RB(A\mid C)\geq 1, \\ RB(A\mid C)-1 & \text{if } RB(A\mid C)\leq 1. \end{cases}$$

The limiting value is ∞, however, when $RB_{\Xi_1}(\xi_1\mid\xi_2)>1$ and so we reject this as a possible measure of evidence. A similar comment applies to the class of measures they consider based on Z. It seems that Z is being proposed because of a confounding of belief with evidence that is better handled by a separate measure of the strength of the evidence; see Section 4.4.

It would appear that the Bayes factor is not included but, as shown in the following section, $L(C,A)$ is a 1 to 1 increasing function of the Bayes factor and thus equivalent. Furthermore, the limiting value of the Bayes factor equals $RB_{\Xi_1}(\xi_1\mid\xi_2)$ and so is identical to the relative belief ratio in the context of continuous probability.

In Popper (1959) there is discussion of the concept of *corroboration* of a theory. This seems very close to what is called the evidence here. Popper makes the important distinction between belief and corroboration and asserts that probability is not a measure of corroboration. Popper (1983) proposes

$$\frac{RB(A\mid C)-1}{(1-P(A))RB(A\mid C)+1}$$

as a measure of corroboration and, for fixed $P(A)$, this is a 1 to 1 increasing function of $RB(A\mid C)$. The limiting value of this is given by

$$\frac{RB_{\Xi_1}(\xi_1\mid\xi_2)-1}{RB_{\Xi_1}(\xi_1\mid\xi_2)+1},$$

which is a 1 to 1 increasing function of $RB_{\Xi_1}(\xi_1 \mid \xi_2)$ and so equivalent to the relative belief ratio.

In Rescher (1958) the *degree of evidential support* is defined by

$$des(A,C) = \begin{cases} P(C)(RB(A|C) - 1) & \text{if } RB(A|C) \leq 1, \\ \frac{P(A)}{1-P(A)}P(C)(RB(A|C) - 1) & \text{if } RB(A|C) \geq 1. \end{cases}$$

This measure also does not behave appropriately in the limit and modifying it so that it does just leads us to the relative belief ratio.

The relative belief ratio $RB(A|C)$ has been discussed previously as a measure of the evidence that A is true given having observed C. Some authors cite Keynes (1921) and refer to it as the *Keynes association factor* but the specific reference to this in published work by Keynes is unclear. Finch (1960) references Carnap (1950) for introducing this ratio. Finch's R measure is

$$R(C,A) = \frac{P(A|C) - P(A)}{P(A)} = RB(A|C) - 1$$

and it is referred to as the power of the observation of C to support A. Good (1960a) introduces *Good's information function* as $I(A : C) = \log RB(C|A)$ and this is clearly equivalent to the relative belief ratio. The function $I(A : C)$ is discussed further in Section 4.3.2. In all these references, however, there is little or no development of the properties of the relative belief ratio or any real attempt to build a theory of inference based on this quantity. More seriously, there is no calibration provided of the values assumed by these measures.

Based on our study, it is concluded that the essence of all the measures of evidence discussed here seems to be expressed by the relative belief ratio. If one wants to include terms beyond $RB(A|C)$ in such a measure, then a good argument for this is needed and there doesn't appear to be one. In fact, many of these measures seem quite arbitrary. Perhaps only the Bayes factor can serve as a worthy rival and so it is now discussed more fully.

4.3.1   The Bayes Factor

The Bayes factor is a very common measure of evidence and seems to be one of the oldest in use; see Wrinch and Jeffreys (1921). Some discussion of the Bayes factor has been provided in Section 3.5.6.

The Bayes factor measures the difference between $P(A)$ and $P(A|C)$ by comparing the odds ratios based on these quantities. In general, when $0 < P(A) < 1$, and $P(C) > 0$, the Bayes factor equals

$$BF(A|C) = \frac{P(A|C)/(1 - P(A|C))}{P(A)/(1 - P(A))}.$$

When $BF(A|C) > 1$ there is evidence in favor of A, when $BF(A|C) < 1$ there is evidence against A and when $BF(A|C) = 1$ there is no evidence either way.

The following result gives the relationship between the relative belief ratio and the Bayes factor and follows immediately from the results in Section 4.2.1.

Proposition 4.3.1 *When* $0 < P(A) < 1$, *then*

$$BF(A|C) = \frac{RB(A|C)}{RB(A^c|C)} = \frac{(1-P(A))RB(A|C)}{1-P(A)RB(A|C)} = \frac{RB(C|A)}{RB(C|A^c)} = \frac{P(C|A)}{P(C|A^c)},$$

$$RB(A|C) = \frac{BF(A|C)}{1-P(A)+P(A)BF(A|C)},$$

and

(i) $BF(A|C) < 1$ *iff* $RB(A|C) < 1$ *and in that case* $BF(A|C) \leq RB(A|C)$ *with equality iff* $RB(A|C) = BF(A|C) = 0$,

(ii) $BF(A|C) > 1$ *iff* $RB(A|C) > 1$ *and in that case* $BF(A|C) \geq RB(A|C)$ *with equality iff* $RB(A|C) = BF(A|C) = 1$.

From (i) and (ii) it is seen that the Bayes factor satisfies the principle of evidence. The relative belief ratio and Bayes factor give measures of evidence in the same direction but $BF(A|C)$ is not a 1 to 1 increasing function of $RB(A|C)$. The Bayes factor gives more extreme values, but the discussion of calibration of these quantities demonstrates that this is irrelevant.

As with the relative belief ratio, it is possible that $BF(A|C) > BF(B|C)$ when $A \subset B$. In Lavine and Schervish (1999) it is argued that this is incoherent. While this behavior is not appropriate for a measure of belief, as already discussed, it is completely appropriate for a measure of evidence. Interestingly, as can be seen from Proposition 4.3.1, it is difficult to find simple formulas, as in Proposition 4.2.4, for the behavior of the Bayes factor under unions and intersections.

The Bayes factor can also be defined via a limit.

Proposition 4.3.2 *When* $N_\varepsilon(\xi_1)$ *is a sequence of neighborhoods shrinking nicely to* ξ_1 *and* $RB_{\Xi_1}(\xi_1|\xi_2)$ *exists, then*

$$BF_{\Xi_1}(\xi_1|\xi_2) = RB_{\Xi_1}(\xi_1|\xi_2)\frac{1-P_{\Xi_1}(\{\xi_1\})}{1-RB_{\Xi_1}(\xi_1|\xi_2)P_{\Xi_1}(\{\xi_1\})}$$

and so $BF_{\Xi_1}(\xi_1|\xi_2) = RB_{\Xi_1}(\xi_1|\xi_2)$ *whenever* $P_{\Xi_1}(\{\xi_1\}) = 0$.

Proof. Directly from the definition of the Bayes factor for $N_\varepsilon(\xi_1)$, it follows that

$$\lim_{\varepsilon \to 0} BF(N_\varepsilon(\xi_1)|\xi_2) = \lim_{\varepsilon \to 0} \frac{P_\Xi(N_\varepsilon(\xi_1)|\xi_2)}{P_{\Xi_1}(N_\varepsilon(\xi_1))}\frac{1-P_{\Xi_1}(N_\varepsilon(\xi_1))}{1-P_\Xi(N_\varepsilon(\xi_1)|\xi_2)}.$$

∎

Note that this is exactly as stated in Proposition 4.3.1 but it is proven that, when $P_{\Xi_1}(\{\xi_1\}) = 0$, the relative belief ratio and the limiting Bayes factor are the same. It is also worth recalling here Lemma 3.5.6, which established the equality of the Bayes factor with a related relative belief ratio, but this does not follow from Proposition 4.3.2.

Another interesting consequence of Proposition 4.3.2 is that the limiting value of $BF(N_\varepsilon^c(\xi_1)|\xi_2)$ equals $1/RB_{\Xi_1}(\xi_1|\xi_2)$ whenever $P_{\Xi_1}(\{\xi_1\}) = 0$. So $1/RB_{\Xi_1}(\xi_1|\xi_2)$ has an interpretation in terms of the evidence in favor of the complement $\{\xi_1\}^c$ in this context. In general, for sets A with $P(A)$ suitably small, $BF(A\,|\,C) \approx RB(A\,|\,C)$ and $BF(A^c\,|\,C) \approx 1/RB(A\,|\,C)$. So in this context, Proposition 4.2.6 (iii) also says that the average value of the evidence in favor of the complement of a value of the variable of interest is neutral when averaging using the posterior.

As discussed in Section 3.5.6, when $P_{\Xi_1}(\{\xi_1\}) = 0$, it is not uncommon to see the probability measure P_{Ξ_1} modified to be the mixture $P_{\Xi_1,\xi_1} = p\delta_{\xi_1} + (1-p)P_{\Xi_1}$ where δ_{ξ_1} is the probability measure degenerate at ξ_1 and $p \in (0,1)$, in order that the Bayes factor can be defined using the basic definition. But this is not a valid reason to modify our initial probability assignments and, moreover, it is unnecessary. Hereafter this is referred to as Jeffreys' definition of the Bayes factor in the continuous case and further discussed in Section 4.5.

The *degree of factual support*, as a measure of evidence, is defined in Kemeny and Oppenheim (1952) as

$$F(A\,|\,C) = \frac{P(C\,|\,A) - P(C\,|\,A^c)}{P(C\,|\,A) + P(C\,|\,A^c)}.$$

Using the symmetry of the relative belief ratio, then

$$F(A\,|\,C) = \frac{RB(C\,|\,A)P(C) - RB(C\,|\,A^c)P(C)}{RB(C\,|\,A)P(C) + RB(C\,|\,A^c)P(C)} = \frac{BF(A\,|\,C) - 1}{BF(A\,|\,C) + 1}$$

and this is a 1 to 1 increasing function of $BF(A\,|\,C)$. For the limiting case, when $P_{\Xi_1}(\{\xi_1\}) = 0$, then $F_{\Xi_1}(\xi_1\,|\,\xi_2) = (RB_{\Xi_1}(\xi_1\,|\,\xi_2) - 1)/(RB_{\Xi_1}(\xi_1\,|\,\xi_2) + 1)$, which is a 1 to 1 increasing function of $RB_{\Xi_1}(\xi_1\,|\,\xi_2)$. Also, using Corollary (4.2.2), the previously defined measure $L(C,A)$ of Table 4.1 is seen to satisfy

$$L(C,A) = \frac{RB(A\,|\,C) - RB(A^c\,|\,C)}{RB(A\,|\,C) + RB(A^c\,|\,C)} = \frac{BF(A\,|\,C) - 1}{BF(A\,|\,C) + 1},$$

which is also a 1 to 1, increasing function of the Bayes factor.

In general, the mathematics of relative belief ratios is much simpler than that of Bayes factors. Furthermore, it seems that a theory of inference based on the relative belief ratio is more tractable than other choices of measures of evidence. Finally, Proposition 4.3.2 shows that, when defining the relative belief ratio and the Bayes factor appropriately as limits in the case when $P(A) = 0$, the same answer is obtained.

4.3.2   Good's Information and Weight of Evidence

As already mentioned, this treatment is not the first to suggest $RB(A\,|\,C)$ as a measure of the evidence that A is true having observed that C is true. Also, Good (1960a, 1965, 1971, 1983) contain discussion of what is called the information concerning A provided by C and defined by $I(A:C) = \log RB(C\,|\,A)$. Proposition 4.2.2 then implies that $I(A:C) = \log RB(A\,|\,C)$. So *Good's information function* is just the log relative belief ratio. Although taking the logarithm is irrelevant, Good seems to have preferred this form because Proposition 4.2.4 (ii) implies that Good's information function possesses a form of additivity, as information is obtained sequentially. On the other hand, taking the logarithm obscures the more important additivity result given by Proposition 4.2.4 (iii).

Good also referred to the log Bayes factor $W(A:C) = \log BF(A|C)$ as the *weight of evidence*, as a measure of the evidence, and distinguished between this and the information. Good (1985) argues for using $W(A:C)$ to measure evidence as opposed to using $I(A:C)$. He did this because, as Proposition 4.3.1 shows,

$$W(A:C) = \log RB(A|C) - \log RB(A^c|C) = I(A:C) - I(A^c:C)$$

is a comparison between the information for A and the information for its complement A^c. This supposes however, that assessing a hypothesis involves a comparison between the hypothesis and its negation and a very different point of view is taken on this in the discussion of calibration in Section 4.4. Also, for the case when $P(A) = 0$, Proposition 4.3.2 shows that there is really no difference between $W(A:C)$ and $I(A:C)$. For us, both the relative belief ratio and the Bayes factor are measuring evidence as change in belief, albeit in different ways in the discrete case, just as $D(C,A) = P(A|C) - P(A)$ is measuring evidence in yet another way.

Good's concern seems generally to give an appropriate definition of evidence in the simple context where $P(A) > 0$ and $P(C) > 0$. While there are some disagreements with his developments, for example, preferring $RB(A|C)$ to $BF(A|C)$, there is much to agree with in his writings on the topic of evidence. Good doesn't seem, however, to have considered a general development of statistical inference based on these ideas. This is carried out in Section 4.5.

4.3.3 Desiderata for a Measure of Evidence

Consider now characterizing a measure of evidence based upon properties that would be appropriate for such a concept. First it is desirable that this measure be applicable to any context where there is a probability measure P on a set Ω and interest is in the truth of A after observing $\omega \in C$, with $P(A) > 0$ and $P(C) > 0$.

The measure of evidence is naturally restricted to be a real-valued function of $(P(A), P(A|C))$ for this simple context, so that it depends only on the beliefs expressed about A, and is in some sense a measure of the difference between $P(A)$ and $P(A|C)$. Let us denote this function by $ev(P(A), P(A|C))$ and suppose that larger values indicate more evidence in favor of A being true. It then seems natural to allow for any strictly increasing, continuous function of ev to also be considered as a valid measure of evidence. So ev corresponds to an equivalence class of evidential functions. Values $ev(0, p)$ with $p > 0$ are not defined, as, in this simple context at least, there can be no evidence obtained in favor of or against A, when $P(A) = 0$.

That ev be continuous in its arguments seems desirable. This is because small changes in $(P(A), P(A|C))$ should result in small changes in the measure of the evidence. Note that this continuity is over all probability models and has nothing to do with the essential discreteness of a particular application. Also, $ev(P(A), P(A|C))$ should be strictly increasing in $P(A|C)$, as evidence for A must increase as $P(A|C)$ increases, with $P(A)$ fixed. Note that $ev(P(A), P(A|C))$ must also strictly decrease as $P(A)$ increases, with $P(A|C)$ fixed, but this is not needed for our argument.

Now consider the measurement of belief. There would appear to be no reason why some other scale, as opposed to the probability scale, cannot be used to measure belief, provided it can be mapped 1 to 1 to the probability scale and it is natural to restrict to 1 to 1, continuous mappings. Furthermore, there must be some invariants of belief that apply to all such scales. For example, if we have probabilities p and q, and the function g maps $[0,1]$ to a new scale, then g must be such that $(g(p),g(q))$ possesses some properties that are essential to (p,q). The most natural property to demand as an invariant of belief is the invariance of relative belief, namely, require that $q/p = g(q)/g(p)$ for all $p \in (0,1]$ and $q \in [0,1]$. So an increase in belief by a factor of r in one scale increases belief by the same factor in another scale. Taking $q = 1$ implies that $g(p) = g(1)p$ where $g(1) \neq 0$.

It is then natural to require that the measurement of evidence also be invariant under such transformations so that the measure of evidence is to a certain extent independent of how belief is measured. So it is required that $ev(rP(A), rP(A|C)) = ev(P(A), P(A|C))$ for every $r \neq 0$ and note that this entails that ev be defined at values outside $[0,1]$. It is then immediate that

$$ev(P(A), P(A|C)) = ev(1, RB(A|C)).$$

This implies that ev is a strictly increasing function of the relative belief ratio. As such, the relative belief ratio is a fundamental measure of evidence, as it is in the equivalence class of evidential functions. In essence $RB(A|C)$ is the maximal invariant under rescalings of the probability scale. So a few intuitively reasonable requirements lead to the relative belief ratio as the measure of evidence.

Notice that $BF(A|C) = RB(A|C)/RB(A^c|C)$ is also invariant under such transformations but it is not a function of $RB(A|C)$. This is not a contradiction, as $BF(A|C)$ is not defined when $P(A) = 1$.

The measure of evidence also needs to be generalizable to continuous contexts so that it reflects properties of the approximation, as discussed in Section 1.4. Let $N_{\Xi_1,\delta}(\xi_1)$ and $N_{\Xi_2,\varepsilon}(\xi_2)$ converge nicely (see Appendix A) to $\xi_1 = \Xi_1(\omega)$ and $\xi_2 = \Xi_2(\omega)$, as $\delta \to 0$ and $\varepsilon \to 0$, respectively. Then, by continuity and invariance,

$$ev(P_{\Xi_1}(N_{\Xi_1,\delta}(\xi_1)), P_{\Xi}(N_{\Xi_1,\delta}(\xi_1)|N_{\Xi_2,\varepsilon}(\xi_2)))$$
$$= ev\left(1, P_{\Xi}(N_{\Xi_1,\delta}(\xi_1)|N_{\Xi_2,\varepsilon}(\xi_2))/P_{\Xi_1}(N_{\Xi_1,\delta}(\xi_1))\right)$$
$$\to ev(1, RB_{\Xi_1}(\xi_1|\xi_2)),$$

whenever $RB_{\Xi_1}(\xi_1|\xi_2)$ is defined as in Definition 4.2.2. In general, for the continuous case, the invariance under rescalings implies invariance under change of support measure and under reparameterization. Both properties can be considered as essential for any valid measure of evidence.

Undoubtedly other properties can be given that imply the relative belief ratio as the natural measure of evidence. For example, Good (1965) provides an axiomatic development of $I(A:C)$. While such developments are satisfying, it is primarily the properties of the theory of inference derived from the relative belief ratio that provide the greatest conviction about the appropriateness of this as the measure of the evidence.

4.4 Measuring the Strength of the Evidence

It is not uncommon for a Bayes factor to be quoted as the evidence for a hypothesis being true or false and its size to be interpreted as the strength of that evidence. For example, a Bayes factor or relative belief ratio of 200 might seem to be very strong evidence in favor of a hypothesis, as the odds in favor or probability, respectively, have increased by a factor of 200. Indeed, there are proposed scales for the Bayes factor that seem to originate with Jeffreys and are discussed in Kass and Raftery (1995) which aim at assessing when there is weak, moderate or strong evidence, etc.

Expression (4.1) shows, however, that in the discrete case, the possible values for the relative belief ratio are highly dependent on the prior probabilities chosen. While the extremes are associated with situations where the truth value of the event in question is known categorically, this still leaves open the interpretation of the strength of intermediate values.

In the continuous case, Example 3.5.2 (the Jeffreys–Lindley paradox) demonstrates the futility of trying to devise a universal scale on which to measure evidence. For, by choosing the prior to be sufficiently diffuse, a Bayes factor as large as desired can be obtained. This comment also applies to the relative belief ratio, as, by Lemma 3.5.6, it is equal to the Bayes factor in that context. So the evidence is being biased in favor of the hypothesis in question simply because the measures of evidence are getting larger as less information is placed into the problem by the prior. An explicit measure of this bias is obtained in general in Section 4.6 and in particular for the context of Example 3.5.2 in Example 4.6.1.

In Example 3.5.2 it is argued that, as a measure of evidence, the Bayes factor/relative belief ratio is behaving correctly and so the problem is not with the measure of evidence. Many authors point to the prior as being the problem and indeed this seems to be the case. But this in itself does not resolve the problem, as it does not tell us how to avoid the bias. Some authors have suggested allowing the prior to depend on the data but this is not a valid solution, as it leads to inferences that violate a basic principle of inference, namely, the principle of conditional probability. Furthermore, there doesn't appear to be another principle that justifies this in some meaningful way.

In this section a general method for calibrating a relative belief ratio in a given problem is presented. This calibration is seen to be context dependent and results from taking a somewhat different point of view about hypothesis assessment. In particular, it is not considered as a hypothesis testing problem where the goal is to *decide* whether a hypothesis is true or false. Rather, when considering a hypothesized value for a parameter of interest, the evidence for this value is compared with the evidence for *each* of the other possible values for that parameter. This is seen to resolve one particular aspect of the Jeffreys–Lindley paradox, as now the classical p-value arises as a measure of the strength of the evidence and not as a measure of the evidence itself. Full resolution of the difficulties associated with this example will await Section 4.6, however, where measuring and controlling the bias introduced into an analysis by a prior is examined.

4.4.1   The Strength of the Evidence

Measuring the strength of the evidence, as given by a relative belief ratio, is some-what more involved than just measuring the evidence itself. Consider the Bayesian context of Example 4.2.4 and the relative belief ratio $RB_\Psi(\psi_0\,|\,x)$, which provides the evidence for the hypothesis $H_0 = \Psi^{-1}\{\psi_0\}$. A natural question then is to ask if a specific value $RB_\Psi(\psi_0\,|\,x) > 1$ is strong evidence in favor, moderate evidence in favor or weak evidence in favor of H_0. Similarly, whenever $RB_\Psi(\psi_0\,|\,x) < 1$, and there is evidence against H_0, the same question concerning strength can be asked. This is the essential issue of calibration and, it has been argued here, there does not appear to be a universal scale which can be applied for this purpose.

What is available, however, are all the values $RB_\Psi(\psi\,|\,x)$ for $\psi \in \Psi$. A natural way to calibrate $RB_\Psi(\psi_0\,|\,x)$ then is to compare this value with the other values of $RB_\Psi(\psi\,|\,x)$. As the posterior $\Pi_\Psi(\cdot\,|\,x)$ expresses beliefs about the true value of ψ, this comparison is to be done by comparing $RB_\Psi(\psi_0\,|\,x)$ to the posterior distribution of $RB_\Psi(\cdot\,|\,x)$. In essence, then, we want to compute a posterior probability that assesses the extent to which we believe in the evidence obtained and this will be the measure of the strength of the evidence. Consider the following motivating ideas.

1. *Evidence in favor:* If $RB_\Psi(\psi_0\,|\,x) > 1$ and a high posterior probability is assigned to the true value having a relative belief ratio bigger than $RB_\Psi(\psi_0\,|\,x)$, then $RB_\Psi(\psi_0\,|\,x)$ is interpreted as weak evidence in favor of H_0. If $RB_\Psi(\psi_0\,|\,x) > 1$ and a low posterior probability is assigned to the true value having a relative belief ratio bigger than $RB_\Psi(\psi_0\,|\,x)$, then $RB_\Psi(\psi_0\,|\,x)$ is interpreted as strong evidence in favor of H_0.

2. *Evidence against:* If $RB_\Psi(\psi_0\,|\,x) < 1$ and a high posterior probability is assigned to the true value having a relative belief ratio bigger than $RB_\Psi(\psi_0\,|\,x)$, then $RB_\Psi(\psi_0\,|\,x)$ is interpreted as strong evidence against H_0. If $RB_\Psi(\psi_0\,|\,x) < 1$ and a low posterior probability is assigned to the true value having a relative belief ratio bigger than $RB_\Psi(\psi_0\,|\,x)$, then $RB_\Psi(\psi_0\,|\,x)$ is interpreted as weak evidence against H_0.

To see the motivation behind these points consider the simplest situation where there are only two possible values of Ψ so $\#(\Psi) = 2$. Recall from Proposition 4.2.3 (i) that, if $RB_\Psi(\psi_0\,|\,x) > 1$, then the other value of ψ has relative belief ratio less than 1 and conversely. Then the posterior probability

$$\Pi_\Psi(RB_\Psi(\psi\,|\,x) = RB_\Psi(\psi_0\,|\,x)\,|\,x) \tag{4.4}$$

can be used to locate $RB_\Psi(\psi_0\,|\,x)$ in the posterior distribution of $RB_\Psi(\cdot\,|\,x)$. For example, when $RB_\Psi(\psi_0\,|\,x) > 1$ and (4.4) is small, then there is evidence in favor of H_0 but our belief that ψ_0 is the true value is weak while, if $RB_\Psi(\psi_0\,|\,x) < 1$ with (4.4) small, then there is evidence against H_0 and our belief that ψ_0 is not the true value is strong. Similarly, when $RB_\Psi(\psi_0\,|\,x) > 1$ and (4.4) is large, then there is evidence in favor of H_0 and our belief that ψ_0 is the true value is strong while, if $RB_\Psi(\psi_0\,|\,x) < 1$ with (4.4) is large, then there is evidence against H_0 and our belief that ψ_0 is not the true value is weak. Notice that the interpretation of the value (4.4) as a measure of the strength of the evidence presented by $RB_\Psi(\psi_0\,|\,x)$ is different depending on whether there is evidence for H_0 or evidence against H_0.

While (4.4) seems like a reasonable measure of the strength of the evidence when there are only a few possible values of ψ, it is clearly inadequate when there are many possible values, all with small prior probabilities, so that (4.4) is small. For example, in the continuous context it will generally be the case that (4.4) is equal to 0. As such, it is necessary to look at other ways of locating $RB_\Psi(\psi_0 \mid x)$ in the posterior distribution of $RB_\Psi(\cdot \mid x)$. The following definition of the strength of the evidence seems more appropriate in such contexts.

Definition 4.4.1 *The strength of the evidence* $RB_\Psi(\psi_0 \mid x)$ *is given by*

$$\Pi_\Psi(RB_\Psi(\psi \mid x) \le RB_\Psi(\psi_0 \mid x) \mid x), \qquad (4.5)$$

namely, the posterior probability that the true value of ψ *has a relative belief ratio no larger than that obtained for* ψ_0.

While (4.5) looks like a p-value, it is important to note that its interpretation is quite different depending on whether $RB_\Psi(\psi_0 \mid x)$ is less than or greater than 1. For when $RB_\Psi(\psi_0 \mid x) < 1$, so there is evidence against the value ψ_0, then a small value of (4.5) indicates a large belief that the true value of ψ is in the set $\{\psi : RB_\Psi(\psi \mid x) > RB_\Psi(\psi_0 \mid x)\}$ and so there is strong evidence against ψ_0. If (4.5) is big when $RB_\Psi(\psi_0 \mid x) < 1$, then there is only weak evidence against ψ_0. When $RB_\Psi(\psi_0 \mid x) > 1$, so there is evidence in favor of the value ψ_0, then a small value of (4.5) indicates a large belief that the true value of ψ is in the set $\{\psi : RB_\Psi(\psi \mid x) > RB_\Psi(\psi_0 \mid x)\}$ and so there is weak evidence in favor of ψ_0. If (4.5) is big when $RB_\Psi(\psi_0 \mid x) > 1$, then there is a large belief that the true value is in the set $\{\psi : RB_\Psi(\psi \mid x) \le RB_\Psi(\psi_0 \mid x)\}$ and so strong evidence in favor of ψ_0. Notice that ψ_0 is the value with the most evidence in the set $\{\psi : RB_\Psi(\psi \mid x) \le RB_\Psi(\psi_0 \mid x)\}$ and so is the natural choice, from within this set, to estimate the true value of ψ. Estimation problems are discussed further in Sections 4.5 and 4.7.

The value of $\Pi_\Psi(RB_\Psi(\psi \mid x) \le RB_\Psi(\psi_0 \mid x) \mid x)$ seems to nicely capture the ideas expressed in points 1 and 2 concerning measuring the strength of the evidence by locating $RB_\Psi(\psi_0 \mid x)$ in the posterior distribution of $RB_\Psi(\cdot \mid x)$. In contexts where there are only a few possible values of ψ, however, one might prefer using (4.4) to measure the strength, although there is no reason why both can't be quoted, or for that matter, additional posterior probabilities quoted that are felt to be appropriate. The following example indicates why (4.4) might be preferred to (4.5) when the cardinality of Ψ is small.

Example 4.4.1 *Measuring Strength in the Binary Case.*

Suppose that $\Psi = \{0, 1\}$, so

$$\Pi_\Psi(RB_\Psi(\psi \mid x) \le RB_\Psi(0 \mid x) \mid x) = \begin{cases} 1 & \text{if } RB_\Psi(0 \mid x) \ge RB_\Psi(1 \mid x), \\ \Pi_\Psi(\{0\} \mid x) & \text{if } RB_\Psi(0 \mid x) < RB_\Psi(0 \mid x), \end{cases}$$

and a similar result is obtained for $RB_\Psi(1 \mid x)$. By Proposition 4.2.3 (i), $RB_\Psi(0 \mid x) = RB_\Psi(1 \mid x)$ if and only if their common value is 1, and so observing x provides no evidence about the value of ψ. When $RB_\Psi(0 \mid x) > RB_\Psi(1 \mid x)$, then Proposition 4.2.3 (i) implies that $RB_\Psi(0 \,|\, x) > 1$ and so (4.5) indicates that we have the strongest possible evidence in favor of $\psi = 0$. This doesn't seem sensible, as this extreme value arises because there are only two possible values. In this context (4.4) seems like a more appropriate assessment of strength. ∎

The size of (4.4) or (4.5) is not evidence one way or the other, but rather these quantities measure the strength of the evidence given by the relative belief ratio. As will be seen, there are great benefits that arise due to separating the measurement of evidence from the measurement of its strength. In particular, it will be seen that this is somewhat analogous to recording an estimate of a quantity together with an assessment of the accuracy of the estimate. Just as recording an estimate without an assessment of its accuracy, such as a standard error, could be regarded as being an incomplete inference, recording a measure of evidence without some assessment of its strength leaves a significant part of the inference story untold.

The following inequalities play a key role in measuring the strength of evidence.

Proposition 4.4.1 *Suppose that (4.2) holds everywhere, then*

$$\Pi_\Psi(RB_\Psi(\psi \,|\, x) = RB_\Psi(\psi_0 \,|\, x) \,|\, x) \leq \Pi_\Psi(RB_\Psi(\psi \,|\, x) \leq RB_\Psi(\psi_0 \,|\, x) \,|\, x)$$
$$\leq RB_\Psi(\psi_0 \,|\, x) \,|\, x).$$

Proof. The first inequality is obvious. The second inequality follows from Markov's inequality as

$$\Pi_\Psi(1/RB_\Psi(\psi \,|\, x) \geq 1/RB_\Psi(\psi_0 \,|\, x) \,|\, x) \leq \frac{E_{\Pi_\Psi(\cdot \,|\, x)}(1/RB_\Psi(\psi \,|\, x))}{1/RB_\Psi(\psi_0 \,|\, x)} = RB_\Psi(\psi_0 \,|\, x)$$

since $E_{\Pi_\Psi(\cdot \,|\, x)}(1/RB_\Psi(\psi \,|\, x)) = 1$. ∎

From Proposition 4.4.1, when $RB_\Psi(\psi_0 \,|\, x)$ is very small, then there is strong evidence against ψ_0 and when $RB_\Psi(\psi_0 \,|\, x) < 1$ and $\Pi_\Psi(RB_\Psi(\psi \,|\, x) = RB_\Psi(\psi_0 \,|\, x) \,|\, x)$ is large, then there is weak evidence against ψ_0. When $RB_\Psi(\psi_0 \,|\, x) > 1$ and $\Pi_\Psi(RB_\Psi(\psi \,|\, x) = RB_\Psi(\psi_0 \,|\, x) \,|\, x)$ is large, then there is strong evidence in favor of ψ_0.

As the next example demonstrates, it is possible that $RB_\Psi(\psi_0 \,|\, x) > 1$ is large with (4.5) small.

Example 4.4.2 *Relative Belief Inferences and the Jeffreys–Lindley Paradox.*

In Example 3.5.2 a contradiction was exhibited between classical frequentist and Bayesian inferences based on the Bayes factor alone. From Lemma 3.5.6 the relative belief ratio $RB(\mu \,|\, x)$ in that example equals the Bayes factor computed using the mixture prior. Therefore,

$$RB(0 \,|\, x) = \sqrt{1 + n\tau_0^2} \exp\left\{ -\frac{n\bar{x}^2}{2} \frac{n\tau_0^2}{1 + n\tau_0^2} \right\} \tag{4.6}$$

and so considering the relative belief ratio by itself does not resolve the paradox. But now also consider the strength of the evidence provided by (4.6). An easy calculation gives

$$\Pi(RB(\mu\,|\,x) \leq RB(0\,|\,x)\,|\,x) = 1 - \Phi((1+1/n\tau_0^2)^{1/2}(|\sqrt{n}\bar{x}| + (n\tau_0^2+1)^{-1}\sqrt{n}\bar{x})) +$$
$$\Phi((1+1/n\tau_0^2)^{1/2}(-|\sqrt{n}\bar{x}| + (n\tau_0^2+1)^{-1}\sqrt{n}\bar{x}))\qquad(4.7)$$

and from this conclude that

$$\Pi(RB(\mu\,|\,x) \leq RB(0\,|\,x)\,|\,x) \to 2(1 - \Phi(|\sqrt{n}\bar{x}|))$$

as $\tau_0^2 \to \infty$. So the classical p-value serves here as a measure of the strength of the evidence given by $RB(0\,|\,x)$ when using a very diffuse prior. This implies that a very large relative belief ratio, or equivalently Bayes factor, is not necessarily strong evidence in favor of the hypothesis.

As a specific numerical example, suppose that $n = 50$, $\tau_0^2 = 400$ and $\sqrt{n}\bar{x} = 1.96$. Then $RB(0\,|\,x) = 20.72$ and Jeffreys' scale (see Kass and Raftery (1995)) says that this is strong evidence in favor of $H_0 = \{0\}$. But (4.7) equals 0.05 and, as such, 20.72 is clearly not strong evidence in favor of H_0 as there is a large posterior probability that the true value has a larger relative belief ratio. In this case the value of μ maximizing $RB(\mu\,|\,x)$ equals $\mu_{LRSE}(x) = \bar{x} = 0.28$ with $RB(\mu_{LRSE}(x)\,|\,x) = 141.40$. As discussed in Section 4.5, $\mu_{LRSE}(x)$ is the estimate of μ implied by the evidence. Note that $\mu_{LRSE}(x) = 0.28$ cannot be interpreted as being close to 0 independent of the application context. If, however, the application dictates that a value of 0.28 is practically speaking close enough to 0 to be treated as 0, then it certainly seems reasonable to proceed as if H_0 is correct and this is supported by the value of the relative belief ratio.

This example clearly illustrates the essence of the problem with the interpretation of large values of a relative belief ratio or Bayes factor as strong evidence in favor of H_0. For, as the prior becomes more diffuse via $\tau_0^2 \to \infty$, the evidence in favor of H_0 becomes arbitrarily large. So it seems possible to bias the evidence a priori in favor of H_0 by choosing τ_0^2 very large and indeed it is shown that this is the case in Example 4.6.1. As noted in Example 3.5.2, the relative belief ratio is behaving correctly in these circumstances but now there is a check on this effect via the computation of the strength of the evidence.

Frequentists may have difficulty accepting that a small value of $2(1 - \Phi(|\sqrt{n}\bar{x}|))$ is not necessarily evidence against H_0. It is to be noted, however, that the frequentist treatment is never really explicit about how evidence is to be measured. For example, large values of a p-value are never considered as evidence in favor of H_0. Our treatment here, however, is based on a precise definition of how evidence is to be measured.

It is not claimed that what has been presented so far in this example completely resolves the paradox. With a change in interpretation for the p-value, however, it is seen that the Bayes factor and the p-value are no longer contradictory. There is still a need, however, to pin down how to choose the value of τ_0^2. Of course, the correct answer is through elicitation, and we will discuss this in Chapter 5, but this does not answer the question of whether or not a prior is inducing bias. The measurement of this bias is discussed in Section 4.6 and, in particular, see Example 4.6.1. ∎

The following result adds further support to interpreting a large value of (4.5) as strong evidence in favor of ψ_0, when $RB_\Psi(\psi_0 \,|\, x) > 1$.

Proposition 4.4.2 *When (4.2) holds and* $RB_\Psi(\psi_0 \,|\, x) > 1$, *then* $RB(RB_\Psi(\psi \,|\, x) < RB_\Psi(\psi_0 \,|\, x)) < RB_\Psi(\psi_0 \,|\, x)$.

Proof. We have that

$$\Pi_\Psi(RB_\Psi(\psi \,|\, x) < RB_\Psi(\psi_0 \,|\, x) \,|\, x)$$

$$= \int_{\{\psi : RB_\Psi(\psi \,|\, x) < RB_\Psi(\psi_0 \,|\, x)\}} \pi_\Psi(\psi \,|\, x)\, \nu_\Psi(d\psi)$$

$$= \int_{\{\psi : RB_\Psi(\psi \,|\, x) < RB_\Psi(\psi_0 \,|\, x)\}} RB_\Psi(\psi \,|\, x) \pi_\Psi(\psi)\, \nu_\Psi(d\psi)$$

$$\leq RB_\Psi(\psi_0 \,|\, x) \int_{\{\psi : RB_\Psi(\psi \,|\, x) < RB_\Psi(\psi_0 \,|\, x)\}} \pi_\Psi(\psi)\, \nu_\Psi(d\psi)$$

$$= RB_\Psi(\psi_0 \,|\, x)\Pi_\Psi(RB_\Psi(\psi \,|\, x) < RB_\Psi(\psi_0 \,|\, x))$$

and equality occurs if and only if $\Pi_\Psi(RB_\Psi(\psi \,|\, x) < RB_\Psi(\psi_0 \,|\, x)) = 0$ since $RB_\Psi(\psi_0 \,|\, x) - RB_\Psi(\psi \,|\, x) > 0$ on $\{\psi : RB_\Psi(\psi \,|\, x) < RB_\Psi(\psi_0 \,|\, x)\}$. So equality implies $\Pi_\Psi(RB_\Psi(\psi \,|\, x) < RB_\Psi(\psi_0 \,|\, x) \,|\, x) = 0$ and thus

$$1 = \Pi_\Psi(RB_\Psi(\psi \,|\, x) \geq RB_\Psi(\psi_0 \,|\, x) \,|\, x)$$

$$= \int_{\{\psi : RB_\Psi(\psi \,|\, x) \geq RB_\Psi(\psi_0 \,|\, x)\}} RB_\Psi(\psi \,|\, x) \pi_\Psi(\psi)\, \nu_\Psi(d\psi)$$

$$\geq RB_\Psi(\psi_0 \,|\, x) > 1$$

and this is a contradiction. ∎

So the evidence that the true value is in $\{\psi : RB_\Psi(\psi \,|\, x) < RB_\Psi(\psi_0 \,|\, x)\}$ is strictly less than the evidence that ψ_0 is the true value, when we have evidence in favor of ψ_0 being true. By a similar argument we also have the following result, which supports the interpretation of the strength for evidence against.

Proposition 4.4.3 *When (4.2) holds and* $RB_\Psi(\psi_0 \,|\, x) < 1$, *then* $RB(RB_\Psi(\psi \,|\, x) > RB_\Psi(\psi_0 \,|\, x)) > RB_\Psi(\psi_0 \,|\, x)$.

4.5   Inference Based on Relative Belief Ratios

In the previous sections of this chapter the relative belief ratio $RB_\Psi(\psi \,|\, x)$ has been introduced as the relevant measure of the evidence that ψ is the true value of the parameter of interest and furthermore, it has been proposed that the strength of this evidence, either in favor or against, be assessed by comparing the observed value $RB_\Psi(\psi \,|\, x)$ against the posterior distribution of $RB_\Psi(\cdot \,|\, x)$.

Our goal now is to provide solutions to the two inference problems raised in Section 1.8, namely, (i) provide an estimate of the true value of ψ together with an assessment of the accuracy of the estimate and (ii) provide a measure of the evidence that the hypothesis $H_0 = \Psi^{-1}\{\psi_0\}$ is either true or false together with an assessment of the strength of this evidence. Since the solution to (ii) has essentially already been discussed, we start with that problem together with some additional aspects.

As a fundamental ingredient for determining the inferences, the relative belief ratio provides a total ordering on Ψ.

The *Relative Belief Preference Ordering:* ψ_1 is not strictly preferred to ψ_2, denoted $\psi_1 \preceq \psi_2$, whenever $RB_\Psi(\psi_1 \,|\, x) \leq RB_\Psi(\psi_2 \,|\, x)$.

So ψ_1 is not strictly preferred to ψ_2 whenever the evidence for ψ_1 being the true value is not strictly greater than the evidence for ψ_2 being the true value. The relative belief preference ordering is similar to the likelihood preference ordering and in fact they agree when $\Psi(\theta) = \theta$. The relative belief preference ordering is defined for all parameters, however, and this is not the case for the likelihood preference ordering.

4.5.1   Hypothesis Assessment

Consider the hypothesis that $\Psi(\theta) = \psi_0$, or equivalently, it is proposed that the true value of θ is in $H_0 = \Psi^{-1}\{\psi_0\}$. The evidence that the hypothesis is true is given by $RB_\Psi(\psi_0 \,|\, x)$ and possible measures of the strength of this evidence are given by (4.4) or (4.5).

In Evans (1997) the quantity (4.5) was referred to as the *observed relative surprise* (ORS) and its usage there was like a p-value. Certainly this looks like a p-value but, as already discussed, its interpretation is quite different. It is incorrect to conclude that there is evidence against ψ_0 when (4.5) is small and $RB_\Psi(\psi_0 \,|\, x) > 1$, as this violates the principle of evidence. Accordingly, (4.5) is no longer recommended as a measure of surprise.

Aitkin (2010), following Dempster (1973), uses $\Pi(L(\theta \,|\, x) \leq L(\theta_0 \,|\, x) \,|\, x)$ for assessment of a point hypothesis for the full model parameter θ, where $L(\cdot \,|\, x)$ is the likelihood, and small values mean evidence against. For the full parameter θ (see Example 4.5.3) it is the case that $\Pi(RB(\theta \,|\, x) \leq RB(\theta_0 \,|\, x) \,|\, x) = \Pi(L(\theta \,|\, x) \leq L(\theta_0 \,|\, x) \,|\, x)$. As mentioned, this violates the principle of evidence whenever $RB(\theta_0 \,|\, x) > 1$ and, in any case, it seems more appropriate to use this quantity as part of assessing the strength of the evidence. For marginal parameter $\psi = \Psi(\theta)$, Aitkin (2010) uses $\Pi_\Psi(L^\Psi(\psi \,|\, x) \leq L^\Psi(\psi_0 \,|\, x) \,|\, x)$ for assessment of hypotheses about ψ, where $L^\Psi(\cdot \,|\, x)$ is the profile likelihood, and this suffers from the additional difficulties associated with profile likelihoods. Other than this, however, there is considerable agreement with the intention underlying Aitkin (2010), namely, to base inference on the measurement of statistical evidence.

It can sometimes arise with statistical problems that a hypothesis is simply stated as a subset $H_0 \subset \Theta$, namely, there is no Ψ and specific value ψ_0 provided such that $H_0 = \Psi^{-1}\{\psi_0\}$, so that Ψ *generates* H_0. This introduces an ambiguity into the problem, as typically there will be many Ψ such that $H_0 = \Psi^{-1}\{\psi_0\}$ for some ψ_0. So in such a context it is necessary to choose a Ψ that generates H_0.

For a specific Ψ,

$$RB_\Psi(\psi_0 \,|\, x) = \frac{\pi_\Psi(\psi_0 \,|\, x)}{\pi_\Psi(\psi_0)} = \frac{\int_{H_0} \pi(\theta \,|\, x) J_\Psi(\theta) \, \mu_{H_0}(d\theta)}{\int_{H_0} \pi(\theta) J_\Psi(\theta) \, \mu_{H_0}(d\theta)} \tag{4.8}$$

and from this it is seen that $RB_\Psi(\psi_0 \,|\, x)$ is invariant to the choice of Ψ whenever $J_\Psi(\theta)$ is constant for every $\theta \in H_0$ and for every Ψ generating H_0. This is the case when Π is discrete but in the continuous case generally this does not hold and so the measure of the evidence and its strength depends on the choice made of Ψ and the volume distortions it introduces. This issue is caused in essence by a failure to fully specify the statistical problem and is similar to the Borel paradox discussed in Example 2.1.2.

One general principle that can be invoked when possible is to choose Ψ so that $J_\Psi(\theta)$ is constant for every $\theta \in \Psi^{-1}\{\psi\}$ for each ψ. In such a case $RB_\Psi(\psi \,|\, x)$ is not affected by volume distortions induced by Ψ and so neither is the evidence nor any measure of its strength. Consider an example.

Example 4.5.1 *Testing Equality of Means.*

Suppose that x is a sample from an $N(\mu_1, \sigma^2)$ distribution independent of a sample y from an $N(\mu_2, \sigma^2)$ distribution where $(\mu_1, \mu_2, \sigma^2) \in R \times R \times R^+$ is unknown and the hypothesis of interest is $H_0 : \mu_1 = \mu_2$. A simple choice for Ψ is $\Psi(\mu_1, \mu_2, \sigma^2) = \mu_1 - \mu_2$ as $H_0 = \Psi^{-1}\{0\}$. This choice of Ψ has $J_\Psi(\mu_1, \mu_2, \sigma^2) = 1/\sqrt{2}$ and so never induces volume distortion into the inference. Any 1 to 1 smooth function of Ψ also has this property when using relative belief.

There are many other Ψ transformations that generate $H_0 : \mu_1 = \mu_2$. For example, for $p > 0$, using $\Psi(\mu_1, \mu_2, \sigma^2) = \mu_1^p - \mu_2^p$, we have $H_0 = \Psi^{-1}\{0\}$. Then $J_\Psi(\mu_1, \mu_2, \sigma^2) = p^{-1}(\mu_1^{2(p-1)} + \mu_2^{2(p-1)})^{-1/2}$, which equals $p^{-1}\mu^{-(p-1)}$ when $\mu_1 = \mu_2 = \mu$ and this is not constant on H_0 when $p \neq 1$. ∎

A general approach to the problem of an unspecified Ψ generating H_0 is to take $\Psi(\theta)$ to be a measure of the distance of θ from H_0, so then $H_0 = \Psi^{-1}\{0\}$. The value $RB_\Psi(0 \,|\, x)$ is then a comparison of the concentration of the posterior distribution of θ about H_0 with the concentration of the prior distribution of θ about H_0. If H_0 is true, then it is to be expected that the posterior concentrates more than the prior about H_0 and so $RB_\Psi(0 \,|\, x)$ should be large. This approach to assessing H_0 is referred to as the *method of concentration*.

For example, in Example 4.5.1 using squared Euclidean distance of (μ_1, μ_2) from $H_0 = L\{1\}$ leads to $\Psi(\mu_1, \mu_2, \sigma^2) = (\mu_1 - \mu_2)^2/2$. In this case $J_\Psi(\mu_1, \mu_2, \sigma^2) = |\mu_1 - \mu_2|^{-1} = \sqrt{2}|\psi|^{-1}$ and so volume distortions play no role in the inferences. In this case, however, there is a singularity induced at H_0 and $RB_\Psi(0 \,|\, x)$ is defined as a limit as in Definition 4.2.2. This approach can be employed to handle general ANOVA type problems, as discussed in Baskurt and Evans (2013, 2015). Also, the method of concentration has been successfully used in a variety on inference problems with contingency tables, as discussed in Evans, Gilula, Guttman and Swartz (1997) and Evans, Gilula and Guttman (2012).

As a general principle, however, it is necessarily preferred that a Ψ generating H_0 be specified as part of the inference problem. It is worth noting, however, that this is again an issue that arises due to treating continuous models as being true, as with discrete contexts it does not arise. As such, an appropriate discretization, as advocated in Section 4.7.1, avoids this problem.

4.5.2   Estimation

The total order on Ψ induced by $RB_\Psi(\cdot\,|x)$ leads necessarily to the estimate

$$\psi_{LRSE}(x) = \arg\sup_{\psi} RB_\Psi(\psi\,|x)$$

as it maximizes the evidence. It is subsequently shown that $RB_\Psi(\psi_{LRSE}(x)\,|x) > 1$, so indeed there is evidence in favor of $\psi_{LRSE}(x)$ being the true value. In Evans (1997) the estimator ψ_{LRSE} is referred to as the least relative surprise estimator of ψ, as $\psi_0 = \psi_{LRSE}(x)$ clearly maximizes (4.5) as a function of ψ_0. A better name would be the *maximum relative belief estimator* and at times this terminology is used in the text.

Undoubtedly one can construct situations where $\psi_{LRSE}(x)$ does not exist, but this possibility is largely ignored, as it does not seem to occur in standard statistical contexts. It can also happen that $\psi_{LRSE}(x)$ is not unique and it seems better to deal with these situations as they arise rather than trying to develop a theory that allows for this. So, unless stated otherwise, it is assumed hereafter that in any statistical problem under discussion, $\psi_{LRSE}(x)$ exists and is unique.

It is also important that an estimate be accompanied by a measure of its accuracy. For this a value $\gamma \in [0,1]$ is chosen and a γ-*credible region* $C_{\Psi,\gamma}(x)$ containing $\psi_{LRSE}(x)$ is quoted where the "size" of $C_{\Psi,\gamma}(x)$ is the assessment of the accuracy. The concept of size is application dependent but could be taken to be the volume of the set $C_{\Psi,\gamma}(x)$ when Ψ is Euclidean, or the cardinality of $C_{\Psi,\gamma}(x)$ in the case of finite sets. The value γ will typically be something reasonably large, such as $\gamma = 0.95$, but other choices are possible and even a range of values can be selected and the relevant sets quoted.

The total ordering induced by $RB_\Psi(\cdot\,|x)$ immediately dictates the form that $C_{\Psi,\gamma}(x)$ must take, namely, $C_{\Psi,\gamma}(x) = \{\psi : RB_\Psi(\psi\,|x) \geq k\}$ for some k. If $C_{\Psi,\gamma}(x)$ did not take this form, there would be values $\psi_1 \in C_{\Psi,\gamma}(x)$, $\psi_2 \notin C_{\Psi,\gamma}(x)$ but $RB_\Psi(\psi_1\,|x) \leq RB_\Psi(\psi_2\,|x)$, which contradicts the principle of evidence. Therefore, a γ-*relative belief region* for ψ is given by

$$\begin{aligned}
C_{\Psi,\gamma}(x) &= \{\psi : G_\Psi(RB_\Psi(\psi\,|x)\,|x) \geq 1 - \gamma\} \\
&= \{\psi : RB_\Psi(\psi\,|x) \geq G_\Psi^{-1}(1 - \gamma\,|x)\}
\end{aligned} \qquad (4.9)$$

where $G_\Psi(\cdot\,|x)$ is the posterior cdf of $RB_\Psi(\cdot\,|x)$ and $G_\Psi^{-1}(\cdot\,|x)$ is the associated quantile function. In Evans (1997) the set $C_{\Psi,\gamma}(x)$ was referred to as a γ-*relative surprise region*.

Proposition 3.5.1 can be applied to this situation with $\omega = \psi$ and $h = RB_\Psi(\cdot\,|x)$. The following summarizes what this result says about relative belief regions and the strength (4.5).

Properties of Relative Belief (RB) Regions and the Strength of the Evidence:
(i) says that RB regions are nested and always contain $\psi_{LRSE}(x)$,
(ii) says that a γ-RB region is indeed a γ-credible region as it contains at least γ of the posterior probability for ψ,
(iii) shows an equivalence between the definition of the strength (4.5) and RB credible regions, namely,

$$\Pi_\Psi(RB_\Psi(\psi\,|\,x) \leq RB_\Psi(\psi_0\,|\,x)\,|\,x) = 1 - \inf\{\gamma : \psi_0 \in C_{\Psi,\gamma}(x)\},$$

and, since $G_\Psi(RB_\Psi(\psi_0\,|\,x)\,|\,x) = \Pi_\Psi(RB_\Psi(\psi\,|\,x) \leq RB_\Psi(\psi_0\,|\,x)\,|\,x)$, then (4.9) says that $C_{\Psi,\gamma}(x)$ is composed of those ψ_0 values for which the strength of evidence for ψ_0 is at least $1 - \gamma$,
(iv) says that RB regions are always continuous from above and
(v) says that RB regions are also continuous from below at any γ such that $G_\Psi(\cdot\,|\,x)$ never assumes the value $1 - \gamma$ or does so with probability 0.

Properties (iv) and (v) imply that γ-RB regions are continuous in γ whenever the posterior distribution of $RB_\Psi(\cdot\,|\,x)$ is continuous. Therefore, however size is measured to assess accuracy, $C_{\Psi,\gamma}(x)$ is the smallest relative belief region containing at least γ of the posterior probability.

Property (iii) demonstrates a relationship between the two measures of the reliability of the basic estimation and hypothesis assessment inferences and, as such, reinforces our view that (4.5) is an appropriate measure of the strength of the evidence. Notice that, putting $str(\psi_0) = \Pi_\Psi(RB_\Psi(\psi\,|\,x) \leq RB_\Psi(\psi_0\,|\,x)\,|\,x)$, by property (iv), the smallest relative belief region containing ψ_0 is $C_{\Psi,1-str(\psi_0)}(x)$. So, if $RB_\Psi(\psi_0\,|\,x) > 1$ and $str(\psi_0)$ is large, then $C_{\Psi,1-str(\psi_0)}(x)$ is small in the sense that not too many points beyond $\psi_{LRSE}(x)$ are needed to cover ψ_0. If $RB_\Psi(\psi_0\,|\,x) < 1$ and $str(\psi_0)$ is small, then $C_{\Psi,1-str(\psi_0)}(x)$ is large in the sense that many points beyond $\psi_{LRSE}(x)$ are needed to cover ψ_0. There is no reason why the "size" of the set $C_{\Psi,1-str(\psi_0)}(x)$, using the relevant application measure of size, shouldn't also play a role in our assessment of strength.

It is clear that for larger values of γ, the set $C_{\Psi,\gamma}(x)$ may contain values ψ for which $RB_\Psi(\psi\,|\,x)\,|\,x) < 1$. This does not contradict our usage of this set, however, as it is being used to assess the accuracy of $\psi_{LRSE}(x)$ and not as a presentation of a set of "plausible" values for ψ. A more appropriate inference for the latter situation would be a set of the form $Pl_{\Psi,q}(x) = \{\psi : RB_\Psi(\psi\,|\,x) > q\}$ for some $q \geq 1$, as then there is evidence in favor of each of the values in $Pl_{\Psi,q}(x)$. The value q is a lower bound on the amount the probability has increased from a priori to a posteriori. The set $Pl_{\Psi,q}(x)$ will be referred to as the *q-plausible region* for ψ and the value $\Pi_\Psi(Pl_{\Psi,q}(x)\,|\,x)$ as the *plausibility* of $Pl_{\Psi,q}(x)$. Clearly $\psi_{LRSE}(x) \in Pl_{\Psi,q}(x)$ whenever $Pl_{\Psi,q}(x) \neq \phi$ or, equivalently, whenever $0 \leq q \leq RB_\Psi(\psi_{LRSE}(x)\,|\,x)\,|\,x)$ and note that $Pl_{\Psi,1}(x) \neq \phi$.

Estimation also plays a role in the assessment of the hypothesis $H_0 = \Psi^{-1}\{\psi_0\}$. For suppose it is concluded that there is evidence against H_0. As is well known, this can happen because there is a large amount of data and a very small deviation from H_0 has been detected that may not be meaningful in the application. A natural solution to this problem is to look at $\psi_{LRSE}(x)$ to see if it differs meaningfully from

ψ_0. This requires that, as part of the application, a meaningful difference δ has been determined so that, if $|\psi_0 - \psi_{true}| < \delta$, then ψ_0 is also considered to be effectively true. For example, if the parameter ψ is the mean of a response and the response variable is measured in centimeters, then a deviation from the hypothesized value of less than one-half of a centimeter does not seem meaningful.

This issue is clearly connected with the use of continuous models and a similar concern arises with the interpretation of the strength of the evidence as measured by (4.5). Further discussion on this is provided in Example 4.7.1. Overall the conclusion is that it is essential in an application that the approximate nature of continuous models be recognized and prescribing such a δ value plays a key role in this.

4.5.3 Prediction Inferences

Suppose after observing x our interest is in predicting a value $y \in \mathcal{Y}$ where y has model given by $\{g_\lambda(\cdot \mid x) : \lambda \in \Lambda\}$ with respect to support measure $\mu_\mathcal{Y}$ on $\mathcal{Y}, \lambda = \Lambda(\theta)$ and $\lambda_{true} = \Lambda(\theta_{true})$. This allows for the possibility that the distribution of y depends on x and that θ may not index the set of possible distributions for y.

The joint density of (θ, x, y) is $\pi(\theta) f_\theta(x) g_{\Lambda(\theta)}(y \mid x)$, the posterior predictive density of y is $q(y \mid x) = \int_\Theta \pi(\theta \mid x) g_{\Lambda(\theta)}(y \mid x) \nu(d\theta)$ and the prior predictive density of y is $q(y) = \int_\Theta \int_\mathcal{X} \pi(\theta) f_\theta(x) g_{\Lambda(\theta)}(y \mid x) \mu(dx) \nu(d\theta)$. Therefore, the relative belief ratio for a future value y is $RB_\mathcal{Y}(y \mid x) = q(y \mid x)/q(y)$.

Inferences for the unknown value of y then proceed just as for a parameter but now using the relative belief ratio $RB_\mathcal{Y}(y \mid x)$, as this measures the change in belief in y from a priori to a posteriori. For example, $y_{LRSE}(x) = \arg\sup RB_\mathcal{Y}(y \mid x)$ is the posterior prediction of a future value of y and a γ-relative belief prediction region for y can be formed to assess the accuracy of this prediction. Furthermore, the evidence for a specific future value y_0 is $RB_\mathcal{Y}(y_0 \mid x)$ and $Q_\mathcal{Y}(RB_\mathcal{Y}(y \mid x) \leq RB_\mathcal{Y}(y_0 \mid x) \mid x)$ can be used as part of measuring the strength of the evidence in favor of or against y_0. So there are really no new concepts that need to be introduced when considering prediction problems.

4.5.4 Examples

A number of examples are presented here of applying relative belief inferences. These examples are chosen to demonstrate particular aspects of the inferences.

Example 4.5.2 *Models With Two Distributions.*

Suppose $\{f_\theta : \theta \in \Theta\}$ where $\Theta = \{\theta_1, \theta_2\}$ and prior π is positive at each θ. Then

$$RB(\theta \mid x) = \frac{f_\theta(x)}{m(x)} = \frac{f_\theta(x)}{\pi(\theta_1) f_{\theta_1}(x) + \pi(\theta_2) f_{\theta_2}(x)} = \frac{f_\theta(x)/f_{\theta_2}(x)}{\pi(\theta_1) f_{\theta_1}(x)/f_{\theta_2}(x) + \pi(\theta_2)},$$

so the evidence for θ_1 is

$$RB(\theta_1 \mid x) = \frac{f_{\theta_1}(x)/f_{\theta_2}(x)}{\pi(\theta_1) f_{\theta_1}(x)/f_{\theta_2}(x) + \pi(\theta_2)},$$

which is a strictly increasing function of the likelihood ratio $f_{\theta_1}(x)/f_{\theta_2}(x)$. Also, $RB(\theta_1\,|\,x) > 1$ if and only if $RB(\theta_2\,|\,x) < 1$ and $RB(\theta_1\,|\,x) = 1$ if and only if $RB(\theta_2\,|\,x) = 1$. Notice that $RB(\theta_1\,|\,x)/RB(\theta_2\,|\,x) = f_{\theta_1}(x)/f_{\theta_2}(x)$, so the relative evidence for θ_1 compared to θ_2 is indeed given by the likelihood ratio and this agrees with the developments in Section 3.2. As $f_{\theta_1}(x)/f_{\theta_2}(x) \to \infty$, then $RB(\theta_1\,|\,x) \to 1/\pi(\theta_1)$, the maximum possible value, while $RB(\theta_2\,|\,x) \to 0$, the minimum possible value.

Suppose that $RB(\theta_1\,|\,x) > 1$, so that $\theta_{LRSE}(x) = \theta_1$. Therefore, $\theta_1 \in C_\gamma(x)$ for every $\gamma \in [0, 1]$. The posterior probability of θ_1 equals

$$\pi(\theta_1\,|\,x) = \frac{\pi(\theta_1)f_{\theta_1}(x)/f_{\theta_2}(x)}{\pi(\theta_1)f_{\theta_1}(x)/f_{\theta_2}(x) + \pi(\theta_2)}$$

and $\pi(\theta_1\,|\,x) \to 1$ as $f_{\theta_1}(x)/f_{\theta_2}(x) \to \infty$. Therefore, for $\gamma < 1$ and $f_{\theta_1}(x)/f_{\theta_2}(x)$ large enough, $C_\gamma(x) = \{\theta_1\}$. Furthermore, the strength of the evidence in favor of $H_0 = \{\theta_1\}$, as measured by (4.4), is $\pi(\theta_1\,|\,x)$ and so the evidence becomes stronger the larger the likelihood ratio $f_{\theta_1}(x)/f_{\theta_2}(x)$. When $\pi(\theta_1) \approx 0$, then $RB(\theta_1\,|\,x) \approx f_{\theta_1}(x)/f_{\theta_2}(x)$ and the strength of the evidence is a balance between $\pi(\theta_1)$ and $f_{\theta_1}(x)/f_{\theta_2}(x)$.

When $\pi(\theta_1) = \pi(\theta_2)$, then effectively relative belief and likelihood inferences are the same. The relative belief approach, however, gives a clearer interpretation for the choice of γ in assessing the accuracy of the estimate, and the strength of the evidence is calibrated by a probability as opposed to just quoting the likelihood ratio. ∎

The following example establishes a more general partial equivalence with likelihood inferences.

Example 4.5.3 *Likelihood Inferences for the Model Parameter.*

Suppose that interest is in inference about the full model parameter θ. In this case, whenever (4.3) holds, $RB(\theta\,|\,x) = f_\theta(x)/m(x)$, which is proportional to any likelihood function for θ. Notice, however, that $RB(\cdot\,|\,x)$ cannot be multiplied by an arbitrary positive constant, as this would destroy its interpretation as a measure of the change in belief and, as such, as the evidence. It is the case, however, that $\theta_{LRSE}(x) = \theta_{MLE}(x)$ and $C_\gamma(x)$ is a likelihood region for θ for each γ. Notice, however, that there is now a much stronger basis for the choice of γ than with pure likelihood theory, as γ is a probability. Also, $RB(\theta\,|\,x)$ gives a measure of the evidence for an individual θ while the likelihood $L(\theta\,|\,x)$ only leads to a measure of relative evidence between two values through the likelihood ratio.

The evidence for the hypothesis $H_0 = \{\theta_0\}$ is given by $RB(\theta_0\,|\,x)$ and the strength, as measured by (4.5), is $\Pi(RB(\theta\,|\,x) \leq RB(\theta_0\,|\,x)\,|\,x) = \Pi(L(\theta\,|\,x) \leq L(\theta_0\,|\,x)\,|\,x)$. This is the posterior probability that the true value has a likelihood no greater than the hypothesized value.

The biggest failing of the pure likelihood approach is its treatment of marginal parameters $\psi = \Psi(\theta)$, as the basic motivating idea behind the likelihood approach does not lead to inferences for such quantities unless they are 1 to 1 reparameterizations. So in a real sense, the relative belief approach builds on the basic ideas in likelihood theory by filling in the gaps in a logical, consistent way. It is easy to see that, for a general marginal parameter $\psi = \Psi(\theta)$, the same discussion applies with $RB(\theta \,|\, x)$ replaced by $RB_\Psi(\psi \,|\, x)$ and $L(\theta \,|\, x)$ replaced by the integrated likelihood, which is proportional to the conditional prior predictive given that $\psi = \Psi(\theta)$, namely,

$$m(x \,|\, \psi) = \int_{\Psi^{-1}\{\psi\}} f_\theta(x) \pi(\theta \,|\, \psi) \, v_{\Psi^{-1}\{\psi\}}(d\theta).$$

It is worth noting here that (4.3) does not strictly apply in a commonly discussed situation. For suppose the prior is given by a mixture prior $\Pi = p\delta_0 + (1-p)\Pi_0$ where δ_0 is degenerate at θ_0 and Π_0 is a continuous prior with density π_0. Then the posterior distribution of θ is

$$\Pi(\cdot \,|\, x) = \frac{p f_{\theta_0}(x)}{p f_{\theta_0}(x) + (1-p) m_0(x)} \delta_0 + \frac{(1-p) m_0(x)}{p f_{\theta_0}(x) + (1-p) m_0(x)} \Pi_0(\cdot \,|\, x),$$

where $m_0(x) = \int_\Theta f_\theta(x) \Pi_0(d\theta)$. Remembering, however, that the relative belief ratio is defined as the limit $RB(\theta \,|\, x) = \lim_{\varepsilon \to 0} \Pi(N_\varepsilon(\theta) \,|\, x) / \Pi(N_\varepsilon(\theta))$ for neighborhoods $N_\varepsilon(\theta)$ shrinking nicely to θ, we get $RB(\theta \,|\, x) = f_\theta(x) / (p f_{\theta_0}(x) + (1-p) m_0(x))$. ∎

The following, relatively simple example, illustrates quite nicely the difference between a decision, loss-based approach to a statistical problem and one based on measuring evidence alone.

Example 4.5.4 *Prosecutor's Fallacy.*

In general, the prosecutor's fallacy seems to refer to any kind of error in probabilistic reasoning made by a prosecutor when arguing for the conviction of a defendant. The paper by Thompson and Schumann (1987) is a good reference and this example is concerned with its relevance to measuring statistical evidence.

Suppose a population is split into two classes where a proportion ε are guilty of a crime and a proportion $1 - \varepsilon$ are not guilty. Suppose further that a particular trait is held by a proportion ψ_1 of those innocent and a proportion ψ_2 of those who are guilty. The overall proportion in the population with the trait is then $(1 - \varepsilon)\psi_1 + \varepsilon\psi_2$ and this will be small whenever ε and ψ_1 are small. The values ε and ψ_1 being small corresponds to the proportion of guilty being very small and the trait being very rare in the population. The prosecutor notes that the defendant has the trait and, because $(1 - \varepsilon)\psi_1 + \varepsilon\psi_2$ is very small, concludes that the defendant is guilty. Actually, as recorded in Thompson and Schumann (1987), it seems that the prosecutor in question actually quoted $1 - \{(1 - \varepsilon)\psi_1 + \varepsilon\psi_2\}$ as the probability of guilt! In any case, our concern here is the reasoning concerning the smallness of $(1 - \varepsilon)\psi_1 + \varepsilon\psi_2$ and what it implies about the guilt of the defendant.

Treating ε as the prior probability that the defendant is guilty, without observing whether or not they have the trait, it is seen that the posterior probability that the defendant is guilty, given that they have the trait, is

$$P(\text{"guilty"} \,|\, \text{"defendant has the trait"}) = \frac{\varepsilon\psi_2}{(1 - \varepsilon)\psi_1 + \varepsilon\psi_2},$$

and this converges to 0 as $\varepsilon \to 0$. This suggests that, when ε is small, the prosecutor's reasoning is fallacious. The relative belief ratio for guilt equals

$$RB(\text{"guilty"} \mid \text{"defendant has the trait"}) = \frac{\psi_2}{(1-\varepsilon)\psi_1 + \varepsilon\psi_2}$$

and the relative belief ratio for innocence is

$$RB(\text{"innocent"} \mid \text{"defendant has the trait"}) = \frac{\psi_1}{(1-\varepsilon)\psi_1 + \varepsilon\psi_2}.$$

Now $RB(\text{"guilty"} \mid \text{"defendant has the trait"}) > 1$ if and only if $\psi_2 > \psi_1$ and this occurs if and only if $RB(\text{"innocent"} \mid \text{"defendant has the trait"}) < 1$. If the trait is at all useful in terms of determining guilt, it is sensible to suppose $\psi_2 > \psi_1$ and, under these circumstances, it is certainly reasonable to say we have evidence in favor of guilt, as the probability of guilt has increased from a priori to a posteriori.

The question now is: does relative belief commit a prosecutor's fallacy when $\psi_2 > \psi_1$? It might seem so as evidence of guilt is always found when the trait is observed. Recall, however, that there are two parts to a relative belief inference whether estimation or hypothesis assessment, namely, we must also assess the accuracy of the inference. Consider assessing the strength of the evidence $RB(\text{"guilty"} \mid \text{"defendant has the trait"})$ that the hypothesis $H_0 :$ "guilty" is true. As there are only two possible values, the posterior probability $P(\text{"guilty"} \mid \text{"defendant has the trait"})$ is a suitable measure of strength of this evidence and, as previously noted, this converges to 0 as $\varepsilon \to 0$. Therefore, even though there is evidence in favor of guilt, it is only weak evidence and it *may be* that it is not appropriate to convict based upon evidence in favor of guilt that is considered weak, given the consequences of this to an innocent person.

The situation is more complicated than this, however, and exposes a clear distinction between taking a decision-based approach and an evidential one. For consider the problem where ε corresponds to the proportion of individuals infected with a deadly infectious disease and ψ_1, ψ_2 correspond to the probabilities of a test for infection being positive in the noninfected and infected populations, respectively. Note that common terminology is to refer to $1 - \psi_1$ as the *specificity* and ψ_2 as the *sensitivity*. A good test will of course have high specificity and sensitivity, implying $\psi_2 > \psi_1$, and so this is exactly the same situation, as, for a patient with a positive test, relative belief will record that there is evidence the patient is infected. Even if this is weak evidence, however, it *may be* foolhardy to simply ignore the evidence and treat the person as not infected.

A standard Bayesian decision approach in this problem is MAP, which, for ε small enough, will declare the defendant innocent and the patient noninfected. In the former case this seems fine but not in the latter case. It would seem that categorical statements are not what is wanted from a statistical procedure in such problems. Undoubtedly decisions will ultimately be made and these decisions may, for good reasons, ignore what the evidence says. But the additional criteria that come into play in making decisions are not statistical in nature. It is surely reasonable to ultimately consider these additional characteristics of a problem but, from a theory of statistics, a natural requirement is a clear statement of the evidence and its strength.

It is also reasonable to consider what happens when ε is fixed and small and $\psi_1 \to 0$. In these circumstances $P(\text{"guilty"} \mid \text{"defendant has the trait"}) \to 1$ and the evidence of guilt becomes strong. This is surely correct, as when $\psi_1 = 0$ it is known categorically that the defendant is guilty. So the prosecutor is not necessarily wrong to consider the defendant guilty. The basis for the prosecutor's reasoning can of course be called into question. This needs to be based on an appropriate theory of evidence so that the reasoning is correct. ∎

The following example was also considered in Chapter 3 and called into question the interpretation of a likelihood ratio.

Example 4.5.5 *Relative Belief Inferences for Example 3.2.1.*

Suppose now there is a prior π_k on Θ_k, where the dependence on k indicates that the prior must depend on the number of symbols. Note also that $\#(\Theta_k) \sim k^M$ as $k \to \infty$. It is immediate that $\theta_{LRSE}(x) = \theta_{MLE}(x)$ so it might appear that nothing has changed. But now also consider what the relative belief ratios indicate. The posterior density of θ is given by

$$
\pi_k(\theta \mid x) = \begin{cases} \frac{\{1/(k+1)+\delta\}\pi_k(x)}{m_k(x)} & \theta = x, \\ \frac{\{1/(k+1)-\delta/k\}\pi_k(x'(x))}{m_k(x)} & \theta = x'(x), \\ 0 & \text{otherwise,} \end{cases}
$$

where $m_k(x) = \{1/(k+1)+\delta\}\pi_k(x) + \{1/(k+1)-\delta/k\}\pi_k(x'(x))$. This implies that

$$
RB(x \mid x) = \frac{1/(k+1)+\delta}{\{1/(k+1)+\delta\}\pi_k(x) + \{1/(k+1)-\delta/k\}\pi_k(x'(x))},
$$

$$
RB(x'(x) \mid x) = \frac{1/(k+1)-\delta/k}{\{1/(k+1)+\delta\}\pi_k(x) + \{1/(k+1)-\delta/k\}\pi_k(x'(x))},
$$

and the dependence of the prior on k must also be taken into account as $k \to \infty$. For example, if we take π_k to be the uniform prior on Θ_k, then $RB(x \mid x) \to \infty$ and $RB(x'(x) \mid x) \to \infty$ as $k \to \infty$ for every M. So evidence for $\theta = x$ does not correspond to evidence against $\theta = x'(x)$.

Also, $\pi(x \mid x) \to 1$ and $\pi(x'(x) \mid x) \to 0$ as $k \to \infty$ so the posterior can also be viewed as providing a misleading measure of the evidence in support of $\theta = x$. This also implies that, for all k large enough and any γ, the γ-relative belief region is $C_\gamma(x) = \{x\}$. For any $q > 1$ and for k large enough, however, the q-plausibility region is $PL_q(x) = \{x, x'(x)\}$ with plausibility equal to 1. This at least indicates that the value $x'(x)$ is plausible.

So under realistic assumptions about the prior, the relative belief ratio leads to more reasonable inferences based on the evidence than the likelihood ratio. Still these inferences do not strictly agree with the natural frequentist inference. Recall that the likelihood ratio $RB(x'(x) \mid x)/RB(x \mid x)$ converges to 0 no matter how we choose the prior.

Another relevant comment for this example concerns the sample size. Suppose instead a sample (x_1, x_2) is observed. If (x_1, x_2) is observed with $0 < l(x_i) < M$, then

$$
f_\theta(x_1, x_2) = \begin{cases} [1/(k+1)+\delta]^2 & x_1 = x_2 = \theta \\ [1/(k+1)+\delta][1/(k+1)-\delta/k] & x_1 = \theta, x_2 = \theta a_i \\ [1/(k+1)-\delta/k]^2 & x_1 = \theta a_i, x_2 = \theta a_j \\ 0 & \text{otherwise.} \end{cases}
$$

When $x_1 = x_2$, this is basically the same situation as when a single value is observed but, when $x_1 \neq x_2$, which occurs with probability approaching $1 - \delta^2$ as $k \to \infty$, then $\theta_{LRSE}(x) = \theta_{MLE}(x) = \theta$ and inference is categorical and correct. The situation improves even more as sample size grows. ■

The next example is a context where the Bayes factor, as defined by Jeffreys, appears to behave anomalously. This is compared with the behavior of the relative belief ratio.

Example 4.5.6 *Information Inconsistency for Bayes Factors (continued).*

Recall that Example 3.5.3 demonstrated that, for a sample of n from the $N(\mu, \sigma^2)$ normal model, with the mixture prior $\Pi = p\delta_{\mu_0} + (1 - p)\Pi_0$ given by $\mu \,|\, \sigma^2 \sim p\delta_{\mu_0} + (1 - p)N(\mu_0, \tau_0^2\sigma^2), 1/\sigma^2 \sim \text{gamma}_{\text{rate}}(\alpha_0, \beta_0), \Psi(\mu) = \mu$, and $H_0 = \{\mu_0\}$, the Bayes factor as defined by Jeffreys did not converge to 0 as $\sqrt{n}|\bar{x} - \mu_0| \to \infty$. Now examine the relative belief ratio for this problem based on the continuous prior Π_0, namely, $\mu \,|\, \sigma^2 \sim N(\mu_0, \tau_0^2\sigma^2), 1/\sigma^2 \sim \text{gamma}_{\text{rate}}(\alpha_0, \beta_0)$. Note that $RB_\Psi(\mu_0 \,|\, x)$ will be different from Jeffreys' Bayes factor since we are not dealing with the full model parameter and so Lemma 3.5.6 does not apply.

A straightforward calculation shows that the prior distribution of μ is given by

$$\mu \sim \mu_0 + \sqrt{\tau_0^2\beta_0/\alpha_0}\, t_{2\alpha_0},$$

where $t_\lambda \sim \text{Student}(\lambda)$, while the posterior distribution is given by

$$\mu \,|\, x \sim \mu(x) + \sqrt{(n + 1/\tau_0^2)^{-1}\beta(x)/(n + 2\alpha_0)}\, t_{n+2\alpha_0},$$

where

$$\mu(x) = (n + 1/\tau_0^2)^{-1}(n\bar{x} + \mu_0/\tau_0^2) = (n + 1/\tau_0^2)^{-1}n(\bar{x} - \mu_0) + \mu_0,$$
$$\beta(x) = (n + 1/\tau_0^2)^{-1}[n(\bar{x} - \mu_0)^2/\tau_0^2] + s^2 + 2\beta_0.$$

Therefore,

$$RB_\Psi(\mu \,|\, x) = \frac{\pi_\Psi(\mu \,|\, x)}{\pi_\Psi(\mu)}$$

$$= \left\{ \frac{\Gamma\left(\frac{n+2\alpha_0}{2}\right)}{\Gamma(\frac{1}{2})\Gamma\left(\frac{n-1+2\alpha_0}{2}\right)} \Big/ \frac{\Gamma\left(\frac{1+2\alpha_0}{2}\right)}{\Gamma(\frac{1}{2})\Gamma\left(\frac{2\alpha_0}{2}\right)} \right\} \left\{ \frac{\sqrt{\tau_0^2\beta_0/\alpha_0}}{\sqrt{(n + 1/\tau_0^2)^{-1}\beta(x)/(n + 2\alpha_0)}} \right\} \times$$

$$\left\{ \frac{(1 + (n + 2\alpha_0)(\mu - \mu(x))^2/(n + 1/\tau_0^2)^{-1}\beta(x))^{-\frac{n+2\alpha_0}{2}}}{(1 + \alpha_0(\mu - \mu_0)^2/\tau_0^2\beta_0)^{-\frac{1+2\alpha_0}{2}}} \right\}. \qquad (4.10)$$

It is of interest that $RB_\Psi(\mu \,|\, x) \to \infty$ as $\tau_0^2 \to \infty$, so the Jeffreys–Lindley paradox applies here as well.

Our concern now, however, is with $RB_\Psi(\mu_0 \,|\, x)$ as $\sqrt{n}|\bar{x} - \mu_0| \to \infty$ with s^2 constant. It is the case that $RB_\Psi(\mu_0 \,|\, x) \to 0$ since, with $\mu = \mu_0$, the first factor in (4.10) is constant, the second factor goes to 0 and the third converges to a constant as

$$\frac{(\mu_0 - \mu(x))^2}{(n + 1/\tau_0^2)^{-1}\beta(x)} = n\frac{n(\bar{x} - \mu_0)^2}{n(\bar{x} - \mu_0)^2/\tau_0^2 + (n + 1/\tau_0^2)s^2 + 2(n + 1/\tau_0^2)\beta_0} \to n\tau_0^2$$

when $\sqrt{n}|\bar{x} - \mu_0| \to \infty$. So the relative belief ratio does not exhibit information inconsistency in this example.

It should be noted that $RB_\Psi(\mu_0 \,|\, x)$ is quite different from $BF(H_0 \,|\, x)$ obtained in Example 3.5.3. For us, the relative belief ratio based on Π_0 is more appropriate, as it treats the continuous model and prior as approximations. The Jeffreys Bayes factor attempts to treat the continuous model as if it were the truth, and then compensates for the problems introduced by the prior Π_0 assigning 0 mass to the point null by requiring a somewhat arbitrary modification of the prior. ∎

4.6 Measuring the Bias in the Evidence

Considering the a priori behavior of the relative belief ratio has a number of useful implications for inference. One of the virtues of basing the inferences on a measure of the evidence is that it is possible to consider the effects on the evidence of the choice of prior before the data is collected. As such, the relative belief ratio can be used as a design tool as part of a statistical investigation. For example, in a well-designed statistical investigation, the amount of data collected, the model and the prior are all chosen as informatively as possible.

First recall Proposition 4.2.7 and write $RB_\Psi(\psi_0 \,|\, x) = m(x \,|\, \psi_0)/m(x) = m_T(T(x) \,|\, \psi_0)/m_T(T(x))$ where T is a minimal sufficient statistic. Following a similar idea found in Royall (2000) for pure likelihood inferences, consider the prior probability of getting misleading evidence. Discussion of these ideas in a Bayesian context appears in Baskurt and Evans (2013).

First consider the possibility of a small value of $RB_\Psi(\psi_0 \,|\, x)$ when $H_0 = \Psi^{-1}\{\psi_0\}$ is true, as we know that this would be misleading evidence.

Proposition 4.6.1 *The prior probability that $RB_\Psi(\psi_0 \,|\, x) \leq q$, given that H_0 is true, is bounded above by q, namely, $M_T(m_T(t \,|\, \psi_0)/m_T(t) \leq q \,|\, \psi_0) \leq q$.*

Proof. Using Markov's inequality,

$$M_T\left(\frac{m_T(t)}{m_T(t \,|\, \psi_0)} \geq \frac{1}{q} \,\middle|\, \psi_0\right) \leq q E_{M_T(\cdot \,|\, \psi_0)}\left(\frac{m_T(t)}{m_T(t \,|\, \psi_0)}\right) = q.$$

∎

So this result tells us that, a priori, the relative belief ratio for H_0 is unlikely to be small when H_0 is true.

Another type of misleading evidence arises when $RB_\Psi(\psi_0 \,|\, x)$ is large when H_0 is false, namely, when $\psi_0 \neq \psi_{true}$. The following result addresses this.

Proposition 4.6.2 *The prior probability that $RB_\Psi(\psi_0 \,|\, x) \geq q$, given that H_0 is false, satisfies*

$$M_T(m_T(t \,|\, \psi_0)/m_T(t) \geq q \,|\, \psi_{true}) \leq E_{M_T(\cdot \,|\, \psi_{true})}(m_T(t \,|\, \psi_0)/m_T(t))/q$$
$$= E_{M_T(\cdot \,|\, \psi_0)}(m_T(t \,|\, \psi_{true})/m_T(t))/q.$$

Proof. The inequality follows from Markov's inequality and the equality follows from

$$E_{M_T(\cdot\,|\,\psi_{true})}\left(\frac{m_T(t\,|\,\psi_0)}{m_T(t)}\right) = \int_{\mathcal{T}} \frac{m_T(t\,|\,\psi_0)}{m_T(t)} m_T(t\,|\,\psi_{true})\,\mu_{\mathcal{T}}(dt)$$

and Proposition 4.2.7. ∎

From Proposition 4.6.2, observe that, when q is large enough, the prior probability of getting evidence at least as large as q in favor of ψ_0 when it is false will be small. How large q has to be to make the upper bound small depends upon the value of $E_{M_T(\cdot\,|\,\psi_{true})}(m_T(t\,|\,\psi_0)/m_T(t))$. It might be expected that this expectation decreases as ψ_{true} moves away from ψ_0, at least for values of ψ_0 that are reasonable for the prior. In Section 4.7.2 it is proved that, under conditions, $M_T(m_T(t\,|\,\psi_0)/m_T(t) \geq q\,|\,\psi_{true})$ can be made small by choice of sample size.

Now consider the a priori behavior of the ratio $RB_\Psi(\psi\,|\,x)$ when ψ is a generalized false value, as discussed in Evans and Shakhatreh (2008). For this calculate the prior probability that $RB_\Psi(\psi\,|\,x) \geq q$ when $\theta \sim \Pi(\cdot\,|\,\psi_{true}), x \sim P_\theta$ and $\psi \sim \Pi_\Psi$ independently of (θ,x). So here ψ is a false value in the generalized sense that it has no connection with the true value of the parameter and the data. From Markov's inequality

$$\Pi_\Psi(RB_\Psi(\psi\,|\,x) \geq q) \leq 1/q,$$

since $E_{\Pi_\Psi}(RB_\Psi(\psi\,|\,x)) = 1$ for every value of x. Therefore, this inequality also holds when we average both sides using the probability measure given by $\theta \sim \Pi(\cdot\,|\,\psi_{true}), x \sim P_\theta$. These results reinforce the interpretation that large values of $RB_\Psi(\psi_0\,|\,x)$ are evidence in favor of H_0.

Examples 3.5.2 and 4.4.2 illustrate the need to be concerned with the bias introduced into a problem by the choices made. While it might be believed that a prior is noninformative when it is diffuse, in fact it might be that bias in favor of any particular hypothesis is being introduced. This seems to be a general phenomenon associated with diffuse priors. A precise definition of bias is now provided.

Definition 4.6.1 *The bias against* $H_0 = \Psi^{-1}\{\psi_0\}$ *is given by*

$$M_T(m_T(t\,|\,\psi_0)/m_T(t) \leq 1\,|\,\psi_0) \tag{4.11}$$

and the bias in favor of H_0 *is given by*

$$M_T(m_T(t\,|\,\psi_0)/m_T(t) \leq 1\,|\,\psi_*) \tag{4.12}$$

for values of $\psi_* \neq \psi_0$.

If (4.11) is large, then this says that there is a priori little chance of detecting evidence in favor of H_0 when H_0 is true. For bias in favor, select values $\psi_* \neq \psi_0$ that represent practically significant deviations from ψ_0, say $\psi_0 \pm \delta$ when $\psi \in R^1$, and δ represents a meaningful difference. If (4.12) is small, then this indicates that the prior is biasing the evidence in favor of ψ_0.

The concept of bias is connected with the idea of a *severe test*, as discussed in Mayo and Spanos (2006). For suppose that (4.11) is large and the observed data

Table 4.2 *Values of* $M_T(RB(0|x) \leq 1 | \mu)$ *for various* τ_0^2 *and* μ *in Example 4.6.1 when* $n = 50$.

τ_0^2	0.04	0.10	0.20	0.40	1.00	2.00	400.00
$\mu = 0.0$	0.20	0.14	0.10	0.07	0.05	0.03	0.00
$\mu = 0.1$	0.31	0.24	0.19	0.15	0.10	0.08	0.01
$\mu = 0.2$	0.56	0.48	0.42	0.35	0.28	0.23	0.04
$\mu = 0.3$	0.79	0.74	0.69	0.63	0.55	0.48	0.15

leads to concluding that there is evidence against H_0. In such a case the evidence is of dubious quality. Similarly, if (4.12) is small and we obtain evidence in favor of H_0, then again the evidence is of dubious quality, as it can rightly be said that a priori it is expected that evidence in favor of H_0 will arise, even when H_0 is meaningfully false. Clearly, it is preferable to have (4.11) small and (4.12) large.

The concept of bias is illustrated via an example.

Example 4.6.1 *Bias and the Jeffreys–Lindley Paradox.*

Consider the role of bias in the problem discussed in Examples 3.52 and 4.42. The conditional prior predictive $M_T(\cdot | \mu)$ is given by $\bar{x} | \mu \sim N(\mu, 1/n)$. Putting $a_n = \{\max(0, (1 + 1/n\tau_0^2) \log((1 + n\tau_0^2)))\}^{1/2}$, then

$$M_T(RB(0|x) \leq 1 | \mu) = 1 - \Phi(a_n - \sqrt{n}\mu) + \Phi(-a_n - \sqrt{n}\mu). \qquad (4.13)$$

As $\tau_0^2 \to \infty$, (4.13) converges to 0 for any μ. This is what is wanted when $\mu = 0$ but otherwise it reflects bias in favor of H_0 when $\mu \neq 0$. We have recorded several values of (4.13) in Table 4.2 when $n = 50$. Note that the first row gives values of (4.11) and the remaining rows give values of (4.12), for various settings of τ_0^2.

As the first row demonstrates, it is only when τ_0^2 is small that there is any bias against H_0 and it isn't severe, as the prior probability of getting evidence against H_0 is not large. From the remaining rows observe that there is bias in favor of H_0 when τ_0^2 is large. For example, when $\mu = 0.2$ and $\tau_0^2 = 400$, the prior probability of obtaining evidence in favor of $\mu = 0$ is $1.00 - 0.04 = 0.96$ and so obtaining evidence in favor of the hypothesis is not surprising. Of course, this interpretation also depends on the value $\mu = 0.2$ being a meaningful deviation from H_0 in the application. If it is not considered as a difference that matters, then this bias can be ignored. We need to consider the bias in favor at values of μ that represent practically meaningful differences. Notice that (4.13) converges to 1 as $\mu \to \pm\infty$.

The specific numerical example in Example 4.4.2 corresponds to the values $n = 50$ and $\tau_0^2 = 400$. So there is no a priori bias against H_0 but some bias in favor of H_0. Recall that $RB(0|x) = 20.72$ is only weak evidence in favor *of* H_0 since (4.7) equals 0.05. Also $\mu_{LRSE}(x) = 0.28$ and $M_T(m_T(t|0)/m_T(t) \leq 1 | 0.28) = 0.12$, which suggests that there is a priori bias in favor of H_0 at values like $\mu = 0.28$. So it is plausible to suspect that weak evidence in favor of H_0 has been obtained because of the bias entailed in the prior, at least if a value like $\mu = 0.28$ is considered as being practically different from 0.

It should also be noted that, as $n \to \infty$, then (4.13) converges to 0 when $\mu \neq 0$ and converges to 1 when $\mu = 0$. So in a situation where the sample size can be chosen, for a given prior, the value n can be selected to make (4.13) suitably small at several relevant values of $\mu \neq 0$ and also make (4.13) suitably large when $\mu = 0$.

For us this is the final step in resolving the Jeffreys–Lindley paradox. In Example 4.4.2 it is demonstrated how the classical p-value and the relative belief ratio (also the Bayes factor here whether considered as a limit or using Jeffreys definition) both play a role in the assessment of the hypothesis. These roles are somewhat different, however, and the p-value no longer plays the primary role. In this example it has been shown explicitly how to quantify the bias that a diffuse prior puts into the problem. Certainly the bias numbers should also be reported as part of any assessment of the hypothesis, as these may lead to doubts concerning the validity of the assessment. ■

One might imagine using calculations such as those represented in Table 4.2 to select a prior, which in this case corresponds to choosing a value of τ_0^2, to try and avoid bias as much as possible. This is not what we are proposing. Rather, priors should be elicited and then, for the selected prior, calculations of bias performed to see if there is an issue. The appropriate control mechanism is of course the sample size. If bias is a problem, then there is a need to increase the amount of data collected until it is effectively controlled. See Proposition 4.7.2 for the relevant convergence results.

In many contexts the ability to control the amount of data collected is not available. Still the biases should be calculated post hoc, as this seems preferable to just ignoring the possibility that such biasing can occur. The following paradox is frequently discussed in the literature on inductive inference and illustrates the value of the concept of bias in resolving anomalies.

Example 4.6.2 *Hempel's (the Raven) Paradox.*

Consider the universe of all objects and the statement $A = $ "if an object is a crow, then it is black" or equivalently "all crows are black." Suppose a black crow is observed so $C = $ "a black crow is observed." It is natural to expect that $RB(A \,|\, C) > 1$ and so this observation produces evidence in favor of A. On the other hand, consider the contrapositive of A, namely, $B = $ "if an object is not black, then it is not a crow" or equivalently "all nonblack objects are not crows." The paradox supposedly arises due to the fact that observing a nonblack object that isn't a crow, such as a white handkerchief, wouldn't necessarily be viewed as evidence in favor of A even though it is in favor of B.

This paradox, and its resolution, can be viewed as being a statistical problem, as opposed to being a problem in pure probabilistic logic, since the statement A is really a hypothesis about a probability distribution. For suppose the universe of all objects is partitioned into four sets as follows:

$L_1 = $ all black crows,
$L_2 = $ all nonblack crows,
$L_3 = $ all black objects that aren't crows,
$L_4 = $ all nonblack objects that aren't crows.

Let θ_i be the proportion of the objects in the universe in L_i and so, for a randomly selected object, $P(L_i) = \theta_i$. Let $x = (x_1, x_2, x_3, x_4)$ be the observation where x_i is the indicator for L_i. So for a randomly selected object $x \sim \text{multinomial}(1, \theta_1, \theta_2, \theta_3, \theta_4)$.

Then the statement A is clearly equivalent to the hypothesis that $\theta_1 + \theta_2 = \theta_1$, or equivalently, $H_0 = \Psi^{-1}\{0\}$ where $\psi = \Psi(\theta_1, \theta_2, \theta_3, \theta_4) = \theta_2$. The contrapositive "all nonblack objects are not crows" is equivalent to the hypothesis that $\theta_2 + \theta_4 = \theta_4$, or equivalently, $H_0 = \Psi^{-1}\{0\}$ where again $\psi = \Psi(\theta_1, \theta_2, \theta_3, \theta_4) = \theta_2$.

For the prior take $\theta = (\theta_1, \theta_2, \theta_3, \theta_4) \sim \text{Dirichlet}(\delta_1, \delta_2, \delta_3, \delta_4)$ for some $(\delta_1, \delta_2, \delta_3, \delta_4)$. This implies that a priori $\psi \sim \text{beta}(\delta_2, \delta_1 + \delta_3 + \delta_4)$. After observing x, the posterior is $\theta \,|\, x \sim \text{Dirichlet}(x_1 + \delta_1, x_2 + \delta_2, x_3 + \delta_3, x_4 + \delta_4)$ and so $\psi \sim \text{beta}(x_2 + \delta_2, x_1 + x_3 + x_4 + \delta_1 + \delta_3 + \delta_4)$. The relative belief ratio for assessing a value ψ_0 is thus

$$RB_\Psi(\psi_0 \,|\, x_1, x_2, x_3, x_4)$$

$$= \frac{\frac{\Gamma(x_1+x_2+x_3+x_4+\delta_1+\delta_2+\delta_3+\delta_4)}{\Gamma(x_2+\delta_2)\Gamma(x_1+x_3+x_4+\delta_1+\delta_3+\delta_4)} \psi_0^{x_2+\delta_2-1}(1-\psi_0)^{x_1+x_3+x_4+\delta_1+\delta_3+\delta_4-1}}{\frac{\Gamma(\delta_1+\delta_2+\delta_3+\delta_4)}{\Gamma(\delta_2)\Gamma(\delta_1+\delta_3+\delta_4)} \psi_0^{\delta_2-1}(1-\psi_0)^{\delta_1+\delta_3+\delta_4-1}}$$

$$= \frac{\Gamma(\delta_1+\delta_3+\delta_4)}{\Gamma(\delta_1+\delta_2+\delta_3+\delta_4)} \frac{\Gamma(x_1+x_2+x_3+x_4+\delta_1+\delta_2+\delta_3+\delta_4)}{\Gamma(x_1+x_3+x_4+\delta_1+\delta_3+\delta_4)} \times$$

$$\frac{\Gamma(\delta_2)}{\Gamma(x_2+\delta_2)} \psi_0^{x_2}(1-\psi_0)^{x_1+x_3+x_4}.$$

Consider now the situations discussed in Hempel's paradox. Notice that, if $x = (0, 1, 0, 0)$ is observed, then $RB_\Psi(0 \,|\, x_1, x_2, x_3, x_4) = 0$ and, as expected, there is categorical evidence against H_0. Now suppose that $x_2 = 0$, so that

$$RB_\Psi(0 \,|\, x_1, 0, x_3, x_4) = \frac{\Gamma(\delta_1+\delta_3+\delta_4)}{\Gamma(\delta_1+\delta_2+\delta_3+\delta_4)} \frac{\Gamma(1+\delta_1+\delta_2+\delta_3+\delta_4)}{\Gamma(1+\delta_1+\delta_3+\delta_4)}$$

$$= \frac{\delta_1+\delta_2+\delta_3+\delta_4}{\delta_1+\delta_3+\delta_4} = \left(1 + \frac{\delta_2}{\delta_1+\delta_3+\delta_4}\right)$$

and consider the actual value of $RB_\Psi(0 \,|\, x_1, x_2, x_3, x_4)$ under realistic choices of the δ_i. Certainly it does not make sense to choose $\delta_1 = \delta_2 = \delta_3 = \delta_4 = 1$, as this implies our prior beliefs about each category of object are the same. A plausible constraint for the hyperparameters is that they satisfy $0 < \delta_2 \ll \delta_1 \ll \delta_3 < \delta_4$, as these describe reasonable beliefs about the next object observed. With such an assignment, it is seen that, while the observations $x = (1, 0, 0, 0), (0, 0, 1, 0)$ and $(0, 0, 0, 1)$ are all equivalent in terms of confirming the nonexistence of a nonblack crow, this evidence seems largely irrelevant, as our beliefs in the hypothesis barely change. This is similar to a conclusion found in Good (1960b).

It is clear, however, that the strength of the evidence, as measured by (4.5), is 1 and so there is the strongest possible evidence in favor of H_0 within the context of this experiment. The question then is, how can the small change in beliefs be reconciled with the strength measure? Computing the bias in favor of H_0 reveals that the evidence in favor of H_0 is actually of limited value. For, as noted, $RB_\Psi(0 \,|\, x_1, x_2, x_3, x_4) = 1 + \delta_2/(\delta_1 + \delta_3 + \delta_4)$ when $x_2 = 0$ and equals 0 otherwise and the a priori distribution of x_2 is Bernoulli(ψ_*) when $\Psi(\theta_1, \theta_2, \theta_3, \theta_4) = \psi_*$. Therefore, the bias in favor of H_0 is given by

$$M(RB_\Psi(0 \,|\, x_1, x_2, x_3, x_4) \leq 1 \,|\, \psi_*) = \psi_*$$

and there is extreme bias in favor whenever ψ_* is small, as surely any realistic value must be, as this is the proportion of nonblack crows in the universe of all objects. For example, if we take $\psi_* = 0.0001$, surely far too large, then the prior probability of observing evidence in favor of H_0 is 0.9999. So, while there is indeed evidence obtained in favor of H_0, the observation of a single object outside of category L_2 is in reality meaningless in terms of verifying H_0 because of bias.

Notice too that as $\delta_1 + \delta_3 + \delta_4 \to \infty$, which will happen under Bayesian updating with increasing sample size when only objects are observed that are not in category L_2, then $RB_\Psi(0 \mid x_1, x_2, x_3, x_4) \to 1$, so that further observations are increasingly irrelevant. Accordingly, we conclude that relative belief behaves appropriately here and the concept of a priori bias adds substantial value toward understanding the implications of the example. ■

It is also interesting to consider the bias in the Prosecutor's Fallacy.

Example 4.6.3 *Prosecutor's Fallacy.*
Suppose again that $\psi_2 > \psi_1$. Then it is the case that

$$RB(\text{"guilty"} \mid \text{"defendant has the trait"}) = \frac{\psi_2}{(1-\varepsilon)\psi_1 + \varepsilon\psi_2} > 1$$

while

$$RB(\text{"guilty"} \mid \text{"defendant doesn't have the trait"}) = \frac{1 - \psi_2}{(1-\varepsilon)(1-\psi_1) + \varepsilon(1-\psi_2)} < 1.$$

Therefore, the bias in favor of "guilty" is just the marginal probability of "defendant doesn't have the trait" given "the defendant is innocent" and this equals $1 - \psi_1$. It is seen from this that there is no a priori bias in favor of guilt whenever ψ_1 is small enough.

The bias against "guilty" is the conditional probability of "defendant doesn't have the trait" given "the defendant is guilty," which equals $1 - \psi_2$. Therefore, there is no bias against guilt provided that ψ_2 is large.

In the context of testing for a disease, this says that there is no bias one way or the other provided the test for the disease is a good one, namely, records an infected person as diseased with high probability and records an uninfected person as nondiseased with high probability. ■

When $H_0 = \{\theta_0\}$, the bias against H_0 is given by $P_{\theta_0, T}(f_{\theta_0, T}(t)/m_T(t) \leq 1)$ and the bias in favor of H_0 is given by $P_{\theta_*, T}(f_{\theta_0, T}(t)/m_T(t) \leq 1)$ for $\theta_* \neq \theta_0$. These probabilities look very much like the kinds of probabilities that one would compute in a frequentist analysis for the size and power of a test, although these are more specific in form. Therefore, in a generalized sense of frequentism, such as discussed in Section 3.5.9, frequentist ideas enter our analysis in terms of determining whether or not an analysis is producing evidence that isn't a foregone conclusion and so more credible. This is not to be interpreted as saying that inference itself, however, is to follow the principle of frequentism. Based on the ingredients of data, model and prior, the principles of conditional probability and evidence together with the relative belief preference ordering lead us unambiguously to inferences.

4.7 Properties of Relative Belief Inferences

While many proposals for inference methods have intuitive support, in the end it is the properties of the methods that add conviction, one way or the other, concerning their validity. In fact, if a proposed methodology seems to have very few properties that really distinguish it as performing well, that in itself could be held against the proposal. Sound methodology should give rise to a reasonably rich theory and not be based just on intuition.

The goal in this section is to consider the properties of relative belief inferences. Some of these have already been discussed, such as their invariance and relationship to the likelihood and the Bayes factor, but as will now be shown, they possess a large number of additional good characteristics and in fact a variety of optimal properties.

4.7.1 Consistency

Perhaps the most basic property that any inference methodology should have is that it be consistent. By this is meant that, as the amount of data collected is increased, the inference converges, in some well-defined sense, to the right answer. If an estimator or hypothesis assessment method does not do this, then it is appropriate to say that it is not valid.

The convergence properties, and more generally the asymptotic properties, of inference methods constitute a vast and somewhat complicated subject, requiring a considerable amount of sophisticated mathematical analysis. In part, this is caused by the infinite and continuity aspects of the models considered. In this text the position has been taken that all aspects of models are in reality finite and that models containing infinities are serving as approximations. As such, if an inference method performs well in the finite context, but has anomalous behavior in the infinite case, then it is fair to say that the bad behavior is caused by the infinities inherent in the model and is not to be regarded as an indication that the approach is incorrect. As an example of this, consider the lack of invariance for MAP inferences in the continuous case. While the invariance of relative belief inferences is a convenient property, it is not the primary reason to prefer relative belief over MAP. Rather, it is considerations concerned with evidence, as MAP inferences are not based on a definition of evidence expressed as change in belief.

Accordingly, the discussion here concerning consistency is restricted to finite models. Actually, it is the finiteness of the parameter space that plays the key role. For practical computations it is subsequently argued that a finite discretization of the parameter space be employed, even with models having continuous parameter spaces. So this restriction really does correspond to the practical implementation of relative belief methodology. The convergence results proved here can be generalized to continuous models but this requires taking into account mathematical considerations that really have little to do with statistics and often the imposition of conditions that are unverifiable in specific applications. There is a need, however, to develop such results.

Our discussion of consistency is also restricted to situations where the data $\tilde{x}_n = (x_1, \ldots, x_n)$ arises as an i.i.d., sample from a distribution $f_{\theta_{true}} \in \{f_\theta : \theta \in \Theta\}$ and all convergence results in this section refer to this process. The notation f_θ is used both for the density of the whole data $\tilde{x}_n$ and for an individual sample value, allowing the context to differentiate.

So suppose now that Θ is finite and the prior π satisfies $\pi(\theta) > 0$ for each $\theta \in \Theta$. Then, using the Savage–Dickey representation for a relative belief ratio, and letting the notation for the prior predictives be explicit about the dependence on n,

$$RB_\Psi(\psi_0 \mid \tilde{x}_n) = \frac{m_n(\tilde{x}_n \mid \psi_0)}{m_n(\tilde{x}_n)}$$

$$= \frac{1}{\pi_\Psi(\psi_0)} \frac{\sum_{\theta \in \Psi^{-1}\{\psi_0\}} \pi(\theta) \exp\left\{ n \left[-\frac{1}{n} \sum_{i=1}^n \log(f_{\theta_{true}}(x_i)/f_\theta(x_i)) \right] \right\}}{\sum_{\theta \in \Theta} \pi(\theta) \exp\left\{ n \left[-\frac{1}{n} \sum_{i=1}^n \log(f_{\theta_{true}}(x_i)/f_\theta(x_i)) \right] \right\}} \tag{4.14}$$

as the $f_{\theta_{true}}$ terms cancel out. Let T_n denote any minimal sufficient statistic. The following result is used repeatedly in our developments.

Proposition 4.7.1 *As $n \to \infty$, the prior predictive satisfies*

$$\frac{m_n(\tilde{x}_n)}{m_n(\tilde{x}_n \mid \theta_{true})} = \frac{m_{T_n}(T_n(\tilde{x}_n))}{m_{T_n}(T_n(\tilde{x}_n) \mid \theta_{true})} \xrightarrow{a.s.} \pi(\theta_{true}).$$

Proof. By the strong law of large numbers,

$$\frac{1}{n} \sum_{i=1}^n \log\left(\frac{f_{\theta_{true}}(x_i)}{f_\theta(x_i)} \right) \xrightarrow{a.s.} E_{P_{\theta_{true}}}\left(\log\left(\frac{f_{\theta_{true}}(X)}{f_\theta(X)} \right) \right) = KL(P_{\theta_{true}}, P_\theta)$$

as $n \to \infty$, where $KL(P_{\theta_{true}}, P_\theta)$ is the Kullback–Leibler divergence of P_θ from $P_{\theta_{true}}$. Now $KL(P_{\theta_{true}}, P_\theta) \geq 0$ with equality if and only if $P_{\theta_{true}} = P_\theta$. Therefore, the result follows from

$$\frac{m_n(\tilde{x}_n)}{m_n(\tilde{x}_n \mid \theta_{true})} = \pi(\theta_{true}) + \sum_{\theta \in \Theta \setminus \{\theta_{true}\}} \pi(\theta) \exp\left\{ n \left[-\frac{1}{n} \sum_{i=1}^n \log\left(\frac{f_{\theta_{true}}(x_i)}{f_\theta(x_i)} \right) \right] \right\}$$

and the fact that, recalling that Θ is finite, there exists $\varepsilon > 0$ and N_ε such that, for all $n > N_\varepsilon$, $n^{-1} \sum_{i=1}^n \log(f_{\theta_{true}}(x_i)/f_\theta(x_i)) > \varepsilon$, whenever $\theta \neq \theta_{true}$. ∎

This leads immediately to a key consistency result for the relative belief ratio.

Corollary 4.7.1 *As $n \to \infty$, then (i) $RB_\Psi(\psi_0 \mid \tilde{x}_n) \xrightarrow{a.s.} 0$ when $\psi_0 \neq \Psi(\theta_{true})$ and (ii) $RB_\Psi(\psi_0 \mid \tilde{x}_n) \xrightarrow{a.s.} 1/\pi_\Psi(\psi_0)$ when $\psi_0 = \Psi(\theta_{true})$.*

Proof. (i) When $\psi_0 \neq \Psi(\theta_{true})$, the numerator in (4.14) satisfies

$$\sum_{\theta \in \Psi^{-1}\{\psi_0\}} \pi(\theta) \exp\left\{ n \left[-\frac{1}{n} \sum_{i=1}^n \log(f_{\theta_{true}}(x_i)/f_\theta(x_i)) \right] \right\} \xrightarrow{a.s.} 0$$

as $n \to \infty$ since $\theta_{true} \notin \Psi^{-1}\{\psi_0\}$ and $m(\tilde{x}_n)/m(\tilde{x}_n \mid \theta_{true})$ converges to a nonzero quantity by Proposition 4.7.1. Therefore, $RB_\Psi(\psi_0 \mid \tilde{x}_n) \xrightarrow{a.s.} 0$ as $n \to \infty$, when $\psi_0 \neq \Psi(\theta_{true})$.

(ii) When $\psi_0 = \Psi(\theta_{true})$, the numerator in (4.14) equals

$$\pi(\theta_{true}) + \sum_{\theta \in \Psi^{-1}\{\psi_0\}\backslash\{\theta_{true}\}} \pi(\theta)\exp\left\{n\left[-\frac{1}{n}\sum_{i=1}^{n}\log(f_{\theta_{true}}(x_i)/f_\theta(x_i))\right]\right\}$$

and this converges to $\pi(\theta_{true})$ just as in Proposition 4.7.1. Combining this with (4.14) and Proposition 4.7.1 establishes that $RB_\Psi(\psi_0\,|\,\tilde{x}_n) \overset{a.s.}{\to} 1/\pi_\Psi(\psi_0)$ as $n \to \infty$, when $\psi_0 = \Psi(\theta_{true})$. ∎

This corollary implies that $RB_\Psi(\psi_0\,|\,\tilde{x}_n)$ is consistent. For (i) says that $RB_\Psi(\psi_0\,|\,\tilde{x}_n)$ converges to categorical evidence against ψ_0 when ψ_0 is not the true value of ψ. Also, using Proposition 4.2.1 (iii), (ii) says that $RB_\Psi(\psi_0\,|\,\tilde{x}_n)$ converges to categorical evidence in favor of ψ_0 when ψ_0 is the true value of ψ.

Now consider the strength of the evidence given by $RB_\Psi(\psi_0\,|\,\tilde{x}_n)$. First there is the following result.

Proposition 4.7.2 *As $n \to \infty$, then (i) $\pi_\Psi(\psi_0\,|\,\tilde{x}_n) \overset{a.s.}{\to} 0$ when $\psi_0 \neq \Psi(\theta_{true})$ and (ii) $\pi_\Psi(\psi_0\,|\,\tilde{x}_n) \overset{a.s.}{\to} 1$ when $\psi_0 = \Psi(\theta_{true})$.*

Proof. (i) Clearly $\pi_\Psi(\psi_0\,|\,\tilde{x}_n) \leq RB_\Psi(\psi_0\,|\,\tilde{x}_n)$ and so the result follows by Corollary 4.7.1 (i).

(ii) Since $\pi_\Psi(\Psi(\theta_{true})\,|\,x) = 1 - \sum_{\psi \in \Psi\backslash\{\Psi(\theta_{true})\}}\pi_\Psi(\psi\,|\,\tilde{x}_n)$, this result follows from part (i) and the finiteness of Ψ. ∎

This together with Corollary 4.7.1 gives the consistency of the measure of strength.

Corollary 4.7.2 *As $n \to \infty$, then (i) $\Pi_\Psi(RB_\Psi(\psi\,|\,\tilde{x}_n) \leq RB_\Psi(\psi_0\,|\,\tilde{x}_n)\,|\,\tilde{x}_n) \overset{a.s.}{\to} 0$ when $\psi_0 \neq \Psi(\theta_{true})$ and (ii) $\Pi_\Psi(RB_\Psi(\psi\,|\,\tilde{x}_n) \leq RB_\Psi(\psi_0\,|\,\tilde{x}_n)\,|\,\tilde{x}_n) \overset{a.s.}{\to} 1$ when $\psi_0 = \Psi(\theta_{true})$.*

Proof. (i) This follows from Corollary 4.7.1 (i) and Proposition 4.4.1 since $\Pi_\Psi(RB_\Psi(\psi\,|\,\tilde{x}_n) \leq RB_\Psi(\psi_0\,|\,\tilde{x}_n)\,|\,\tilde{x}_n) \leq RB_\Psi(\psi_0\,|\,\tilde{x}_n)$.

(ii) This follows from Proposition 4.7.2 (ii) and Proposition 4.4.1 since $\pi_\Psi(\psi_0\,|\,\tilde{x}_n) \leq \Pi_\Psi(RB_\Psi(\psi\,|\,\tilde{x}_n) \leq RB_\Psi(\psi_0\,|\,\tilde{x}_n)\,|\,\tilde{x}_n)$. ∎

So indeed the strength of the evidence goes to the maximum possible value when ψ_0 is a false value and also when ψ_0 is the true value.

Now consider the consistency properties of $\psi_{LRSE}(\tilde{x}_n)$ and $C_{\Psi,\gamma}(\tilde{x}_n)$. For this consider the metric d on $\Psi = \{\psi_1, \ldots, \psi_m\}$ given by $d(\psi_i, \psi_j) = 0$ when $i = j$ and $d(\psi_i, \psi_j) = 1$, otherwise. Convergence of sequence $\psi(\tilde{x}_n)$ to ψ^* with respect to this metric means that there exists an N such that for all $n \geq N$ then $\psi(\tilde{x}_n) = \psi^*$.

Proposition 4.7.3 *As $n \to \infty$, then (i) $\psi_{LRSE}(\tilde{x}_n) \overset{a.s.}{\to} \Psi(\theta_{true})$ and (ii) when $\gamma < 1$, then $\psi(\tilde{x}_n) \overset{a.s.}{\to} \Psi(\theta_{true})$ for any sequence $\psi(\tilde{x}_n) \in C_{\Psi,\gamma}(\tilde{x}_n)$.*

Proof. (i) Necessarily $RB_\Psi(\Psi(\theta_{true})\,|\,\tilde{x}_n) \leq RB_\Psi(\psi_{LRSE}(\tilde{x}_n)\,|\,\tilde{x}_n)$ and so Corollary 4.7.1 (ii) implies that $RB_\Psi(\psi_{LRSE}(\tilde{x}_n)\,|\,\tilde{x}_n) \overset{a.s.}{\to} 1/\pi_\Psi(\psi_{true})$. Suppose there is a subsequence n_m such that $\psi_{LRSE}(\tilde{x}_{n_m}) \overset{a.s.}{\to} \psi_0 \neq \Psi(\theta_{true})$ as $m \to \infty$. Then by Corollary 4.7.1 (i) $RB_\Psi(\psi_{LRSE}(\tilde{x}_{n_m})\,|\,\tilde{x}_{n_m}) \overset{a.s.}{\to} 0$, which contradicts the first statement, so

$\psi_{LRSE}(\tilde{x}_{n_m}) \overset{a.s.}{\to} \Psi(\theta_{true})$. Since every subsequence of $\tilde{x}_n$ has a convergent subsequence, and every convergent subsequence converges to $\Psi(\theta_{true})$, the result follows.

(ii) Note that $\psi_{LRSE}(\tilde{x}_n) \in C_{\Psi,\gamma}(\tilde{x}_n)$ for each n, so that there exists a sequence $\psi(\tilde{x}_n) \in C_{\Psi,\gamma}(\tilde{x}_n)$ for each n. Since $\psi_{LRSE}(\tilde{x}_n) \overset{a.s.}{\to} \Psi(\theta_{true})$, it must be the case that $\pi_\Psi(\psi_{LRSE}(\tilde{x}_n) \mid \tilde{x}_n) \overset{a.s.}{\to} 1$ by Corollary 4.7.1 (ii). Therefore, for $\gamma < 1$ and all n large enough, $C_{\Psi,\gamma}(\tilde{x}_n) = \{\psi_{LRSE}(\tilde{x}_n)\}$, which establishes the result. ∎

Corollary 4.7.2 is a satisfying result for the interpretation of the strength of evidence. The following example illustrates, however, the need to always remember that continuous models serve as approximations.

Example 4.7.1 *The Distribution of the Strength with Continuous Models.*

Consider the context of Example 4.4.2 and the distribution of the strength of the evidence, namely,

$$\Pi(RB(\mu \mid x) \le RB(0 \mid x) \mid x) = 1 - \Phi((1 + 1/n\tau_0^2)^{1/2}(|\sqrt{n}\bar{x}| + (n\tau_0^2 + 1)^{-1}\sqrt{n}\bar{x}))$$
$$+ \Phi((1 + 1/n\tau_0^2)^{1/2}(-|\sqrt{n}\bar{x}| + (n\tau_0^2 + 1)^{-1}\sqrt{n}\bar{x})).$$

Clearly, when $\mu_{true} \ne 0$, then $\Pi(RB(\mu \mid x) \le RB(0 \mid x) \mid x) \overset{a.s}{\to} 0$ as $n \to \infty$. But now consider what happens when $\mu_{true} = 0$, as then the strength converges almost surely to the random variable

$$U = 1 - \Phi((|Z|) + \Phi((-|Z|) = 2(1 - \Phi((|Z|)),$$

where $Z \sim N(0,1)$. So $U \sim U(0,1)$ when $\mu = 0$ and the strength of the evidence does not behave properly.

Corollary 4.7.2 (ii) establishes, however, that this behavior is an artefact of continuity alone. As such, in applications where continuous models are used, this has to be taken into account, at least when assessing the strength of the evidence for a relative belief ratio that is greater than 1. The problem is easily solved, however, by discretizing the parameter space, as then Corollary 4.7.2 (ii) applies.

It is easy to see why this phenomenon occurs when $\mu_{true} = 0$. The posterior distribution of μ is given by $\mu \mid x \sim N((n + 1/\tau_0^2)^{-1}n\bar{x}, (n + 1/\tau_0^2)^{-1})$ and for large n this is approximated by the $N(\bar{x}, 1/n)$ distribution. So when computing the strength of the evidence and n is large, we are trying to make distinctions among values of μ that are all effectively equal to 0. In fact, it does not make sense to imagine that it can ever be verified that $\mu = 0$ identically. For suppose that the x_i measurements are in centimeters. Then a deviation of $\bar{x}$ from 0 of only a few millimeters is not meaningful, no matter how large the sample size is, as this reflects spurious accuracy. The best one can hope for is that the true mean can be identified as being within a half a centimeter of 0. ∎

The lesson of Example 4.7.1 is that, when assessing the strength of evidence in favor of a hypothesis, the fact that a continuous model is only an approximation cannot be ignored. A simple solution, when the parameter ψ is real valued, is to choose a relevant $\delta > 0$ and an interval $(\psi_{low}, \psi_{up}]$ divided up into subintervals $(\psi_{low}, \psi_{low} + \delta], (\psi_{low} + \delta, \psi_{low} + 2\delta], \ldots$ and then carry out the analysis for the discretized version of the parameter. Choosing the interval $(\psi_{low}, \psi_{up}]$ is relatively straightforward, as this is dictated by $\Pi_{\Psi}((\psi_{low}, \psi_{up}]) \approx 1$, namely, all of the prior probability for ψ is effectively contained in the interval. Choosing δ is more difficult, but in essence the question of when two values of ψ are meaningfully different needs to addressed in any application. If an answer to this cannot be provided, then one can only wonder at the relevance of hypothesis assessment for the parameter. This discussion can be generalized to allow the grid size to vary across $(\psi_{low}, \psi_{up}]$ and also extended to multidimensional parameters by applying it to each coordinate of ψ.

4.7.2   Convergence of Bias Measures

The biases introduced in Section 4.6 can be used when considering the data collection aspect of a statistical analysis to control the amount of bias that a prior puts into an analysis. The following result shows how this can be achieved.

Proposition 4.7.4 As $n \to \infty$, then (i) $M_{T_n}(m_{T_n}(t \mid \psi_0)/m_{T_n}(t) \leq 1 \mid \psi_0) \to 0$ and (ii) $M_{T_n}(m_{T_n}(t \mid \psi_0)/m_{T_n}(t) \leq 1 \mid \psi_*) \to 1$ when $\psi_0 \neq \psi_*$.

Proof. (i) By the theorem of total probability,

$$M_{T_n}\left(\frac{m_{T_n}(t \mid \psi_0)}{m_{T_n}(t)} \leq 1 \;\middle|\; \psi_0\right) = E_{\Pi(\cdot \mid \psi_0)}\left(P_\theta\left(\frac{m_{T_n}(T_n(\tilde{x}_n) \mid \psi_0)}{m_{T_n}(T_n(\tilde{x}_n))} \leq 1\right)\right).$$

Now, considering $\tilde{x}_n$ as being i.i.d. P_θ where $\theta \in \Psi^{-1}\{\psi_0\}$, Corollary 4.7.1 (ii), implies that $m_{T_n}(T_n(\tilde{x}_n) \mid \psi_0)/m_{T_n}(T_n(\tilde{x}_n)) \xrightarrow{a.s.} 1/\pi_\Psi(\psi_0)$. Since $1/\pi_\Psi(\psi_0) > 1$, then $P_\theta(m_{T_n}(t \mid \psi_0)/m_{T_n}(t) \leq 1) \to 0$ as $n \to \infty$ and by bounded convergence this implies the result.

(ii) By the theorem of total probability,

$$M_{T_n}\left(\frac{m_{T_n}(t \mid \psi_0)}{m_{T_n}(t)} \leq 1 \;\middle|\; \psi_*\right) = E_{\Pi(\cdot \mid \psi_*)}\left(P_\theta\left(\frac{m_{T_n}(T_n(\tilde{x}_n) \mid \psi_0)}{m_{T_n}(T_n(\tilde{x}_n))} \leq 1\right)\right)$$

and considering $\tilde{x}_n$ as being i.i.d. P_θ, where $\theta \in \Psi^{-1}\{\psi_*\}$, so $\theta \notin \Psi^{-1}\{\psi_0\}$, Corollary 4.7.1 (i), implies that $m_{T_n}(T_n(\tilde{x}_n) \mid \psi_0)/m_{T_n}(T_n(\tilde{x}_n)) \xrightarrow{a.s.} 0$ and this implies the result by bounded convergence. ∎

So the bias can be controlled by sample size.

The proposition also implies the following concerning obtaining misleading evidence.

Corollary 4.7.3 As $n \to \infty$, then (i) $M_{T_n}(m_{T_n}(t \mid \psi_0)/m_{T_n}(t) \leq q \mid \psi_0) \to 0$ for any $q \leq 1$ and (ii) $M_{T_n}(m_{T_n}(t \mid \psi_0)/m_{T_n}(t) \geq q \mid \psi_*) \to 0$ as $n \to \infty$ when $\psi_0 \neq \psi_*$, for any $q > 1$.

This result tells us that the prior probability of obtaining misleading evidence, either against or in favor of ψ_0 at level q, can be made small by choosing n suitably large.

4.7.3   Optimality of Relative Belief Credible Regions

A basic optimality result is derived here for a relative belief credible region $C_{\Psi,\gamma}(x)$ for a parameter $\psi = \Psi(\theta)$. From this a number of results follow. Recall that $C_{\Psi,\gamma}(x) = \{\psi : RB_\Psi(\psi \,|\, x) \geq G_\Psi^{-1}(1 - \gamma) \,|\, x)\}$ where G_Ψ is the cdf of the posterior distribution of the relative belief ratio $RB_\Psi(\cdot \,|\, x)$. Let $c_\gamma(x) = G_\Psi^{-1}(1 - \gamma)$ denote the $(1 - \gamma)$-quantile of this distribution. The following was proved in Evans, Guttman and Swartz (2006) and the proof is provided in the appendix to this chapter.

Proposition 4.7.5 *The set* $C_{\Psi,\gamma}(x)$ *minimizes* $\Pi_\Psi(C)$ *among all sets* $C \subset \Psi$ *satisfying* $\Pi_\Psi(C \,|\, x) \geq \Pi_\Psi(C_{\Psi,\gamma}(x) \,|\, x)$.

Proposition 4.7.5 says that among all subsets of Ψ that have at least the posterior content of $C_{\Psi,\gamma}(x)$, the set $C_{\Psi,\gamma}(x)$ has the smallest prior content. If $\Pi_\Psi(C_{\Psi,\gamma}(x) \,|\, x) = \gamma$, then $C_{\Psi,\gamma}(x)$ has the smallest prior content among all subsets of Ψ that have posterior content at least γ.

An immediate consequence of Proposition 4.7.5 is the following.

Corollary 4.7.4 *The set* $C_{\Psi,\gamma}(x)$ *maximizes* $RB_\Psi(C \,|\, x)$ *among all sets* $C \subset \Psi$ *satisfying* $\Pi_\Psi(C \,|\, x) = \Pi_\Psi(C_{\Psi,\gamma}(x) \,|\, x)$ *when* $\Pi_\Psi(C_{\Psi,\gamma}(x)) \neq 0$.

So among all subsets of Ψ that have posterior content equal to that of $C_{\Psi,\gamma}(x)$, the set $C_{\Psi,\gamma}(x)$ has the largest relative belief ratio. In other words, the set $C_{\Psi,\gamma}(x)$ has the largest increase in belief from a priori to a posteriori among all exact γ-credible regions for ψ. In a direct sense, it is making the best use of the information in the data. Interestingly, these credible regions also have an optimality property with respect to the Bayes factor.

Corollary 4.7.5 *The set* $C_{\Psi,\gamma}(x)$ *maximizes* $BF_\Psi(C \,|\, x)$ *among all sets* $C \subset \Psi$ *satisfying* $\Pi_\Psi(C \,|\, x) = \Pi_\Psi(C_{\Psi,\gamma}(x) \,|\, x)$ *when* $0 < \Pi_\Psi(C_{\Psi,\gamma}(x)) < 1$.

Proof. This follows from the fact that the function $f(x) = (1 - x)/x$ is decreasing in $x \in [0, 1]$. ∎

Corollaries 4.7.4 and 4.7.5 also lead to an a priori optimality property. For suppose there is another rule that leads to γ-credible region $B_{\Psi,\gamma}(x)$ for ψ and $\Pi_\Psi(B_{\Psi,\gamma}(x) \,|\, x) = \Pi_\Psi(C_{\Psi,\gamma}(x) \,|\, x)$ for every x. So, for example, with a continuous model, then often exact γ-credible regions can be constructed using relative belief or hpd. Recall that M denotes the prior predictive probability measure of the data. The following result is immediate.

Corollary 4.7.6 *If* $\Pi_\Psi(B_{\Psi,\gamma}(x) \,|\, x) = \Pi_\Psi(C_{\Psi,\gamma}(x) \,|\, x)$ *for every* x, *then*

$$E_M(RB_\Psi(C_{\Psi,\gamma}(x) \,|\, x)) \geq E_M(RB_\Psi(B_{\Psi,\gamma}(x) \,|\, x))$$

whenever $\Pi_\Psi(C_{\Psi,\gamma}(x)) > 0$ *with* M-*probability* 1.

This says that the prior mean of the relative belief ratio of the region $B_{\Psi,\gamma}(x)$ is maximized by $C_{\Psi,\gamma}(x)$. A similar result holds for the Bayes factor of the region.

The following shows that belief always increases, from a priori to a posteriori, for relative belief regions. As it turns out, belief can actually decrease for other approaches to constructing credible regions such as hpd regions.

Corollary 4.7.7 *If* $\Pi_\Psi(C_{\Psi,\gamma}(x)) \neq 0$, *then* $RB_\Psi(C_{\Psi,\gamma}(x)\,|\,x) \geq 1$ *and if* $0 < \Pi_\Psi(C_{\Psi,\gamma}(x)) < 1$, *then* $BF_\Psi(C_{\Psi,\gamma}(x)\,|\,x) \geq 1$.

Proof. We have $1 = \Pi_\Psi(\Theta\,|\,x) \geq \Pi_\Psi(C_{\Psi,\gamma}(x)\,|\,x)$ and so $RB_\Psi(C_{\Psi,\gamma}(x)\,|\,x) \geq RB_\Psi(\Theta\,|\,x) = 1$. ∎

The inequalities in Corollary 4.7.7 are often strict, as is typically the case with continuous models.

Proposition 4.7.6 *If*

$$0 < \Pi_\Psi(C_{\Psi,\gamma}(x)) = \Pi_\Psi(\{\psi : RB_\Psi(\psi\,|\,x) > c_\gamma(x)\}) < 1,$$
$$0 < \Pi_\Psi(C_{\Psi,\gamma}(x)\,|\,x) = \Pi_\Psi(\{\psi : RB_\Psi(\psi\,|\,x) > c_\gamma(x)\}\,|\,x) < 1,$$

then $RB_\Psi(C_{\Psi,\gamma}(x)\,|\,x) > 1$ *and* $BF_\Psi(C_{\Psi,\gamma}(x)\,|\,x) > 1$.

Proof. Note that

$$\Pi_\Psi(C_{\Psi,\gamma}(x)\,|\,x) = \int_{\{\psi : RB_\Psi(\psi\,|\,x) > c_\gamma(x)\}} \pi_\Psi(\psi\,|\,x)\,\nu_\Psi(d\psi)$$
$$> c_\gamma(x) \int_{\{\psi : RB_\Psi(\psi\,|\,x) > c_\gamma(x)\}} \pi_\Psi(\psi)\,\nu_\Psi(d\psi) = c_\gamma(x)\Pi_\Psi(C_{\Psi,\gamma}(x))$$

so $RB_\Psi(C_{\Psi,\gamma}(x)\,|\,x) > c_\gamma(x)$ and similarly

$$\Pi_\Psi(C^c_{\Psi,\gamma}(x)\,|\,x) = \Pi_\Psi(\{\psi : RB_\Psi(\psi\,|\,x) < c_\gamma(x)\}\,|\,x) \leq c_\gamma(x)\Pi_\Psi(C^c_{\Psi,\gamma}(x))$$

so $RB^c_\Psi(C_{\Psi,\gamma}(x)\,|\,x) \leq c_\gamma(x)$. Therefore, using Proposition 4.3.1,

$$BF_\Psi(C_{\Psi,\gamma}(x)\,|\,x) = RB_\Psi(C_{\Psi,\gamma}(x)\,|\,x)/RB^c_\Psi(C_{\Psi,\gamma}(x)\,|\,x) > c_\gamma(x)/c_\gamma(x) = 1$$

and $RB_\Psi(C_{\Psi,\gamma}(x)\,|\,x) > 1$. ∎

Since $BF_\Psi(C^c_{\Psi,\gamma}(x)\,|\,x) = 1/BF_\Psi(C_{\Psi,\gamma}(x)\,|\,x)$, under the conditions of the proposition, $BF_\Psi(C^c_{\Psi,\gamma}(x)\,|\,x) < 1$ and $RB_\Psi(C^c_{\Psi,\gamma}(x)\,|\,x) < 1$.

We consider a numerical example illustrating these results.

Example 4.7.2 *Probability of Joint Success.*

Suppose x is observed from a binomial(n, θ_1) distribution, an independent y is observed from a binomial(n, θ_2) distribution, independent uniform priors are placed on θ_1 and θ_2 and interest is in making inference about $\psi = \theta_1\theta_2$. This is the probability of simultaneous success from tossing two coins where the coins have probability of heads equal to θ_1 and θ_2, respectively.

Suppose $n = 5$ and $x = 4, y = 1$ is observed. In Table 4.3 some γ-hpd intervals and γ-relative belief (RB) intervals are provided for ψ. Notice that these intervals are quite different. Also the RB intervals always dominate the hpd intervals in the sense that the Bayes factor and relative belief ratio of the RB interval are always greater than the corresponding quantities for the hpd interval, confirming Corollaries 4.7.5 and 4.7.4. The estimate determined by the hpd criterion is $\psi_{MAP}(x) = 0.122$ while $\psi_{LRSE}(x) = 0.186$.

Table 4.3 *Comparison of relative belief and hpd intervals.*

γ	hpd	RB (hpd)	BF (hpd)	rb	RB (rb)	BF (rb)
.95	(.008, .447)	1.25	5.99	(.028, .501)	1.32	7.35
.75	(.032, .293)	1.47	2.82	(.071, .361)	1.69	3.35
.50	(.059, .216)	1.57	2.16	(.110, .284)	1.74	2.48
.25	(.089, .163)	1.63	1.84	(.119, .270)	1.76	2.36

In this example, the hpd intervals always have $RB_\Psi\left(C_{\Psi,\gamma}(x)\,|\,x\right) > 1$ and $BF_\Psi\left(C_{\Psi,\gamma}(x)\,|\,x\right) > 1$, but other methods of forming the intervals may not. For example, taking the left tail of the posterior as a γ-credible interval for ψ, then the left tail 0.4-credible interval has relative belief ratio equal to 0.730 and Bayes factor equal to 0.640. ■

There is another interesting consequence of Proposition 4.7.5 and this is concerned with the prior probability of a credible region covering a false value. The definition of this probability is as follows.

Definition 4.7.1 *The prior probability of γ-credible region $B_{\Psi,\gamma}(x)$ covering a false value is given by*

$$E_M(\Pi_\Psi(B_{\Psi,\gamma}(x))) = \int_\Theta \int_\Psi P_\theta(\psi \in B_{\Psi,\gamma}(x))\,\Pi_\Psi(d\psi)\,\Pi(d\theta). \qquad (4.15)$$

Suppose we generate $\theta \sim \Pi, x\,|\,\theta \sim P_\theta$ and, independently of (θ,x), generate $\psi \sim \Pi_\Psi$. So ψ is a false value in the generalized sense that it is only equal to the true value $\Psi(\theta)$ by chance and ψ has no connection with the data. This restricts attention to those false values which are reasonable in light of the prior. It is clear then that (4.15) is the probability that the random region $B_{\Psi,\gamma}(x)$ contains ψ. The prior probability that a γ-*credible region* $B_{\Psi,\gamma}(x)$ covers the true value is given by

$$E_M(\Pi_\Psi\left(B_{\Psi,\gamma}(x)\,|\,x\right)) = \int_\Theta P_\theta(\Psi(\theta) \in B_{\Psi,\gamma}(x))\,\Pi(d\theta)$$

and note that $E_M(\Pi_\Psi\left(B_{\Psi,\gamma}(x)\,|\,x\right)) \geq \gamma$. So the expected prior content of a credible region is equal to the prior probability that the credible region covers a false value and the expected posterior content is equal to the prior probability that the credible region covers the true value, when we average using the prior predictive. Note that "true value" is also being interpreted in a generalized sense here.

The following result establishes that the relative belief region minimizes the probability of covering a false value.

Proposition 4.7.7 *If $\gamma = \Pi_\Psi\left(C_{\Psi,\gamma}(x)\,|\,x\right)$ for every x, then the γ-relative belief region minimizes the prior probability of covering a false value, namely, $E_M(\Pi_\Psi(B_{\Psi,\gamma}(x))) \geq E_M(\Pi_\Psi(C_{\Psi,\gamma}(x)))$ for every other γ-credible region $B_{\Psi,\gamma}(x)$.*

Proof. This follows from Proposition 4.7.5 as $\Pi_\Psi(B_{\Psi,\gamma}(x)\,|\,x) \geq \Pi_\Psi(C_{\Psi,\gamma}(x)\,|\,x)$. ■

A repeated sampling interpretation of this is obtained by considering a sequence (θ_i, x_i, ψ_i), for $i = 1, 2, \ldots,$ of independent values from the joint distribution $\Pi \times P_\theta \times$

Π_{Ψ}. Then Proposition 4.7.7 says that, among all γ-credible regions for $\Psi(\theta)$ formed from $\Pi \times P_{\theta}$, a γ-relative belief region for $\Psi(\theta)$ minimizes the limiting proportion of times that ψ_i is in the region, while the limiting proportion of times $\Psi(\theta_i)$ is in the region equals γ.

There is also an unbiasedness property for relative belief regions.

Proposition 4.7.8 *The prior probability that a relative belief region covers the true value is always greater than or equal to the prior probability that the region covers a false value, namely,*

$$E_M(\Pi_{\Psi}\left(C_{\Psi,\gamma}(x)\,|\,x\right)) \geq E_M(\Pi_{\Psi}(C_{\Psi,\gamma}(x))).$$

Proof. Corollary 4.7.7 implies that $\Pi_{\Psi}\left(C_{\Psi,\gamma}(x)\,|\,x\right) \geq \Pi_{\Psi}(C_{\Psi,\gamma}(x))$ and so the result follows. ∎

It seems natural to refer to this property of relative belief regions as *Bayesian unbiasedness*.

Proposition 4.7.8 can be extended. In fact, the theorem holds for any probability measure M^*, namely, under the conditions of the proposition, $E_{M^*}(\Pi_{\Psi}(B_{\Psi,\gamma}(x))) \geq E_{M^*}(\Pi_{\Psi}(C_{\Psi,\gamma}(x)))$ for every other γ-credible region $B_{\Psi,\gamma}(x)$. If $M^* \neq M$, however, the interpretation that $E_{M^*}(\Pi_{\Psi}(B_{\Psi,\gamma}(x))$ equals the probability of covering a false value or that $E_{M^*}(\Pi_{\Psi}\left(B_{\Psi,\gamma}(x)\,|\,x\right))$ is the probability of covering the true value no longer applies.

The value $E_M(\Pi_{\Psi}(C_{\Psi,\gamma}(x)))$ can be used in a design context as an a priori measure of the accuracy of $C_{\Psi,\gamma}$. Consider the following example.

Example 4.7.3 *Location Normal.*

Consider the context of Example 3.5.2 so $x = (x_1,\ldots,x_n)$ is a sample from the $N(\mu,1)$ distribution and $\mu \sim N(0,\tau_0^2)$. The posterior distribution of μ is $N((n+1/\tau_0^2)^{-1}n\bar{x}, (n+1/\tau_0^2)^{-1})$. The ratio of the posterior density to the prior density is, in this case, proportional to the likelihood $\exp\{-n(\mu - \bar{x})^2/2\}$. Therefore, a γ-relative belief interval for $\mu = \Psi(\mu)$ is a likelihood interval and so takes the form $C_{\Psi,\gamma}(x) = \bar{x} \pm k_{\gamma}(n,\bar{x},\tau_0^2)$ where $k_{\gamma}(n,\bar{x},\tau_0^2) \geq 0$ is computed from

$$\gamma = \Phi\left((n+1/\tau_0^2)^{-1/2}\bar{x}/\tau_0^2 + (n+1/\tau_0^2)^{1/2}k_{\gamma}(n,\bar{x},\tau_0^2)\right) - \Phi\left((n+1/\tau_0^2)^{-1/2}\bar{x}/\tau_0^2 - (n+1/\tau_0^2)^{1/2}k_{\gamma}(n,\bar{x},\tau_0^2)\right).$$

Now $\Pi_{\Psi}(C_{\Psi,\gamma}(x)) = \Phi\left((\bar{x}+k_{\gamma}(n,\bar{x},\tau_0^2))/\tau_0\right) - \Phi\left((\bar{x}-k_{\gamma}(n,\bar{x},\tau_0^2))/\tau_0\right)$ and a priori $\bar{x} \sim N(0, \tau_0^2 + 1/n)$. Therefore,

$$E_M\left(\Pi_{\Psi}(C_{\Psi,\gamma}(x))\right) = \int_{-\infty}^{\infty} \Phi\left(((\tau_0^2+1/n)^{1/2}z+k_{\gamma}(n,(\tau_0^2+1/n)^{1/2}z))/\tau_0\right)\varphi(z) - \Phi\left(((\tau_0^2+1/n)^{1/2}z-k_{\gamma}(n,(\tau_0^2+1/n)^{1/2}z)/\tau_0\right)\varphi(z)\,dz.$$

For example, Table 4.4 gives some values of the prior probability of covering a false value when $\gamma = 0.95$ and $\tau_0^2 = 1$.

Table 4.4 *Values of $E_M(\Pi_\Psi(C_{\Psi,\gamma}(x)))$ in Example 4.7.3.*

n	1	10	25	50
$E_M\left(\Pi_\Psi(C_{\Psi,\gamma}(x))\right)$	0.700	0.322	0.212	0.152

It is easy to see that $E_M\left(\Pi_\Psi(C_{\Psi,\gamma}(x))\right)$ converges to 0 as $n \to \infty$. So by choosing n large enough, this can be made as small as desired and so control the error in our inference. If $\tau_0^2 \to \infty$, then the prior probability of covering a false value generated from the prior also converges to 0. ■

In Section 4.7.5 additional optimality results concerning relative belief credible regions are presented. In Shang, Ng, Sehrawat, Li and Englert (2013) relative belief regions were obtained independently in the context of quantum state estimation. Also, Shalloway (2014) establishes optimality results for relative belief regions based upon Kullback–Leibler divergence.

4.7.4   Optimality of Relative Belief Hypothesis Assessment

Optimality results for hypothesis assessment are typically derived in the context where the statistician categorically accepts or rejects $H_0 = \Psi^{-1}\{\psi_0\}$. A different point of view has been adopted here. Rather, an assessment is produced of the evidence that the hypothesis is true or false. A decision to accept or reject the hypothesis can then be made that incorporates additional criteria like costs and utilities, but the evidence comes first and there is no reason to suppress this in favor of other criteria. Still, it is interesting to see if relative belief can be adapted to produce optimal tests in some sense. It is shown in this section that this is the case.

With classical frequentist hypothesis testing theory there is an equivalence between tests that reject with size (probability of rejecting when H_0 is true) α and $(1 - \alpha)$-confidence regions. In that case, an optimality property for the confidence region in terms of minimizing the probability of covering false values is equivalent to the test being most powerful (maximizing the probability of rejecting when H_0 is false). It might seem then that we could simply proceed in the same way here, namely, choose a γ and accept H_0 whenever $\psi_0 \in C_{\Psi,\gamma}(x)$, leading to *acceptance region* $\{x : \psi_0 \in C_{\Psi,\gamma}(x)\}$. This approach, however, is unsatisfactory. Recalling the relationship between the strength of the evidence and credible regions, namely,

$$\Pi_\Psi(RB_\Psi(\psi \,|\, x) \le RB_\Psi(\psi_0 \,|\, x) \,|\, x) = 1 - \inf\{\gamma : \psi_0 \in C_{\Psi,\gamma}(x)\},$$

it is seen that a decision to accept H_0 when $\psi_0 \in C_{\Psi,\gamma}(x)$ is equivalent to accepting H_0 when $\Pi_\Psi(RB_\Psi(\psi \,|\, x) \le RB_\Psi(\psi_0 \,|\, x) \,|\, x) \ge 1 - \gamma$. So in effect a decision to accept H_0 is being made based on the strength of the evidence rather than the evidence itself. Recall Example 3.5.2, where it is shown that the evidence can be in favor of H_0, as reflected in a large value of $RB_\Psi(\psi_0 \,|\, x)$, while the strength can be very small. Note too that it is possible that $\{x : \psi_0 \in C_{\Psi,\gamma}(x)\}$ contains values such that $RB_\Psi(\psi_0 \,|\, x) < 1$ and to accept ψ_0 in that case violates the principle of evidence.

In fact, it doesn't seem that the procedure of accepting H_0 if and only if $\psi_0 \in C_{\Psi,\gamma}(x)$ has any particularly good properties with respect to the specific value ψ_0. A more natural approach to this problem is to consider the *relative belief acceptance region* $A(\psi_0) = \{x : RB_\Psi(\psi_0 \,|\, x) > 1\}$. Let $\gamma(\psi_0) = M(\{x : RB_\Psi(\psi_0 \,|\, x) > 1\} \,|\, \psi_0)$ be the prior probability of accepting H_0 given that it is true. Note that this is the complement of the bias against H_0. The following result shows that $A(\psi_0)$ is the optimal acceptance region among all those acceptance regions whose conditional prior content given that H_0 is true is at least $\gamma(\psi_0)$. The proof is provided in the appendix to this chapter.

Proposition 4.7.9 *Among all acceptance regions* $A \subset \mathscr{X}$ *satisfying* $M(A \,|\, \psi_0) \geq \gamma(\psi_0)$, *the relative belief acceptance region* $A(\psi_0)$ *minimizes* $M(A)$.

Notice that

$$
\begin{aligned}
M(A) &= E_{\Pi_\Psi}(M(A \,|\, \psi)) \\
&= E_{\Pi_\Psi}(I_{\{\psi_0\}^c}(\psi) M(A \,|\, \psi)) + \Pi_\Psi(\{\psi_0\}) M(A \,|\, \psi_0) \\
&\geq E_{\Pi_\Psi}(I_{\{\psi_0\}^c}(\psi) M(A(\psi_0) \,|\, \psi)) + \Pi_\Psi(\{\psi_0\}) M(A(\psi_0) \,|\, \psi_0).
\end{aligned}
$$

Therefore, when $\Pi_\Psi(\{\psi_0\}) = 0$, the set $A(\psi_0)$ minimizes $E_{\Pi_\Psi}(I_{\{\psi_0\}^c}(\psi) M(A(\psi_0) \,|\, \psi))$, the prior probability that H_0 is accepted when it is false, among all acceptance regions A satisfying $M(A \,|\, \psi_0) \geq M(A(\psi_0) \,|\, \psi_0)$. Furthermore, when $\Pi_\Psi(\{\psi_0\}) > 0$, the set $A(\psi_0)$ minimizes the prior probability that H_0 is accepted when it is false, among all acceptance regions A satisfying $M(A \,|\, \psi_0) = M(A(\psi_0) \,|\, \psi_0)$.

The value of $M(A(\psi_0) \,|\, \psi_0)$ is a characteristic of the ingredients of the problem but note that, under i.i.d. sampling, Proposition 4.7.4 (i) implies that this converges to 1 as sample size increases. So $M(A(\psi_0) \,|\, \psi_0)$ can be used as a design tool.

The proof of Proposition 4.7.9 is virtually identical to the proof of Proposition 4.7.5. Furthermore, Proposition 4.7.9 can be generalized to prove that, among all acceptance regions $A \subset \mathscr{X}$ satisfying $M(A \,|\, \psi_0) \geq M(\{x : RB_\Psi(\psi_0 \,|\, x) \geq q\} \,|\, \psi_0)$, the region $\{x : RB_\Psi(\psi_0 \,|\, x) \geq q\}$ minimizes $M(A)$. This in turn proves the following where $R(\psi_0) = \{x : RB_\Psi(\psi_0 \,|\, x) < 1\}$ is the *relative belief rejection region*.

Corollary 4.7.8 *Among all rejections regions* $R \subset \mathscr{X}$ *satisfying* $M(R \,|\, \psi_0) \leq M(R(\psi_0) \,|\, \psi_0)$, *the region* $R(\psi_0)$ *maximizes* $M(R)$.

As such, $R(\psi_0)$ is the optimal rejection region for H_0. Note that in general $R(\psi_0) \neq A^c(\psi_0)$, as when $RB_\Psi(\psi_0 \,|\, x) = 1$, it does not make sense to accept or reject H_0.

As with Proposition 4.7.5, there are a number of interesting consequences of Proposition 4.7.9 and these are summarized as follows. The proofs are virtually identical.

Corollary 4.7.9 *(i) If* $M(A) \neq 0$, *then* $A(\psi_0)$ *maximizes*

$$
RB(A \,|\, \psi_0) = M(A \,|\, \psi_0)/M(A)
$$

among all $A \subset \mathscr{X}$ *satisfying* $M(A \,|\, \psi_0) = M(A(\psi_0) \,|\, \psi_0)$.

(ii) If $0 < M(A) < 1$, then $A(\psi_0)$ maximizes

$$BF(A \mid \psi_0) = M(A \mid \psi_0)M(A^c)/M(A^c \mid \psi_0)M(A)$$

among all $A \subset \mathcal{X}$ satisfying $M(A \mid \psi_0) = M(A(\psi_0) \mid \psi_0)$.
(iii) If $M(A(\psi_0)) \neq 0$, then $RB(A(\psi_0) \mid \psi_0) \geq 1$ and if $0 < M(A(\psi_0)) < 1$, then $BF(A(\psi_0) \mid \psi_0) \geq 1$.

Note that $RB(A \mid \psi_0) = M(A \mid \psi_0)/M(A)$ is an a priori relative belief ratio for the set $A \subset \mathcal{X}$ after being told that ψ_0 is the true value. So the set $A(\psi_0)$ has the maximum increase in belief that the data x will be in the set, among all sets with the same conditional prior probability.

Hypothesis testing has been a somewhat controversial topic in Bayesian statistics. Certainly the Jeffreys–Lindley paradox is an indication of this. Many Bayesians, and especially Berger and Sellke (1987), Berger and Delampady (1987) and Johnson (2013), have argued effectively against a *p*-value approach to this problem, instead basing testing on the Bayes factor. Their developments, however, are based on Jeffreys' definition of the Bayes factor and, as already discussed, our preference is to approach all our inference problems using the relative belief ratio. At least a part of the motivation for this lies with a desire to avoid the prescription that, with continuous models, the Bayes factor *must* be based on the mixture prior as in Jeffreys' definition. Of course, a statistician may choose to use such a prior, and we are not claiming that this is inappropriate, as this depends upon the application. It can be argued, however, that defining objects via limits in continuous problems is more suitable, as this recognizes the approximation role that such models and priors are playing. This approach avoids the need for the mixture prior and the resulting inferences have better properties.

4.7.5   Optimality of Relative Belief Estimation

The problem of estimating $\psi = \Psi(\theta)$ is solved in the relative belief formulation for inference by the estimator $\psi_{LRSE}(x) = \arg\sup_\psi RB_\Psi(\psi \mid x)$, together with an assessment of its accuracy as given by the γ-relative belief region $C_{\Psi,\gamma}(x)$. These inferences are forced on us by the relative belief preference ordering. Furthermore, $\psi_{LRSE}(x)$ is optimal in the sense that it maximizes the evidence in favor. Since $\psi_{LRSE}(x) \in C_{\Psi,\gamma}(x)$ for every γ, Corollary 4.7.7 suggests that $RB_\Psi(\psi_{LRSE}(x) \mid x) > 1$ and this is in fact the case.

Proposition 4.7.10 *It is always the case that $RB_\Psi(\psi_{LRSE}(x) \mid x) > 1$.*

Proof. Suppose $RB_\Psi(\psi_{LRSE}(x) \mid x) < 1$. Then $\pi_\Psi(\psi \mid x) < \pi_\Psi(\psi)$ for every $\psi \in \Psi$ and integrating both sides with respect to ψ gives the contradiction $1 < 1$, so $RB_\Psi(\psi_{LRSE}(x) \mid x) \geq 1$.

Then $RB_\Psi(\psi_{LRSE}(x) \mid x) = 1$ iff $\pi_\Psi(\psi_{LRSE}(x) \mid x) = \pi_\Psi(\psi_{LRSE}(x))$ and the uniqueness of $\psi_{LRSE}(x)$ implies $\pi_\Psi(\psi \mid x) < \pi_\Psi(\psi)$ for all other values of ψ (as it is always assumed that $\pi_\Psi(\psi) > 0$ for all values of ψ). In the case where $\Pi_\Psi(\{\psi_{LRSE}(x)\}) = 0$ this again gives the contradiction $1 < 1$ and when $\Pi_\Psi(\{\psi_{LRSE}(x)\}) > 0$, then $\pi_\Psi(\psi_{LRSE}(x)\,|\,x) = 1 - \Pi_\Psi(\psi \neq \psi_{LRSE}(x)\,|\,x) > 1 - \Pi_\Psi(\psi \neq \psi_{LRSE}(x)) = \pi_\Psi(\psi_{LRSE}(x))$, which is a contradiction to the assumed equality. ∎

It is interesting to consider if ψ_{LRSE} has any additional optimal properties with respect to other measures of accuracy. For this, a loss function is introduced that, in the spirit of Robert (1996) and Bernardo (2005), is intrinsic to the problem in the sense that the loss depends only on the prior and does not require further specification. As such, when checking the prior (see Chapter 5), this is also checking the suitability of the loss.

The results in this section were developed in Evans and Jang (2011c). The intention here is to establish additional optimality properties for ψ_{LRSE} and not to argue that a decision-theoretic formulation is the appropriate context for a discussion of estimation problems. In essence, ψ_{LRSE} is justified by the relative belief preference ordering, which is in turn based on the evidence. It will be demonstrated, however, that a loss-based approach leads to insights concerning the statistical behavior of the estimator.

4.7.5.1 Finite Ψ

The loss is defined first for situations where Ψ is finite.

Definition 4.7.2 *When the set Ψ is finite, the relative belief loss function for estimating $\psi = \Psi(\theta)$ is the map $L_{RB} : \Theta \times \Psi \to [0,\infty)$ given by*

$$L_{RB}(\theta, \psi) = \frac{I_{\{\theta:\Psi(\theta)\neq\psi\}}(\theta)}{\pi_\Psi(\Psi(\theta))}.$$

So there is no loss when ψ equals the true value and the loss is proportional to the reciprocal of the prior evaluated at the true value whenever ψ doesn't equal the true value. This loss has the property that it penalizes errors more severely whenever the truth is in the tails of the prior. As such, the loss is attempting to compensate for a prior that will lead to prior-data conflict. This should lead to estimators that are less dependent on the prior and this is indeed the case; see Section 4.7.6.

The following establishes the optimality of ψ_{LRSE} with respect to L_{RB}.

Proposition 4.7.11 *Suppose that $\pi_\Psi(\psi) > 0$ for every $\psi \in \Psi$ and that Ψ is finite. Then ψ_{LRSE} is a Bayes rule for the loss function L_{RB}.*

Proof. For estimator δ, the posterior risk is given by

$$r(\delta\,|\,x) = \int_\Theta \frac{I_{\{\theta:\Psi(\theta)\neq\delta(x)\}}(\theta)}{\pi_\Psi(\Psi(\theta))}\pi(\theta\,|\,x)\,v(d\theta) = \int_\Psi \frac{I_{\{\psi:\psi\neq\delta(x)\}}(\psi)}{\pi_\Psi(\psi)}\pi_\Psi(\psi\,|\,x)\,v_\Psi(d\psi)$$

$$= \int_\Psi RB_\Psi(\psi\,|\,x)\,v_\Psi(d\psi) - RB_\Psi(\delta(x)\,|\,x). \qquad (4.16)$$

Since Ψ is finite, the first term in (4.16) is finite and a Bayes rule at x is given by a value $\delta(x)$ that maximizes the second term, which establishes the result. ∎

From (4.16), the Savage–Dickey formula and direct calculation, the prior risk of δ is

$$r(\delta) = \#(\Psi) - E_M(RB_\Psi(\delta(x)\,|\,x)) = \sum_\psi M(\delta(x) \neq \psi\,|\,\psi). \qquad (4.17)$$

Therefore, finding a Bayes rule with respect to L_{RB} is equivalent to finding δ that maximizes $E_M(\pi_\Psi(\delta(x)\,|\,x)/\pi_\Psi(\delta(x)))$, the prior expected relative belief ratio evaluated at the estimate. Clearly, ψ_{LRSE} is a Bayes rule, as it maximizes the relative belief ratio for each x.

It is interesting to compare (4.17) with what is obtained for the prior risk when using the 0-1 loss $L_{0,1}(\theta, \psi) = I(\Psi(\theta) \neq \psi)$ that produces the MAP estimator $\psi_{MAP}(x) = \arg\sup \pi_\Psi(\psi\,|\,x)$ as a Bayes rule. For 0-1 loss, the prior risk for δ equals

$$\sum_\psi M(\delta(x) \neq \psi\,|\,\psi)\pi_\Psi(\psi), \tag{4.18}$$

the prior probability of making an error. Note that $\psi_{MAP} = \psi_{LRSE}$ when Π_Ψ is uniform and (4.17) is always an upper bound on (4.18). In (4.17) the conditional probabilities of making an error, namely, $M(\delta(x) \neq \psi\,|\,\psi)$, are equally weighted, while in (4.18) these are weighted by their prior probabilities. Consider again the context of the prosecutor's fallacy.

Example 4.7.4 *Prosecutor's Fallacy.*

This has been previously discussed in Examples 4.5.4 and 4.6.3 and in general can be considered as a classification problem with two classes. The innocent class can be labeled by ψ_1 and the guilty class by ψ_2, the respective conditional probabilities that a randomly selected individual from the class has the trait.

The relevant posterior probabilities are given by

$$P(\text{"guilty"}\,|\,\text{"defendant has the trait"}) = \frac{\varepsilon \psi_2}{(1-\varepsilon)\psi_1 + \varepsilon \psi_2},$$

$$P(\text{"guilty"}\,|\,\text{"defendant doesn't have the trait"}) = \frac{\varepsilon(1-\psi_2)}{(1-\varepsilon)(1-\psi_1) + \varepsilon(1-\psi_2)}.$$

When

$$\varepsilon < \min\{\psi_1/(\psi_1 + \psi_2), (1-\psi_1)/(2-\psi_1-\psi_2)\}, \tag{4.19}$$

then $\psi_{MAP}(\text{"defendant has the trait"}) = \psi_1$, since $\varepsilon \psi_2 < (1-\varepsilon)\psi_1$, and $\psi_{MAP}(\text{"defendant doesn't have the trait"}) = \psi_1$, since $\varepsilon(1-\psi_2) < (1-\varepsilon)(1-\psi_1)$. The prior risk of ψ_{MAP} under relative belief loss is equal to $0 + 1 = 1$ and under 0-1 loss it equals $0 \cdot (1-\varepsilon) + 1 \cdot \varepsilon = \varepsilon$. Note too that, for any $\gamma < 1$ and ε small enough, the γ-hpd region for ψ is $H_{\Psi,\gamma}(\text{"defendant has the trait"}) = \{\psi_1\}$, indicating high precision for the estimate when γ is large.

By contrast, whenever $\psi_2 > \psi_1$, which should be the case in any application, then $\psi_{LRSE}(\text{"defendant has the trait"}) = \psi_2$, $\psi_{LRSE}(\text{"defendant doesn't have the trait"}) = \psi_1$, independent of ε. The prior risk under relative belief loss is $\psi_1 + (1-\psi_2) < 1$, while under 0-1 loss it equals $\psi_1(1-\varepsilon) + (1-\psi_2)\varepsilon$, which is greater than ε when (4.19) holds. For ε small enough, the γ-RB region for ψ is $C_{\Psi,\gamma}(\text{"defendant has the trait"}) = \{\psi_1, \psi_2\}$, which indicates no precision for the estimate.

So when ε is small, ψ_{MAP} simply ignores the small class while ψ_{LRSE} does not. Certainly this seems appropriate in the legal context, but it is surely not sensible in the medical application discussed in Example 4.5.4. It is particularly alarming that ψ_{MAP} does this even when ψ_1 is very small. Of course, it is not appropriate to use

ψ_{LRSE} simply as a classifier either. What is needed is a meaningful presentation of the evidence that then can be combined with other particulars of the application to determine appropriate decisions. As discussed in Example 4.5.4, this is what relative belief inferences provide. ∎

4.7.5.2 Countable Ψ

The loss function L_{RS} does not provide meaningful results when Ψ is countably infinite, as (4.17) shows that $r(\delta)$ will be infinite. When countably infinite Ψ is used, however, this is in essence saying that the real parameter space is finite but of large cardinality and the infinite context is being used as an approximation. So it is natural to consider a sequence of finite parameter spaces $\Psi_\eta = \{\psi : \pi_\Psi(\psi) \geq \eta\}$ together with L_{RB} and a renormalized prior, apply Proposition 4.7.11, and then derive the limiting estimator as $\eta \to 0$ as an approximation to what is essentially a finite problem. Another approach is to define a sequence of bounded loss functions

$$L_{RB,\eta}(\theta, \psi) = I_{\{\theta:\Psi(\theta)\neq\psi\}}(\theta)/\max(\eta, \pi_\Psi(\Psi(\theta))), \qquad (4.20)$$

so $L_{RB,\eta} \to L_{RB}$ as $\eta \to 0$, and consider the limiting estimator. The following result, proved in the appendix to this chapter, shows that both approaches lead to ψ_{LRSE}.

Proposition 4.7.12 *Suppose that Ψ is countable, $\pi_\Psi(\psi) > 0$ for every $\psi \in \Psi$ and $\psi_{LRSE}(x)$ is the unique maximizer of $RB_\Psi(\psi \,|\, x)$ for all x. If δ_η is the Bayes rule for the problem with Ψ_η and loss L_{RB}, or for the problem using Ψ and loss $L_{RB,\eta}$, then $\delta_\eta(x) \to \psi_{LRSE}(x)$ as $\eta \to 0$, for every $x \in \mathcal{X}$. For all sufficiently small η, the Bayes rule at x equals $\psi_{LRSE}(x)$.*

In a general estimation problem, an estimator δ is unbiased with respect to a loss function L if $E_{P_\theta}(L(\theta', \delta(x))) \geq E_{P_\theta}(L(\theta, \delta(x)))$ for all $\theta', \theta \in \Theta$. This says that on average $\delta(x)$ is closer to the true value than any other value when we interpret $L(\theta, \delta(x))$ as a measure of distance between $\delta(x)$ and $\Psi(\theta)$. A similar concept can be defined in the Bayesian context.

Definition 4.7.3 *The estimator δ is Bayesian unbiased with respect to loss L when*

$$\int_\Theta \int_\Theta E_\theta(L(\theta', \delta(x)))\,\Pi(d\theta)\,\Pi(d\theta') \geq \int_\Theta E_\theta(L(\theta, \delta(x)))\,\Pi(d\theta) = r(\delta). \quad (4.21)$$

The role of θ' in (4.21) is as a false value, as it is generated from the prior, independent of the true value θ, and with no connection to the data. Therefore, δ is Bayesian unbiased if $\delta(x)$ is on average closer to the true value than a false value.

Consider now the unbiasedness of $\psi_{LRSE}(x)$. For this we consider losses of the form

$$L(\theta, \psi) = I_{\{\theta:\Psi(\theta)\neq\psi\}}(\theta)h(\Psi(\theta)) \qquad (4.22)$$

for some nonnegative function h which satisfies $\int_\Theta h(\Psi(\theta))\,\Pi(d\theta) < \infty$. This class of loss functions includes L_{RB} when Ψ is finite, $L_{RB,\eta}$ and $L_{0,1}$. The following result is proved in the appendix to this chapter.

Proposition 4.7.13 *If Ψ is countable, then $\psi_{LRSE}(x)$ is Bayesian unbiased under (4.22). Moreover, any estimator δ is Bayesian unbiased whenever $\pi_\Psi(\delta(x)|x) \geq \pi_\Psi(\delta(x))$ for all x.*

The relevance of the last statement of Proposition 4.7.13 is that, as established in Proposition 4.7.10, the estimator ψ_{LRSE} always possesses this property and, from the proof of Proposition 4.7.13, it makes sense to refer to such an estimator as being *uniformly (in x) Bayesian unbiased.*

4.7.5.3   General Ψ

When ψ has a continuous prior distribution, again consider this as an approximation to an essentially finite problem. For this assume that the spaces involved are locally Euclidean, mappings are sufficiently smooth and take all support measures to be the analogs of Euclidean volume. Further details on the mathematical requirements underlying these assumptions can be found in Appendix A. The argument provided here applies quite generally but is simplified somewhat by taking all the spaces to be open subsets of Euclidean spaces with Euclidean volume as support measures.

For each $\lambda > 0$ discretize the set Ψ via a countable partition $\{B_\lambda(\psi) : \psi \in \Psi\}$ where $\psi \in B_\lambda(\psi), \Pi_\Psi(B_\lambda(\psi)) > 0$, $\sup_{\psi \in \Psi} \mathrm{diam}(B_\lambda(\psi)) \to 0$ as $\lambda \to 0$. For example, the $B_\lambda(\psi)$ could be equal volume rectangles in R^k. Further, assume that $\Pi_\Psi(B_\lambda(\psi))/\nu_\Psi(B_\lambda(\psi)) \to \pi_\Psi(\psi)$ as $\lambda \to 0$ for every ψ. This will hold whenever π_Ψ is continuous everywhere and $B_\lambda(\psi)$ converges nicely to $\{\psi\}$ as $\lambda \to 0$. Let $\psi_\lambda(\psi) \in B_\lambda(\psi)$ be such that $\psi_\lambda(\psi') = \psi_\lambda(\psi)$ whenever $\psi' \in B_\lambda(\psi)$ and $\Psi_\lambda = \{\psi_\lambda(\psi) : \psi_\lambda(\psi) \in B_\lambda(\psi), \psi \in \Psi\}$ be the discretized version of Ψ. Note that one point is chosen in each $B_\lambda(\psi)$. This will be referred to as a *regular discretization* of Ψ. The discretized prior on Ψ_λ is $\pi_{\Psi,\lambda}(\psi_\lambda(\psi)) = \Pi_\Psi(B_\lambda(\psi))$ and the discretized posterior is $\pi_{\Psi,\lambda}(\psi_\lambda(\psi)|x) = \Pi_\Psi(B_\lambda(\psi)|x)$. We let $\hat{\psi}_\lambda(x)$ denote the maximum relative belief estimate for the problem with parameter space Ψ_λ.

For the discretized problem we consider either truncating the possibly infinite set Ψ_λ to $\Psi_{\lambda,\eta}$, and use the loss L_{RB}, or use Ψ_λ and the loss

$$L_{RB,\lambda,\eta}(\theta, \psi_\lambda(\psi)) = I_{\{\theta : \psi_\lambda(\Psi(\theta)) \neq \psi_\lambda(\psi)\}}(\theta)/\max(\eta, \pi_{\Psi,\lambda}(\psi_\lambda(\Psi(\theta)))). \quad (4.23)$$

Let $\delta_{\lambda,\eta}(x)$ denote a Bayes rule for both versions and require $RB_\Psi(\cdot|x)$ to be bounded away from $RB_\Psi(\psi_{LRSE}(x)|x)$ outside any neighborhood of $\psi_{LRSE}(x)$ to avoid pathologies. The following results are proved in the appendix to this chapter.

Proposition 4.7.14 *If π_Ψ is positive and continuous, $\psi_{LRSE}(x)$ is the unique maximizer of $RB_\Psi(\cdot|x)$ with*

$$\sup_{\{\psi : \|\psi - \psi_{LRSE}(x)\| \geq \varepsilon\}} RB_\Psi(\psi|x) < RB_\Psi(\psi_{LRSE}(x)|x)$$

when $\varepsilon > 0$, and there is a regular discretization of Ψ, then there exists $\eta(\lambda) > 0$ such that a Bayes rule $\delta_{\lambda,\eta(\lambda)}(x) \to \psi_{LRSE}(x)$ as $\lambda \to 0$ for all x.

Corollary 4.7.10 *Furthermore, $\hat{\psi}_\lambda(x)$ converges to $\psi_{LRSE}(x)$ as $\lambda \to 0$.*

By Proposition 4.7.13, the estimator $\hat{\psi}_\lambda$ is uniformly Bayesian unbiased for the discretized problem and so ψ_{LRSE} is the limit of uniformly Bayesian unbiased estimators.

We consider an important example.

Example 4.7.5 *Regression.*

Suppose that $y = X\beta + e$ where $y \in R^n, X \in R^{n \times k}$ is fixed, $\beta \in R^k, e \sim N_n(0, \sigma^2 I)$, place a prior on β and assume that σ^2 is known to simplify the discussion. Then having observed $(X, y), \beta_{LRSE}(x) = b = (X'X)^{-1}X'y$, which in this case is the MLE of β. It is interesting to contrast this with the posterior mode or posterior mean. For example, if $\beta \sim N_k(0, \tau^2 I)$, then the posterior distribution of β is $N_k(\mu_{post}(\beta), \Sigma_{post}(\beta))$ where

$$\mu_{post}(\beta) = \Sigma_{post}(\beta)\sigma^{-2}X'Xb,$$

$$\Sigma_{post}(\beta) = (\tau^{-2}I + \sigma^{-2}X'X)^{-1}$$

and the posterior mean and mode both equal $\mu_{post}(\beta)$. The spectral decomposition $X'X = Q\Lambda Q' = \Sigma_{i=1}^k \lambda_i q_i q_i'$, implies $||\mu_{post}(\beta)|| = ||(I + (\sigma^2/\tau^2)\Lambda^{-1})^{-1}Q'b||$. Since $||b|| = ||Q'b||$ and $1/(1 + \sigma^2/(\tau^2\lambda_i)) < 1$ for each i, it is the case that $\mu_{post}(\beta)$ moves the MLE toward the prior mean 0. This is often cited as a positive attribute of $\mu_{post}(\beta)$. But consider the situation where the true value of β lies in the tails of the prior as it is then wrong to move β toward the prior mean. When τ^2 is chosen large, to avoid the possibility that the true value of β lies in the tails of the prior, then $\mu_{post}(\beta) \approx b$.

So it is not clear that shrinking the MLE is a good thing to do, particularly as this requires giving up invariance. Perlman and Chaudhuri (2012) demonstrate that shrinking toward a point can result in worse performance. The issue of prior-data conflict also plays a role. Following Evans and Moshonov (2006) and Evans and Jang (2011b), there will generally be no prior-data conflict whenever b is not in the tails of the prior and so there will be little benefit from shrinking b in such a situation. These issues will be discussed further in Chapter 5.

Suppose now we want to estimate $\psi = w'\beta$ for some setting w of the predictors. Then, ψ has prior distribution $N(0, \sigma^2_{prior}(\psi)) = N(0, \tau^2 w'w)$, posterior $N(\mu_{post}(\psi), \sigma^2_{post}(\psi)) = N(w'\mu_{post}(\beta), w'\Sigma_{post}(\beta)w)$ and $\psi_{MAP}(y) = \mu_{post}(\psi)$. Note that $\sigma^2_{prior}(\psi) - \sigma^2_{post}(\psi) = w'(\tau^2 I - \Sigma_{post}(\beta))w = \tau^2 w'Q'(I - (I + (\tau^2/\sigma^2)\Lambda)^{-1})Qw > 0$ and so

$$\psi_{LRSE}(y) = (1 - \sigma^2_{post}(\psi)/\sigma^2_{prior}(\psi))^{-1}\mu_{post}(\psi). \tag{4.24}$$

When $\sigma^2_{post}(\psi)$ is much smaller than $\sigma^2_{prior}(\psi)$, then $\psi_{LRSE}(y) \approx \psi_{MAP}(y)$ but $\sigma^2_{prior}(\psi) > \sigma^2_{post}(\psi)$ so $|\psi_{LRSE}(y)| > |\psi_{MAP}(y)|$. In general, $\psi_{LRSE}(y)$ is not equal to $w'b$, the plug-in MLE of ψ, although it is easy to see that $\psi_{LRSE}(y) = w'b$ whenever $w = q_i$. Both $\psi_{MAP}(y) \to w'b$ and $\psi_{LRSE}(y) \to w'b$ as $\tau^2/\sigma^2 \to \infty$. When $\tau^2/\sigma^2 \to 0, \psi_{MAP}(y) \to 0$ and $\psi_{LRSE}(y) \to (||w||^2/\Sigma_{i=1}^k \lambda_i w_i^2)\Sigma_{i=1}^k \lambda_i w_i q_i'b$, which equals $w'b$ when the λ_i are equal. So ψ_{LRSE} does not allow the prior to override the data.

Relative belief predictions are also optimal just as with estimation. So suppose interest is in predicting a response z at the predictor value $w \in R^k$. When $\beta \sim N_k(0, \tau^2 I)$, the prior distribution of z is $z \sim N(0, \sigma^2 + \tau^2 w'w) = N(0, \sigma^2_{prior}(z))$ and the posterior is $N(\mu_{post}(z), \sigma^2_{post}(z))$ with

$$\mu_{post}(z) = w' \mu_{post}(\beta), \quad \sigma^2_{post}(z) = \sigma^2 + w' \Sigma_{post}(\beta) w$$

and

$$z_{LRSE}(y) = (1 - \sigma^2_{post}(z)/\sigma^2_{prior}(z))^{-1} \mu_{post}(z). \tag{4.25}$$

Note that $\sigma^2_{prior}(z) - \sigma^2_{post}(z) = \sigma^2_{prior}(w'\beta) - \sigma^2_{post}(w'\beta) > 0$ and so $|z_{LRSE}(y)| > |\mu_{post}(z)|$ and $z_{LRSE}(y)$ is further from the prior mean than $z_{MAP}(y) = \mu_{post}(z)$. Also, when $\sigma^2_{post}(z)/\sigma^2_{prior}(z)$ is small, then $z_{LRSE}(y)$ and $z_{MAP}(y)$ are very similar. Finally, from (4.24) and (4.25), then $z_{LRSE}(y) = (\sigma^2_{prior}(z)/\sigma^2_{post}(\psi)) w' \psi_{LRSE}(y) = (1 + \sigma^2/\tau^2) \psi_{LRSE}(y)$ and so the relative belief predictor at x is more dispersed than the relative belief estimator of the mean at w. This makes good sense, as it is necessary to take into account the additional variation due to prediction. By contrast, $z_{MAP}(y) = \psi_{MAP}(y)$. ∎

4.7.5.4 Relative Belief Credible Regions and Loss

A loss function L for estimating ψ also leads to credible regions, namely, the γ-*lowest posterior loss credible region*

$$L_\gamma(x) = \{\psi : r(\psi \,|\, x) \le l_\gamma(x)\} \tag{4.26}$$

where $l_\gamma(x) = \inf\{k : \int_{\{\psi_0 : r(\psi_0 | x) \le k\}} \pi_\Psi(\psi \,|\, x) \, \nu_\Psi(d\psi) \ge \gamma\}$; see Section 3.5.5 and Bernardo (2005). Here ψ in (4.26) is interpreted as the decision function that takes the value ψ constantly in x. Clearly, as $\gamma \to 0$, the set $L_\gamma(x)$ converges to the value of a Bayes rule at x.

Consider now the regions that arise from prior-based loss functions.

Proposition 4.7.15 *Suppose that Ψ is finite, $\pi_\Psi(\psi) > 0$ for every $\psi \in \Psi$. Then $C_{\Psi,\gamma}(x)$ is a γ-lowest posterior loss credible region based on loss L_{RB}.*

Proof. From (4.26) and (4.16) the γ-lowest posterior loss credible region is

$$L_\gamma(x) = \left\{ \psi : RB_\Psi(\psi \,|\, x) \ge \int_\Psi RB_\Psi(\zeta \,|\, x) \, \nu_\Psi(d\zeta) - l_\gamma(x) \right\}$$

and $l_\gamma(x) = \inf\{k : \Pi_\Psi(\{\psi : r(\psi \,|\, x) \le k\} \,|\, x) \ge \gamma\}$. As $\int_\Psi RB_\Psi(z \,|\, x) \, \nu_\Psi(dz)$ is constant, we see that $L_\gamma(x) = C_{\Psi,\gamma}(x)$. ∎

When Ψ is countable, consider the finite parameter spaces Ψ_η, or the loss function (4.20), and let $\eta \to 0$ to determine the relevant regions. A γ-lowest posterior loss region $L_{\eta,\gamma}(x)$ is defined as in Proposition 4.7.15. The following result is proved in the appendix to this chapter.

Proposition 4.7.16 *Suppose that* Ψ *is countable,* $\pi_\Psi(\psi) > 0$ *for every* $\psi \in \Psi$ *and* $\Pi_\Psi(C_{\Psi,\gamma}(x)\,|\,x) = \gamma$. *For the parameter space* Ψ_η *or the loss function (4.20), then* $C_{\Psi,\gamma}(x) \subset \liminf_{\eta\to 0} L_{\eta,\gamma}(x)$ *for all* $\gamma' < \gamma$ *and* $\limsup_{\eta\to 0} L_{\eta,\gamma}(x) \subset C_{\Psi,\gamma}(x)$ *for all* $\gamma' > \gamma$.

While Proposition 4.7.16 does not establish the exact convergence $\lim_{\eta\to 0} L_{\eta,\gamma}(x) = C_{\Psi,\gamma}(x)$, it is suspected that this does hold quite generally. Actually the proof also shows that $C_{\Psi,\gamma}(x) \subset \liminf_{\eta\to 0} L_{\eta,\gamma}(x)$ when using the loss function (4.20).

Now consider the continuous case and suppose there is a regular discretization. For $S^* \subset \Psi_\lambda = \{\psi_\lambda(\psi) : \psi_\lambda(\psi) \in B_\lambda(\psi)\}$, namely, S^* is a subset of a discretized version of Ψ, define the *undiscretized* version of S^* to be $S = \cup_{\psi\in S^*} B_\lambda(\psi)$. Let $C^*_{\lambda,\gamma}(x)$ be the γ-relative surprise region for the discretized problem and let $C_{\Psi,\lambda,\gamma}(x)$ be its undiscretized version. In a continuous context, two sets will be considered as equal if they differ only by a set of measure 0 with respect to Π_Ψ. In the appendix to this chapter the following result is proved.

Proposition 4.7.17 *Suppose that* π_Ψ *is positive and continuous, that there is a regular discretization of* Ψ *and that* $RB_\Psi(\cdot\,|\,x)$ *has a continuous posterior distribution. Then* $\lim_{\lambda\to 0} C_{\Psi,\lambda,\gamma}(x) = C_{\Psi,\gamma}(x)$.

This says that a γ-relative belief region for the discretized problem (after undiscretizing) converges to the γ-relative belief region for the original problem.

Proposition 4.7.17 has interest in its own right, but it can also be used to prove that relative belief regions are limits of lowest posterior loss regions. Let $L^*_{\eta,\lambda,\gamma}(x)$ be the γ-lowest posterior loss region obtained for the discretized problem using either the parameter space $\Psi_{\lambda,\eta}$ or the loss function (4.23) and let $L_{\eta,\lambda,\gamma}(x)$ be the undiscretized version. The following result is proved in the appendix to this chapter.

Proposition 4.7.18 *Suppose that* π_Ψ *is positive and continuous, that there is a regular discretization of* Ψ *and that* $RB_\Psi(\cdot\,|\,x)$ *has a continuous posterior distribution. Then*

$$C_{\Psi,\gamma}(x) = \lim_{\lambda\to 0}\liminf_{\eta\to 0} L_{\eta,\lambda,\gamma}(x) = \lim_{\lambda\to 0}\limsup_{\eta\to 0} L_{\eta,\lambda,\gamma}(x).$$

4.7.6 Robustness of Relative Belief Inferences

It is reasonable to ask how sensitive relative belief inferences are to the prior when interest is in $\psi = \Psi(\theta)$. For example, how robust to the prior are $\psi_{LRSE}(x), C_{\Psi,\gamma}(x), RB_\Psi(\psi_0\,|\,x)$ and $\Pi_\Psi(RB_\Psi(\psi\,|\,x) \leq RB_\Psi(\psi_0\,|\,x)\,|\,x)$, as these four objects represent essential relative belief inferences. The focus here is on the ε-contamination approach to perturbing the marginal prior π_Ψ. General discussions of the robustness of Bayesian inferences to the prior can be found in Berger (1994) and Rios Insua and Ruggeri (2000).

By the Savage–Dickey result, $RB_\Psi(\psi\,|\,x) = m(x\,|\,\psi)/m(x)$ and note that $m(x\,|\,\psi) = \int_{\Psi^{-1}\{\psi_0\}} f_\theta(x)\pi(\theta\,|\,\psi)\,\nu_{\Psi^{-1}\{\psi\}}(d\theta)$ does not depend on π_Ψ. Therefore, since

$$\psi_{LRSE}(x) = \arg\sup_\psi RB_\Psi(\psi\,|\,x) = \arg\sup_\psi m(x\,|\,\psi),$$

the relative belief estimate is maximally robust to π_Ψ. Furthermore, $C_{\Psi,\gamma}(x)$ is of the form $\{\psi : m(x\,|\,\psi) \geq k\}$ for some k and thus the form of the relative belief regions for ψ is maximally robust to π_Ψ. The specific region chosen for the assessment of the accuracy of $\psi_{LRSE}(x)$ depends on the posterior and is not independent of π_Ψ but, as will be demonstrated here, the posterior content of $C_{\Psi,\gamma}(x)$ also possesses optimal robustness properties.

Consider now ε-contaminated priors for θ of the form

$$\pi^{\varepsilon,q}(\theta) = \pi(\theta\,|\,\Psi(\theta))[(1-\varepsilon)\pi_\Psi(\Psi(\theta)) + \varepsilon q(\Psi(\theta))]$$

where q is a prior density on Ψ with respect to ν_Ψ. So only the marginal prior of Ψ is being perturbed in π^ε. The posterior density of ψ based on this prior is then

$$\pi_\Psi^{\varepsilon,q}(\psi\,|\,x) = (1-\varepsilon_x)\,\pi_\Psi(\psi\,|\,x) + \varepsilon_x q(\psi\,|\,x)$$

where

$$\pi_\Psi(\psi\,|\,x) = \pi_\Psi(\psi) \int_{\Psi^{-1}\{\psi\}} \pi(\theta\,|\,\psi) \frac{f_\theta(x)}{m(x)}\, \nu_{\Psi^{-1}\{\psi\}}(d\theta) = \pi_\Psi(\psi) \frac{m(x\,|\,\psi)}{m(x)},$$

$$q(\psi\,|\,x) = q(\psi) \int_{\Psi^{-1}\{\psi\}} \pi(\theta\,|\,\psi) \frac{f_\theta(x)}{m_q(x)}\, \nu_{\Psi^{-1}\{\psi\}}(d\theta) = \pi_\Psi(\psi) \frac{m(x\,|\,\psi)}{m_q(x)},$$

$$m(x) = \int_\Psi \pi_\Psi(\psi) m(x\,|\,\psi)\, \nu_\Psi(d\psi),$$

$$m_q(x) = \int_\Psi q(\psi) m(x\,|\,\psi)\, \nu_\Psi(d\psi)$$

and

$$\varepsilon_x = \frac{\varepsilon m_q(x)/m(x)}{1 - (1 - m_q(x)/m(x))\varepsilon}.$$

Note that, if T is a minimal sufficient statistic for the model, then $m_q(x)/m(x) = m_{qT}(T(x))/m_T(T(x))$.

Consider the sensitivity of the posterior content of $C_{\Psi,\gamma}(x)$ to the prior. To assess the robustness of the posterior content of a set $A \subset \Psi$, it makes sense to look at

$$\delta(A) = \Pi_\Psi^{upper}(A\,|\,x) - \Pi_\Psi^{lower}(A\,|\,x)$$

where $\Pi_\Psi^{upper}(A\,|\,x) = \sup_Q \Pi_\Psi^{\varepsilon,Q}(A\,|\,x)$ and $\Pi_\Psi^{lower}(A\,|\,x) = \inf_Q \Pi_\Psi^{\varepsilon,Q}(A\,|\,x)$ where the supremum and infimum are taken over all probability measures on Ψ. For this let $\varepsilon^* = \varepsilon/(1-\varepsilon)$ and

$$r(A) = \sup_{\psi \in A} RB_\Psi(\psi\,|\,x) = \sup_{\psi \in A} m(x\,|\,\psi)/m(x),$$

so $r(\Psi) = RB_\Psi(\psi_{LRSE}(x)\,|\,x)$ and it is assumed that $\psi_{LRSE}(x)$ is unique.

The following is proved in the appendix to this chapter and is a generalization of results found in Huber (1973).

Lemma 4.7.1 *Let Q denote a probability measure on Ψ. For prior measure $\Pi_\Psi^{\varepsilon,Q} = (1-\varepsilon)\Pi_\Psi + \varepsilon Q$ on Ψ and $A \subset \Psi$,*
(i) $\Pi_\Psi^{upper}(A\,|\,x) = (\Pi_\Psi(A\,|\,x) + \varepsilon^* r(A))/(1 + \varepsilon^* r(A))$,
(ii) $\Pi_\Psi^{lower}(A\,|\,x) = \Pi_\Psi(A\,|\,x)/(1 + \varepsilon^* r(A^c))$,
(iii)

$$\delta(A) = \frac{\Pi_\Psi(A\,|\,x)\,\varepsilon^*(r(A^c) - r(A))}{(1 + \varepsilon^* r(A))(1 + \varepsilon^* r(A^c))} + \frac{\varepsilon^* r(A)}{(1 + \varepsilon^* r(A))},$$

(iv) $\delta(A^c) = \delta(A)$.

Observe that always one and only one of $r(A), r(A^c)$ equals $r(\Psi)$. Let $\gamma^*(x) = \Pi_\Psi(C_{\Psi,\gamma}(x)\,|\,x)$ be the exact posterior content of the γ-relative belief region. The following generalizes results found in Wasserman (1989) and de la Horra and Fernandez (1994) and is proved in the appendix to this chapter.

Proposition 4.7.19 *The following hold:*
(i) among all sets $A \subset \Psi$ satisfying $\Pi_\Psi(A\,|\,x) \leq \gamma^*(x)$ and $r(A) = r(\Psi)$, the set $C_{\Psi,\gamma}(x)$ minimizes $\delta(A)$,
(ii) among all sets $A \subset \Psi$ satisfying $\Pi_\Psi(A\,|\,x) \geq \gamma^*(x)$ and $r(A^c) = r(\Psi)$, the set $C_{\Psi,1-\gamma^*(x)}^c(x)$ minimizes $\delta(A)$,
(iii) when $\gamma^*(x) = \gamma \geq 1/2$, then, among all sets $A \subset \Psi$ satisfying $\Pi_\Psi(A\,|\,x) = \gamma$, the set $C_{\Psi,\gamma}(x)$ minimizes $\delta(A)$.

Consider now the meaning of the separate parts of Proposition 4.7.19. For any system of credible regions, say $B_{\Psi,\gamma}(x)$, it makes sense to require that these sets are monotonically increasing in γ and that the smallest set $\lim_{\gamma \searrow 0} B_{\Psi,\gamma}(x)$ contains a single point to be taken as the estimate of ψ. The relative belief regions satisfy this, under our assumptions of a unique maximizer of $RB_\Psi(\cdot\,|\,x)$, and the estimate is $\psi_{LRSE}(x)$. Also, any system of credible regions $B_{\Psi,\gamma}(x)$ that does not shrink to $\psi_{LRSE}(x)$ must violate the relative belief ordering in the sense that, for γ small enough, there will exist $\psi_1 \in B_{\Psi,\gamma}(x)$, $\psi_2 \notin B_{\Psi,\gamma}(x)$ but $RB_\Psi(\psi_2\,|\,x) > RB_\Psi(\psi_1\,|\,x)$.

Effectively (i) is then saying that $C_{\Psi,\gamma}(x)$ is the most robust, with respect to posterior content, among all systems of credible regions $B_{\Psi,\gamma}(x)$ with $\Pi_\Psi(B_{\Psi,\gamma}(x)\,|\,x) \leq \gamma^*(x)$ that lead to the estimate $\psi_{LRSE}(x)$. Note that sets A with $\Pi_\Psi(A\,|\,x) > \gamma^*(x)$ have to be excluded because, for example, the set $A = \Psi$ is always maximally robust with respect to content but does not provide a meaningful assessment of the accuracy of the estimate. Given that ψ_{LRSE} and the form of $C_{\Psi,\gamma}$ are maximally robust, this further supports the claim that relative belief estimation is maximally robust to the choice of the marginal prior.

For (ii), note that the sets $C_{\Psi,1-\gamma}^c(x)$ do not satisfy the criteria for being a system of credible regions. Also the set $C_{\Psi,1-\gamma}^c(x)$ could be considered as a γ-*incredible region* in the sense that it contains those values of ψ the least supported by the evidence and has content no greater than γ. So (ii) says that $C_{\Psi,1-\gamma^*(x)}^c(x)$ is the most robust choice with respect to content among all incredible regions with posterior probability content at least equal to $\gamma^*(x)$.

Part (iii) indicates that, when there are many sets with posterior content exactly equal to $\gamma \geq 1/2$, and this is often true in the continuous case, then $C_{\Psi,\gamma}(x)$ is maximally robust among these sets with respect to posterior content. So overall, relative belief estimation of ψ possesses optimal robustness properties with respect to the marginal prior π_Ψ.

For hypothesis assessment it is necessary to consider the sensitivity of the evidence given by $RB_\Psi(\psi_0 \,|\, x)$. The relative belief ratio for ψ_0 based on $\pi^{\varepsilon, Q}$ equals

$$RB_\Psi^\varepsilon(\psi_0 \,|\, x) = \frac{m(x \,|\, \psi_0)}{(1 - \varepsilon)m(x) + \varepsilon m_q(x)} = \frac{RB_\Psi(\psi_0 \,|\, x)}{1 - \varepsilon(1 - m_q(x)/m(x))}. \qquad (4.27)$$

The *Gâteaux derivative* of $RB_\Psi(\cdot \,|\, x)$ at ψ_0 in the direction q is used to assess the sensitivity of this quantity.

Proposition 4.7.20 *The Gâteaux derivative of $RB_\Psi(\cdot \,|\, x)$ at ψ_0 in the direction q is $RB_\Psi(\psi_0 \,|\, x)\{1 - m_{qT}(T(x))/m_T(T(x))\}$.*

Proof. This follows from (4.27) as

$$\lim_{\varepsilon \to 0} \frac{RB_\Psi^\varepsilon(\psi_0 \,|\, x) - RB_\Psi(\psi_0 \,|\, x)}{\varepsilon} = RB_\Psi(\psi_0 \,|\, x) \lim_{\varepsilon \to 0} \left\{ \frac{1 - m_q(x)/m(x)}{1 - \varepsilon(1 - m_q(x)/m(x))} \right\}.$$

∎

So the robustness of the value $RB_\Psi(\psi_0 \,|\, x)$ depends on the possible values for $m_{qT}(T(x))/m_T(T(x))$. Notice that the absolute value of the Gâteaux derivative is increasing in $m_q(x)/m(x)$. It will be seen in Chapter 5 that relatively small values of $m_T(T(x))$ are associated with prior-data conflict. If the conflict arises through the choice of π_Ψ, rather than the conditional prior of θ given $\Psi(\theta) = \psi$, then it is reasonable to suppose that there is a prior q such that $m_{qT}(T(x))/m_T(T(x))$ is large but not otherwise. So provided there is no serious prior-data conflict, it can be expected that $RB_\Psi(\psi_0 \,|\, x)$ is robust to perturbations in π_Ψ. A simple example illustrates this.

Example 4.7.6 *Location Normal.*

Suppose that $x = (x_1, \ldots, x_n)$ is a sample from the $N(\mu, 1)$ distribution and the prior is $\mu \sim N(\mu_0, \sigma_0^2)$. Then the prior predictive distribution of the minimal sufficient statistic is given by $\bar{x} \sim N(\mu_0, 1/n + \sigma_0^2)$. Prior-data conflict occurs here whenever $\bar{x}$ is the tails of the $N(\mu_0, 1/n + \sigma_0^2)$ distribution and this suggests that the prior $N(\mu_0, \sigma_0^2)$ is not located properly. Certainly it is possible to consider increasing σ_0^2 to deal with this but alternatively the prior can be relocated or a combination of both.

Let q denote the density of the $N(\mu_1, \sigma_0^2)$ distribution so that $\bar{x} \sim N(\mu_1, 1/n + \sigma_0^2)$. This implies that

$$m_q(x)/m(x) = \exp\{-(1/n + \sigma_0^2)^{-1}[(\bar{x} - \mu_1)^2 - (\bar{x} - \mu_0)^2]/2\}$$

and, as a function of μ_1, this is maximized when $\mu_1 = \bar{x}$ with maximal value $\exp\{(1/n + \sigma_0^2)^{-1}(\bar{x} - \mu_0)^2/2\}$, which converges to ∞ as $\bar{x} \to \pm\infty$. Therefore, when there is no prior-data conflict, namely, when $\bar{x}$ is a reasonable value for the $N(\mu_0, 1/n + \sigma_0^2)$ distribution, the relative belief ratio is robust to the choice of prior. ∎

One measure of the strength of the evidence given by $RB_\Psi(\psi_0 \,|\, x)$ is the posterior probability of the set $\{\psi : m(x \,|\, \psi) \leq m(x \,|\, \psi_0)\}$ and this set is maximally robust to π_Ψ. The strength is a posterior probability, however, and this depends on π_Ψ.

Proposition 4.7.21 *The Gâteaux derivative in the direction q of* $\Pi_\Psi(RB_\Psi(\psi\,|\,x) \leq RB_\Psi(\psi_0\,|\,x)\,|\,x))$ *equals*

$$\frac{m_{qT}(T(x))}{m_T(T(x))} \left\{ \begin{array}{l} Q(RB_\Psi(\psi\,|\,x) \leq RB_\Psi(\psi_0\,|\,x)\,|\,x) - \\ \Pi_\Psi(RB_\Psi(\psi\,|\,x) \leq RB_\Psi(\psi_0\,|\,x)\,|\,x) \end{array} \right\}. \tag{4.28}$$

Proof. From the definition of $\Pi_\Psi^{\varepsilon,Q}(\cdot\,|\,x)$,

$$\Pi_\Psi^\varepsilon(RB_\Psi^\varepsilon(\psi\,|\,x) \leq RB_\Psi^\varepsilon(\psi_0\,|\,x)\,|\,x) = (1 - \varepsilon_x)\,\Pi_\Psi(RB_\Psi^\varepsilon(\psi\,|\,x) \leq RB_\Psi^\varepsilon(\psi_0\,|\,x)\,|\,x)$$
$$+ \varepsilon_x Q(RB_\Psi^\varepsilon(\psi_0\,|\,x) \leq RB_\Psi^\varepsilon(\psi_0\,|\,x)\,|\,x)$$

and so

$$\Pi_\Psi^{\varepsilon,Q}(RB_\Psi^\varepsilon(\psi_0\,|\,x) \leq RB_\Psi^\varepsilon(\psi_0\,|\,x)\,|\,x) = \Pi_\Psi(m(x\,|\,\psi) \leq m(x\,|\,\psi_0)\,|\,x) +$$
$$\varepsilon_x \left\{ Q(m(x\,|\,\psi) \leq m(x\,|\,\psi_0)\,|\,x) - \Pi_\Psi(m(x\,|\,\psi) \leq m(x\,|\,\psi_0)\,|\,x) \right\}.$$

This implies that

$$\lim_{\varepsilon \to 0} \frac{\Pi_\Psi^{\varepsilon,Q}(RB_\Psi^\varepsilon(\psi_0\,|\,x) \leq RB_\Psi^\varepsilon(\psi_0\,|\,x)\,|\,x) - \Pi_\Psi(RB_\Psi(\psi\,|\,x) \leq RB_\Psi(\psi_0\,|\,x)\,|\,x)}{\varepsilon}$$
$$= \frac{m_q(x)}{m(x)} \left\{ Q(m(x\,|\,\psi) \leq m(x\,|\,\psi_0)\,|\,x) - \Pi_\Psi(m(x\,|\,\psi) \leq m(x\,|\,\psi_0)\,|\,x) \right\}$$
$$= \frac{m_q(x)}{m(x)} \left\{ \begin{array}{l} Q(RB_\Psi(\psi\,|\,x) \leq RB_\Psi(\psi_0\,|\,x)\,|\,x) - \\ \Pi_\Psi(RB_\Psi(\psi\,|\,x) \leq RB_\Psi(\psi_0\,|\,x)\,|\,x) \end{array} \right\}.$$

∎

So, based on previous comments, the strength of the evidence can be expected to be robust to the marginal prior provided there is no prior-data conflict with π_Ψ. Note that essentially the same proof gives that the Gâteaux derivative of $\Pi_\Psi(RB_\Psi(\psi\,|\,x) = RB_\Psi(\psi_0\,|\,x)\,|\,x)$ equals (4.28) with all the inequalities replaced by equalities. Therefore, the same conclusions apply when using this quantity to assess the strength of the evidence.

Recall that MAP-based inferences implicitly use $\pi_\Psi(\psi_0\,|\,x)$ as a measure of the evidence that ψ_0 is the true value. The following result is immediate.

Proposition 4.7.22 *The Gâteaux derivative of the posterior density of* $\psi = \Psi(\theta)$ *at* ψ_0 *in the direction q is given by* $\{m_{qT}(T(x))/m_T(T(x))\}\{q(\psi_0\,|\,x) - \pi_\Psi(\psi_0\,|\,x)\}$.

This suggests that more than prior-data conflict affects the robustness of MAP-based inferences. For, if ψ_0 is in the tails of the posterior $\pi_\Psi(\cdot\,|\,x)$, then there will be a choice of prior q that makes $q(\psi_0\,|\,x) - \pi_\Psi(\psi_0\,|\,x)$ large. As such, MAP-based inferences are less robust than relative belief inferences. A similar result is obtained for the Bayesian p-value in Evans and Zou (2001). Further developments of these robustness results can be found in Al-Labadi and Evans (2015).

4.8   Concluding Comments

The goal of this chapter has been to develop a theory of inference based upon an explicit method of measuring statistical evidence. Providing a measure of statistical evidence seems to require that the ingredients to a statistical problem include a proper prior, as any sensible measure is based on how the data changes beliefs. While there are a number of ways one could choose to measure change in belief, the relative belief ratio seems preferable from many points of view. It has been shown that this measure, and the theory of statistical inference based on it, possess a number of good properties. The basic intuition behind relative belief, the many good characteristics and its role in resolving a number of thorny issues for inference all lead us to have strong convictions concerning its value.

There is one other characteristic that is worth mentioning. Given that the ingredients for a statistical problem are composed solely of the model, prior and data, there are really only three basic principles that underlie the theory, namely, the principle of conditional probability, the principle of evidence and the relative belief ordering. Once these principles are agreed upon, all the inferences follow necessarily from the principles applied to the ingredients. This is very satisfying, as there are no arbitrary choices required as part of inference. Issues about the perceived accuracy of an estimate or the strength of evidence remain, but these are application dependent and are not statistical in nature.

As previously discussed, our purpose here is to establish a gold standard for inference. The role of such a standard is similar to the gold standard established for data collection. It is an ideal that we strive for in a statistical analysis which may be only approximately attainable. Such a failure, however, does not diminish the role that the gold standard plays in statistics. Without such a standard, there is no way to determine what is right or wrong and why.

Is also our view that any valid theory of statistical inference needs to satisfy the principle of empirical criticism, namely, all ingredients needed for an application of the theory need to be able to be checked against the data to ensure that they are not contradicted by the data. For the ingredients needed for relative belief theory, this is discussed in Chapter 5.

In addition to the sources cited in this chapter, general discussions of the concept of evidence and the role it plays in the philosophy of science can be found in Horwich (1982), Ayer (1972) and Achinstein (2001) and there are many others. The statistical literature on this topic is not as extensive, but certainly Royall (1997) and Thompson (2007) are very explicit in identifying the role of evidence as key to the development of statistical theory. Also, as discussed in Section 3.3.3, Birnbaum's work on statistical evidence has played a significant role in considerations about the foundations of statistics. I. J. Good also dealt extensively with this topic and Good (1985) represents a source of much of his thought on this and other topics.

4.9 Appendix

4.9.1 Proof of Proposition 4.7.5

Let $C \subset \Psi$ be such that $\Pi_\Psi(C\,|\,x) \geq \Pi_\Psi(C_{\Psi,\gamma}(x)\,|\,x)$. Put

$$\Psi_0 = \{\psi : I_{C_{\Psi,\gamma}(x)}(\psi) - I_C(\psi) = 0\},$$
$$\Psi_1 = \{\psi : I_{C_{\Psi,\gamma}(x)}(\psi) - I_C(\psi) < 0\},$$
$$\Psi_2 = \{\psi : I_{C_{\Psi,\gamma}(x)}(\psi) - I_C(\psi) > 0\},$$

and note that $\{\Psi_0, \Psi_1, \Psi_2\}$ is a partition of Ψ. Then

$$\Psi_1 = \{\psi : I_{C_{\Psi,\gamma}(x)}(\psi) - I_C(\psi) < 0, \pi_\Psi(\psi\,|\,x_0)/\pi_\Psi(\psi) \leq c_\gamma(x)\}$$

because $\pi_\Psi(\psi\,|\,x)/\pi_\Psi(\psi) > c_\gamma(x)$ implies $I_{C_{\Psi,\gamma}(x)}(\psi) = 1$, which implies

$$I_{C_{\Psi,\gamma}(x)}(\psi) - I_C(\psi) = 1 - I_C(\psi) \geq 0,$$

a contradiction to the definition of Ψ_1. Also

$$\Psi_2 = \{\psi : I_{C_{\Psi,\gamma}(x)}(\psi) - I_C(\psi) > 0, \pi_\Psi(\psi\,|\,x)/\pi_\Psi(\psi) \geq c_\gamma(x)\}$$

because $\pi_\Psi(\psi\,|\,x)/\pi_\Psi(\psi) < c_\gamma(x)$ implies $I_{C_{\Psi,\gamma}(x)}(\psi) = 0$, which implies

$$I_{C_{\Psi,\gamma}(x)}(\psi) - I_C(\psi) = -I_C(\psi) \leq 0.$$

Therefore,

$$\Pi_\Psi(C_{\Psi,\gamma}(x)) - \Pi_\Psi(C) = \int (I_{C_{\Psi,\gamma}(x)}(\psi) - I_C(\psi))\,\Pi_\Psi(d\psi)$$
$$= \int_{\Psi_1} (I_{C_{\Psi,\gamma}(x)}(\psi) - I_C(\psi))\,\Pi_\Psi(d\psi) + \int_{\Psi_2} (I_{C_{\Psi,\gamma}(x)}(\psi) - I_C(\psi))\,\Pi_\Psi(d\psi).$$

Now note that $\Pi_\Psi(d\psi) = \pi_\Psi(\psi)\,\upsilon_\Psi(d\psi), \Pi_\Psi(d\psi\,|\,x_0) = \pi_\Psi(\psi\,|\,x)\,\upsilon_\Psi(d\psi)$ and because $I_{C_{\Psi,\gamma}(x)}(\psi) - I_C(\psi) < 0$ and $\pi_\Psi(\psi\,|\,x)/\pi_\Psi(\psi) \leq c_\gamma(x)$ when $\psi \in \Psi_1$, then

$$\int_{\Psi_1} (I_{C_{\Psi,\gamma}(x)}(\psi) - I_C(\psi))\,\Pi_\Psi(d\psi) \leq c_\gamma^{-1}(x) \int_{\Psi_1} (I_{C_{\Psi,\gamma}(x)}(\psi) - I_C(\psi))\,\Pi_\Psi(d\psi\,|\,x)$$
$$= c_\gamma^{-1}(x) E_{\Pi_\Psi(\cdot\,|\,x)}(I_{\Psi_1}(I_{C_{\Psi,\gamma}(x)} - I_C)).$$

Similarly, because $I_{C_{\Psi,\gamma}(x)}(\psi) - I_C(\psi) > 0$ and $\pi_\Psi(\psi\,|\,x)/\pi_\Psi(\psi) \geq c_\gamma(x)$ when $\psi \in \Psi_2$, then

$$\int_{\Psi_2} (I_{C_{\Psi,\gamma}(x)}(\psi) - I_C(\psi))\,\Pi_\Psi(d\psi) \leq c_\gamma^{-1}(x) \int_{\Psi_2} (I_{C_{\Psi,\gamma}(x)}(\psi) - I_C(\psi))\,\Pi_\Psi(d\psi\,|\,x)$$
$$= c_\gamma^{-1}(x) E_{\Pi_\Psi(\cdot\,|\,x)}(I_{\Psi_2}(I_{C_{\Psi,\gamma}(x)} - I_C)).$$

Therefore,

$$\Pi_\Psi(C_{\Psi,\gamma}(x)) - \Pi_\Psi(C) = \int (I_{C_{\Psi,\gamma}(x)}(\psi) - I_C(\psi)) \Pi_\Psi(d\psi)$$

$$\leq c_\gamma^{-1}(x) E_{\Pi_\Psi(\cdot | x)}(I_{C_{\Psi,\gamma}(x)} - I_C) = c_\gamma^{-1}(x)(\Pi(C_{\Psi,\gamma}(x) | x) - \Pi(C | x)) \leq 0$$

as it was assumed that $\Pi(C | x) \geq \Pi(C_{\Psi,\gamma}(x) | x)$. Therefore, it is concluded that $\Pi_\Psi(C_{\Psi,\gamma}(x)) \leq \Pi_\Psi(C)$.

4.9.1.1   Proof of Proposition 4.7.9

Let $A \subset \mathcal{X}$ be such that $M(A | \psi_0) \geq \gamma(\psi_0)$. Put

$$\mathcal{X}_0 = \{x : I_{A(\psi_0)}(x) - I_A(x) = 0\},$$
$$\mathcal{X}_1 = \{x : I_{A(\psi_0)}(x) - I_A(x) < 0\},$$
$$\mathcal{X}_2 = \{x : I_{A(\psi_0)}(x) - I_A(x) > 0\},$$

and note that $\{\mathcal{X}_0, \mathcal{X}_1, \mathcal{X}_2\}$ is a partition of $\mathcal{X}$. Then

$$\mathcal{X}_1 = \{\psi : I_{A(\psi_0)}(x) - I_A(x) < 0, m(x | \psi_0)/m(x) \leq 1\}$$

because $m(x | \psi_0)/m(x) > 1$ implies $I_{A(\psi_0)}(x) = 1$, which implies $I_{A(\psi_0)}(x) - I_A(x) = 1 - I_A(x) \geq 0$, a contradiction to the definition of $\mathcal{X}_1$. Also

$$\mathcal{X}_2 = \{\psi : I_{A(\psi_0)}(x) - I_A(x) > 0, m(x | \psi_0)/m(x) > 1\}$$

because $m(x | \psi_0)/m(x) \leq 1$ implies $I_{A(\psi_0)}(x) = 0$, which implies $I_{A(\psi_0)}(x) - I_A(x) = -I_A(x) \leq 0$.

Now

$$M(A(\psi_0)) - M(A) = \int (I_{A(\psi_0)}(x) - I_A(x)) M(dx)$$

$$= \int_{\mathcal{X}_1} (I_{A(\psi_0)}(x) - I_A(x)) M(dx) + \int_{\mathcal{X}_2} (I_{A(\psi_0)}(x) - I_A(x)) M(dx).$$

Note that $M(dx) = m(x) \mu(dx), M(dx | \psi_0) = m(x | \psi_0) \mu(dx)$ and because $I_{A(\psi_0)}(x) - I_A(x) < 0$ and $m(x | \psi_0)/m(x) \leq 1$ when $x \in \mathcal{X}_1$, then

$$\int_{\mathcal{X}_1} (I_{A(\psi_0)}(x) - I_A(x)) M(dx) \leq \int_{\mathcal{X}_1} (I_{A(\psi_0)}(x) - I_A(x)) M(dx | \psi_0)$$

$$= E_{M(\cdot | \psi_0)} \left(I_{\mathcal{X}_1} (I_{A(\psi_0)} - I_A) \right).$$

Similarly, because $I_{A(\psi_0)}(x) - I_A(x) > 0$ and $m(x | \psi_0)/m(x) > 1$ when $x \in \mathcal{X}_2$, then

$$\int_{\mathcal{X}_2} (I_{A(\psi_0)}(x) - I_A(x)) M(dx) \leq \int_{\mathcal{X}_2} (I_{A(\psi_0)}(x) - I_A(x)) M(dx | \psi_0)$$

$$= E_{M(\cdot | \psi_0)} (I_{\mathcal{X}_2} (I_{A(\psi_0)} - I_A)).$$

Therefore,

$$M(A(\psi_0)) - M(A) = \int (I_{A(\psi_0)}(x) - I_A(x)) M(dx)$$

$$\leq E_{M(\cdot\,|\,\psi_0)}(I_{A(\psi_0)} - I_A) = M(A(\psi_0)\,|\,\psi_0) - M(A\,|\,\psi_0) \leq 0$$

as it was assumed that $M(A\,|\,\psi_0) \geq M(A(\psi_0)\,|\,\psi_0)$. Therefore, it is concluded that $M(A(\psi_0)) \leq M(A)$.

4.9.1.2 Proof of Proposition 4.7.12

Consider first the approach where Ψ is replaced by Ψ_η. We then replace Θ by $\Theta_\eta = \{\theta : \pi_\Psi(\Psi(\theta)) \geq \eta\}$ and use the conditional prior $\pi_\eta(\theta) = \pi(\theta)/l_\eta$ on Θ_η where $l_\eta = \sum_{\theta \in \Theta_\eta} \pi(\theta) \to 1$ as $\eta \downarrow 0$. Then an easy calculation shows that the prior and posterior on Ψ_η are, respectively, $\pi_\Psi(\psi)/l_\eta$ and $(m(x)/m_\eta(x))\pi_\Psi(\psi\,|\,x)$ where $m_\eta(x) = \sum_{\theta \in \Theta_\eta} f_\theta(x)\pi(\theta) \to m(x) = \sum_{\theta \in \Theta} f_\theta(x)\pi(\theta)$ as $\eta \downarrow 0$. As such, in the finite problem based on Ψ_η, the relative belief ratio equals $(m(x)l_\eta/m_\eta(x))RB_\Psi(\psi\,|\,x)$. Therefore, when $\eta \leq \pi_\Psi(\psi_{LRSE}(x))$, the LRSE for the finite problem is given by $\psi_{LRSE}(x)$ and, by Proposition 4.7.11, this is the value of the Bayes rule for this problem.

For the approach using the bounded loss $L_{RB,\eta}$,

$$r_\eta(\delta\,|\,x) = \int_\Psi \frac{I_{\{\psi:\psi \neq \delta(x)\}}(\psi)}{\max(\eta, \pi_\Psi(\psi))} \pi_\Psi(\psi\,|\,x)\, \nu_\Psi(d\psi)$$

$$= \int_\Psi \frac{\pi_\Psi(\psi\,|\,x)}{\max(\eta, \pi_\Psi(\psi))} \nu_\Psi(d\psi) - \frac{\pi_\Psi(\delta(x)\,|\,x)}{\max(\eta, \pi_\Psi(\delta(x)))}. \tag{4.29}$$

The first term in (4.29) is constant in $\delta(x)$ and bounded above by $1/\eta$, so the value of a Bayes rule at x is obtained by finding $\delta(x)$ that maximizes the second term. Consider η as fixed and note that

$$\frac{\pi_\Psi(\delta(x)\,|\,x)}{\max(\eta, \pi_\Psi(\delta(x)))} = \begin{cases} \pi_\Psi(\delta(x)\,|\,x)/\eta & \text{if } \eta > \pi_\Psi(\delta(x)) \\ \pi_\Psi(\delta(x)\,|\,x)/\pi_\Psi(\delta(x)) & \text{if } \eta \leq \pi_\Psi(\delta(x)). \end{cases} \tag{4.30}$$

There are at most finitely many values of ψ satisfying $\eta \leq \pi_\Psi(\psi)$ and so $RB_\Psi(\psi\,|\,x)$ assumes a maximum on this set, say at $\psi_\eta(x)$. There are infinitely many values of ψ satisfying $\eta > \pi_\Psi(\psi)$ but clearly there exists $\eta' < \eta$ so that $\{\psi : \eta' < \pi_\Psi(\psi) < \eta\}$ is nonempty and finite. Thus, $\pi_\Psi(\psi\,|\,x)$ assumes its maximum on the set $\{\psi : \pi_\Psi(\psi) < \eta\}$ in the subset $\{\psi : \eta' < \pi_\Psi(\psi) < \eta\}$, say at $\psi'_\eta(x)$. Therefore, a Bayes rule $\delta_\eta(x)$ is given by $\delta_\eta(x) = \psi_\eta(x)$ when $RB_\Psi(\psi_\eta(x)\,|\,x) \geq \pi_\Psi(\psi'_\eta(x)\,|\,x)/\eta$ and $\delta_\eta(x) = \psi'_\eta(x)$ otherwise. If $\eta > \pi_\Psi(\delta(x))$, then $\pi_\Psi(\delta(x)\,|\,x)/\eta < RB_\Psi(\delta(x)\,|\,x) \leq RB_\Psi(\psi_{LRSE}(x)\,|\,x)$. Therefore, whenever $\eta \leq \pi_\Psi(\psi_{LRSE}(x))$, the maximizer of (4.30) is given by $\delta(x) = \psi_{LRSE}(x)$.

4.9.1.3 Proof of Proposition 4.7.13

The prior risk of δ is given by

$$
\begin{aligned}
r(\delta) &= \int_\Theta \int_{\mathcal{X}} L(\theta, \delta(x)) P_\theta(dx)\, \Pi(d\theta) \\
&= \int_\Theta \int_{\mathcal{X}} [h(\Psi(\theta)) - I(\Psi(\theta) = \delta(x)) h(\Psi(\theta))] P_\theta(dx)\, \Pi(d\theta) \\
&= \int_\Theta h(\Psi(\theta))\, \Pi(d\theta) - \int_{\mathcal{X}} \int_\Theta I(\Psi(\theta) = \delta(x)) h(\Psi(\theta))\, \Pi(d\theta \,|\, x) M(dx) \\
&= \int_\Theta h(\Psi(\theta))\, \Pi(d\theta) - \int_{\mathcal{X}} h(\delta(x)) \pi_\Psi(\delta(x)\,|\,x) M(dx)
\end{aligned}
$$

and

$$
\begin{aligned}
&\int_\Theta \int_\Theta \int_{\mathcal{X}} L(\theta', \delta(x)) P_\theta(dx)\, \Pi(d\theta)\, \Pi(d\theta') \\
&= \int_\Theta \int_\Theta \int_{\mathcal{X}} [h(\Psi(\theta')) - I(\Psi(\theta') = \delta(x)) h(\Psi(\theta'))] P_\theta(dx)\, \Pi(d\theta)\, \Pi(d\theta') \\
&= \int_\Theta h(\Psi(\theta))\, \Pi(d\theta) - \int_{\mathcal{X}} h(\delta(x)) \pi_\Psi(\delta(x)) M(dx).
\end{aligned}
$$

Therefore, δ is Bayesian unbiased if and only if

$$
\int_{\mathcal{X}} h(\delta(x)) [\pi_\Psi(\delta(x)\,|\,x) - \pi_\Psi(\delta(x))] M(dx) \geq 0.
$$

Proposition 4.7.10 establishes $RB_\Psi(\psi_{\mathrm{LRSE}}(x)\,|\,x) > 1$ and this concludes the proof.

4.9.1.4 Proof of Proposition 4.7.14 and Corollary 4.7.10

Consider first the approach where we use the loss $L_{RB,\lambda,\eta}$. Just as in Proposition 4.7.12, a Bayes rule $\delta_{\lambda,\eta}(x)$ maximizes $\pi_{\Psi,\lambda}(\delta(x)\,|\,x)/\max(\eta, \pi_{\Psi,\lambda}(\delta(x)))$ for $\delta(x) \in \Psi_\lambda$. Furthermore, as in Proposition 4.7.12, such a rule exists. Now define $\eta(\lambda)$ so that $0 < \eta(\lambda) < \Pi_\Psi(B_\lambda(\psi_{\mathrm{LRSE}}(x)))$. Note that $\eta(\lambda) \to 0$ as $\lambda \to 0$. Therefore, as $\lambda \to 0$,

$$
\frac{\pi_{\Psi,\lambda}(\psi_\lambda(\psi_{\mathrm{LRSE}}(x))\,|\,x)}{\max(\eta(\lambda), \pi_{\Psi,\lambda}(\psi_\lambda(\psi_{\mathrm{LRSE}}(x))))} = \frac{\pi_{\Psi,\lambda}(\psi_\lambda(\psi_{\mathrm{LRSE}}(x))\,|\,x)}{\pi_{\Psi,\lambda}(\psi_\lambda(\psi_{\mathrm{LRSE}}(x)))}
$$
$$
\to RB_\Psi(\psi_{\mathrm{LRSE}}(x)\,|\,x). \tag{4.31}
$$

Let $\varepsilon > 0$. Let λ_0 be such that $\sup_{\psi \in \Psi} \operatorname{diam}(B_\lambda(\psi)) < \varepsilon/2$ for all $\lambda < \lambda_0$. Then for $\lambda < \lambda_0$, and any $\delta(x)$ satisfying $\|\delta(x) - \psi_{\mathrm{LRSE}}(x)\| \geq \varepsilon$,

$$
\begin{aligned}
\frac{\pi_{\Psi,\lambda}(\psi_\lambda(\delta(x))\,|\,x)}{\pi_{\Psi,\lambda}(\psi_\lambda(\delta(x)))} &= \frac{\int_{B_\lambda(\psi_\lambda(\delta(x)))} \pi_\Psi(\psi\,|\,x)\, \nu_\Psi(d\psi)}{\int_{B_\lambda(\psi_\lambda(\delta(x)))} \pi_\Psi(\psi)\, \nu_\Psi(d\psi)} \\
&= \frac{\int_{B_\lambda(\psi_\lambda(\delta(x)))} RB_\Psi(\psi\,|\,x) \pi_\Psi(\psi)\, \nu_\Psi(d\psi)}{\int_{B_\lambda(\psi_\lambda(\delta(x)))} \pi_\Psi(\psi)\, \nu_\Psi(d\psi)} \\
&\leq \sup_{\{\psi: \|\psi - \psi_{\mathrm{LRSE}}(x)\| > \varepsilon/2\}} RB_\Psi(\psi\,|\,x) < RB_\Psi(\psi_{\mathrm{LRSE}}(x)\,|\,x). \tag{4.32}
\end{aligned}
$$

By (4.31) and (4.32) there exists $\lambda_1 < \lambda_0$ such that, for all $\lambda < \lambda_1$,

$$\frac{\pi_{\Psi,\lambda}(\psi_\lambda(\psi_{LRSE}(x))\,|\,x)}{\pi_{\Psi,\lambda}(\psi_\lambda(\psi_{LRSE}(x)))} > \sup_{\{\psi:\|\psi-\psi_{LRSE}(x)\|>\varepsilon/2\}} \frac{\pi_\Psi(\psi\,|\,x)}{\pi_\Psi(\psi)}. \qquad (4.33)$$

Therefore, when $\lambda < \lambda_1$, a Bayes rule $\delta_{\lambda,\eta(\lambda)}(x)$ satisfies

$$\frac{\pi_{\Psi,\lambda}(\delta_{\lambda,\eta(\lambda)}(x)\,|\,x)}{\pi_{\Psi,\lambda}(\delta_{\lambda,\eta(\lambda)}(x))} \geq \frac{\pi_{\Psi,\lambda}(\delta_{\lambda,\eta(\lambda)}(x)\,|\,x)}{\max(\eta(\lambda),\pi_{\Psi,\lambda}(\delta_{\lambda,\eta(\lambda)}(x)))}$$

$$\geq \frac{\pi_{\Psi,\lambda}(\psi_\lambda(\psi_{LRSE}(x))\,|\,x)}{\max(\eta(\lambda),\pi_{\Psi,\lambda}(\psi_\lambda(\psi_{LRSE}(x))))} = \frac{\pi_{\Psi,\lambda}(\psi_\lambda(\psi_{LRSE}(x))\,|\,x)}{\pi_{\Psi,\lambda}(\psi_\lambda(\psi_{LRSE}(x)))}. \qquad (4.34)$$

By (4.32), (4.33) and (4.34), this implies that $\|\delta_{\lambda,\eta(\lambda)} - \psi_{LRSE}(x)\| < \varepsilon$ and the convergence is established.

For the proof of Corollary 4.7.10 the above implies $\pi_{\Psi,\lambda}(\hat{\psi}_\lambda(x)\,|\,x)/\pi_{\Psi,\lambda}(\hat{\psi}_\lambda(x))$ $\geq \pi_{\Psi,\lambda}(\delta_{\lambda,\eta(\lambda)}(x)\,|\,x)/\pi_{\Psi,\lambda}(\delta_{\lambda,\eta(\lambda)}(x))$ and so by (4.32), (4.33) and (4.34) this implies that $\|\hat{\psi}_\lambda(x) - \psi_{LRSE}(x)\| < \varepsilon$ and the convergence of $\hat{\psi}_\lambda(x)$ to $\psi_{LRSE}(x)$ is established.

Now consider the approach using the finite set $\Psi_{\lambda,\eta}$ and the loss L_{RB}. Then, as in the proof of Proposition 4.7.12, we have that $\delta_{\lambda,\eta(\lambda)}(x) = \hat{\psi}_\lambda(x)$ whenever $\eta(\lambda) < \pi_{\Psi,\lambda}(\hat{\psi}_\lambda(x))$ and so this the result follows from Corollary 4.7.10.

4.9.1.5 Proof of Proposition 4.7.16

When using Ψ_η, following the proofs of Propositions 4.7.12 and 4.7.16, a γ-lowest posterior loss region takes the form $L_{\eta,\gamma}(x) = \{\psi \in \Psi_\eta : RB_\Psi(\psi\,|\,x) \geq l_{\eta,\gamma}(x)\}$ where $l_{\eta,\gamma}(x) = \sup\{k : \Pi_{\Psi_\eta}(\{\psi \in \Psi_\eta : RB_\Psi(\psi\,|\,x) \geq k\}\,|\,x) \geq \gamma\}$. For $c > 0$, let $S_c(x) = \{\psi \in \Psi : RB_\Psi(\psi\,|\,x) \geq c\}, S_{\eta,c}(x) = \{\psi \in \Psi_\eta : RB_\Psi(\psi\,|\,x) \geq c\}$ so $S_{\eta,c}(x) \uparrow S_c(x)$ as $\eta \to 0$. Suppose c is such that $\Pi_\Psi(S_c(x)\,|\,x) < \gamma$ so $\Pi_\Psi(S_{\eta,c}(x)\,|\,x) < \gamma$. Since $\Pi_{\Psi_\eta}(S_{\eta,c}(x)\,|\,x) = (m(x)/m_\eta(x))\Pi_\Psi(S_{\eta,c}(x)\,|\,x)$ and $m(x)/m_\eta(x) \to 1$ as $\eta \to 0$, there exists η_0 such that for all $\eta < \eta_0, \Pi_{\Psi_\eta}(S_{\eta,c}(x)\,|\,x) < \gamma$, which implies $S_{\eta,c}(x) \subset L_{\eta,\gamma}(x)$, so $S_c(x) \subset \liminf_{\eta \to 0} L_{\eta,\gamma}(x)$. Therefore, $C_{\Psi,\gamma'}(x) \subset \liminf_{\eta \to 0} L_{\eta,\gamma}(x)$ for all $\gamma' < \gamma$. Now suppose c satisfies $\Pi_\Psi(S_c(x)\,|\,x) > \gamma$. Again there exists η_0 such that for all $\eta < \eta_0, \Pi_{\Psi_\eta}(S_{\eta,c}(x)\,|\,x) > \gamma$, so $L_{\eta,\gamma}(x) \subset S_{\eta,c}(x)$ and $\limsup_{\eta \to 0} L_{\eta,\gamma}(x) \subset S_c(x)$. Therefore, $\limsup_{\eta \to 0} L_{\eta,\gamma}(x) \subset C_{\Psi,\gamma'}(x)$ for all $\gamma' > \gamma$.

When using $L_{RB,\eta}$, following the proof of Proposition 4.7.16, it is seen that a γ-lowest posterior loss region takes the form $L_{\eta,\gamma}(x) = \{\psi : \pi_\Psi(\psi\,|\,x)/\max(\eta, \pi_\Psi(\psi)) \geq l_{\eta,\gamma}(x)\}$ where $l_{\eta,\gamma}(x) = \sup\{k : \Pi_\Psi(\{\psi : \pi_\Psi(\psi\,|\,x)/\max(\eta, \pi_\Psi(\psi)) \geq k\}\,|\,x) \geq \gamma\}$. For $c > 0$, let $S_c(x) = \{RB_\Psi(\psi\,|\,x) \geq c\}, S_{\eta,c}(x) = \{\pi_\Psi(\psi\,|\,x)/\max(\eta, \pi_\Psi(\psi)) \geq c\}$, so $S_{\eta,c}(x) \uparrow S_c(x)$ as $\eta \to 0$. Suppose c is such that $\Pi_\Psi(S_c(x)\,|\,x) \leq \gamma$. Then $\Pi_\Psi(S_{\eta,c}(x)\,|\,x) \leq \gamma$ for all η and so $S_{\eta,c}(x) \subset L_{\eta,\gamma}(x)$. This implies that $S_c(x) \subset \liminf_{\eta \to 0} L_{\eta,\gamma}(x)$ and, since $\Pi_\Psi(C_{\Psi,\gamma}(x)\,|\,x) = \gamma$, this implies that $C_{\Psi,\gamma}(x) \subset \liminf_{\eta \to 0} L_{\eta,\gamma}(x)$. Now suppose c is such that $\Pi_\Psi(S_c(x)\,|\,x) > \gamma$. Then, since $c_\eta \to 1$, there exists η_0 such that for all $\eta < \eta_0$ we have $\Pi_\Psi(S_{\eta,c}(x)\,|\,x) > \gamma$. Since $L_{\eta,\gamma}(x) \subset S_{\eta,c}(x)$, we have that $\limsup_{\eta \to 0} L_{\eta,\gamma}(x) \subset S_c(x)$. Then choosing $c = c_{\gamma'}(x)$ for $\gamma' > \gamma$ implies that $\limsup_{\eta \to 0} L_{\eta,\gamma}(x) \subset C_{\Psi,\gamma'}(x)$.

4.9.1.6   Proof of Proposition 4.7.17

Let $S_c(x) = \{\psi : RB_\Psi(\psi|x) \geq c\}, S_{\lambda,c}(x) = \{\psi : \Pi_\Psi(B_\lambda(\psi)|x)/\Pi_\Psi(B_\lambda(\psi)) \geq c\}$ and note that $\lim_{\lambda \to 0} \Pi_\Psi(B_\lambda(\psi)|x)/\Pi_\Psi(B_\lambda(\psi)) = RB_\Psi(\psi|x)$ for every ψ. If $RB_\Psi(\psi|x) > c$, there exists λ_0 such that for all $\lambda < \lambda_0$, then

$$\Pi_\Psi(B_\lambda(\psi)|x)/\Pi_\Psi(B_\lambda(\psi)) > c$$

and this implies that $\psi \in \liminf_{\lambda \to 0} S_{\lambda,c}(x)$. Now $\Pi_\Psi(RB_\Psi(\psi|x) = c) = 0$ and so $S_c(x) \subset \liminf_{\lambda \to 0} S_{\lambda,c}(x)$ (after possibly deleting a set of Π_Ψ-measure 0 from $S_c(x)$). Now, if $\psi \in \limsup_{\lambda \to 0} S_{\lambda,c}(x)$, then $\Pi_\Psi(B_\lambda(\psi)|x)/\Pi_\Psi(B_\lambda(\psi)) \geq c$ for infinitely many $\lambda \to 0$, which implies $RB_\Psi(\psi|x) \geq c$, and therefore $\psi \in S_c(x)$. Therefore $S_c(x) = \lim_{\lambda \to 0} S_{\lambda,c}(x)$ (up to a set of Π_Ψ-measure 0) so that $\lim_{\lambda \to 0} \Pi_\Psi(S_{\lambda,c}(x) \Delta S_c(x)|x) = 0$ for any c.

Let $c_{\lambda,\gamma}(x) = \sup\{c \geq 0 : \Pi_\Psi(S_{\lambda,c}(x)|x) \geq \gamma\}$, so $S_{c_\gamma(x)}(x) = C_{\Psi,\gamma}(x)$ and $S_{\lambda,c_{\lambda,\gamma}(x)}(x) = C_{\Psi,\lambda,\gamma}(x)$. Therefore,

$$\Pi_\Psi(C_{\Psi,\gamma}(x) \Delta C_{\Psi,\lambda,\gamma}(x)|x) = \Pi_\Psi(S_{c_\gamma(x)}(x) \Delta S_{\lambda,c_{\lambda,\gamma}(x)}(x)|x)$$
$$\leq \Pi_\Psi(S_{c_\gamma(x)}(x) \Delta S_{\lambda,c_\gamma(x)}(x)|x) + \Pi_\Psi(S_{\lambda,c_{\lambda,\gamma}(x)}(x) \Delta S_{\lambda,c_\gamma(x)}(x)|x). \quad (4.35)$$

Since $S_{c_\gamma(x)}(x) = \lim_{\lambda \to 0} S_{\lambda,c_\gamma(x)}(x)$, then $\Pi_\Psi(S_{c_\gamma(x)}(x) \Delta S_{\lambda,c_\gamma(x)}(x)|x) \to 0$ and $\Pi_\Psi(S_{\lambda,c_\gamma(x)}(x)|x) \to \Pi_\Psi(S_{c_\gamma(x)}(x)|x) = \gamma$ as $\lambda \to 0$. Now consider the second term in (4.35). Since $RB_\Psi(\psi|x)$ has a continuous posterior distribution, $\Pi_\Psi(RB_\Psi(\psi|x) \geq c|x)$ is continuous in c. Let $\varepsilon > 0$ and note that for all λ small enough, $\Pi_\Psi(S_{\lambda,c_{\gamma-\varepsilon}(x)}(x)|x) < \gamma$ and $\Pi_\Psi(S_{\lambda,c_{\gamma+\varepsilon}(x)}(x)|x) > \gamma$, which implies that $c_{\gamma+\varepsilon}(x) \leq c_{\lambda,\gamma}(x) \leq c_{\gamma-\varepsilon}(x)$ and therefore $S_{\lambda,c_{\gamma+\varepsilon}(x)}(x) \subset S_{\lambda,c_{\lambda,\gamma}(x)} \subset S_{\lambda,c_{\gamma-\varepsilon}(x)}(x)$. As $S_{\lambda,c_{\lambda,\gamma}(x)}(x) \subset S_{\lambda,c_\gamma(x)}(x)$ or $S_{\lambda,c_{\lambda,\gamma}(x)}(x) \supset S_{\lambda,c_\gamma(x)}(x)$, then $\Pi_\Psi(S_{\lambda,c_{\lambda,\gamma}(x)}(x) \Delta S_{\lambda,c_\gamma(x)}(x)|x) = |\Pi_\Psi(S_{\lambda,c_{\lambda,\gamma}(x)}(x)|x) - \Pi_\Psi(S_{\lambda,c_\gamma(x)}(x)|x)|$. For all λ small enough, $|\Pi_\Psi(S_{\lambda,c_{\lambda,\gamma}(x)}(x)|x) - \Pi_\Psi(S_{\lambda,c_\gamma(x)}(x)|x)|$ is bounded above by $\max\{|\Pi_\Psi(S_{\lambda,c_{\gamma+\varepsilon}(x)}(x)|x) - \Pi_\Psi(S_{\lambda,c_\gamma(x)}(x)|x)|, |\Pi_\Psi(S_{\lambda,c_{\gamma-\varepsilon}(x)}(x)|x) - \Pi_\Psi(S_{\lambda,c_\gamma(x)}(x)|x)|\}$ and this upper bound converges to ε as $\lambda \to 0$. Since ε is arbitrary, we have that the second term in (4.35) goes to 0 as $\lambda \to 0$ and this proves the result.

4.9.1.7   Proof of Proposition 4.7.18

Suppose, without loss of generality, that $0 < \gamma < 1$. Let $\varepsilon > 0$ and $\delta > 0$ satisfy $\gamma + \delta \leq 1$. Put $\gamma'(\lambda,\gamma) = \Pi_\Psi(C_{\Psi,\lambda,\gamma}(x)|x), \gamma''(\lambda,\gamma) = \Pi_\Psi(C_{\gamma+\delta}(x)|x)$ and note that $\gamma'(\lambda,\gamma) \geq \gamma, \gamma''(\lambda,\gamma) \geq \gamma + \delta$. By Proposition 4.7.17, it follows that $C_{\Psi,\lambda,\gamma}(x) \to C_{\Psi,\gamma}(x)$ and $C_{\Psi,\lambda,\gamma+\delta}(x) \to C_{\Psi,\gamma+\delta}(x)$ as $\lambda \to 0$, so $\gamma'(\lambda,\gamma) \to \gamma$ and $\gamma''(\lambda,\gamma) \to \gamma + \delta$ as $\lambda \to 0$. This implies that there is a $\lambda_0(\delta)$ such that, for all $\lambda < \lambda_0(\delta)$, then $\gamma'(\lambda,\gamma) < \gamma''(\lambda,\gamma)$. Therefore, by Proposition 4.7.17, for all $\lambda < \lambda_0(\delta)$,

$$C_{\Psi,\lambda,\gamma}(x) \subset \liminf_{\eta \to 0} L_{\eta,\lambda,\gamma'(\lambda,\gamma)}(x) \subset \limsup_{\eta \to 0} L_{\eta,\lambda,\gamma'(\lambda,\gamma)}(x) \subset C_{\Psi,\lambda,\gamma+\delta}(x). \quad (4.36)$$

From (4.36) and Proposition 4.7.17, $C_{\Psi,\gamma}(x) \subset \liminf_{\lambda \to 0} \liminf_{\eta \to 0} L_{\eta,\lambda,\gamma(\lambda,\gamma)}(x) \subset$ $\limsup_{\lambda \to 0} \limsup_{\eta \to 0} L_{\eta,\lambda,\gamma(\lambda,\gamma)}(x) \subset C_{\Psi,\gamma+\delta}(x)$. Since $\lim_{\delta \to 0} C_{\Psi,\gamma+\delta}(x) = C_{\Psi,\gamma}(x)$, this establishes the result.

4.9.1.8 Proof of Lemma 4.7.1

Note first that

$$Q(A \mid x) = \int_A \frac{m(x \mid \psi) q(\psi)}{m_q(x)} \, v_\Psi(d\psi). \qquad (4.37)$$

Therefore, using (4.37),

$$
\begin{aligned}
\Pi_\Psi^\varepsilon(A \mid x) &= \frac{(1-\varepsilon) m(x) \Pi_\Psi(A \mid x) + \varepsilon m_q(x) Q(A \mid x)}{(1-\varepsilon) m(x) + \varepsilon m_q(x)} \\
&= \frac{\Pi_\Psi(A \mid x) + \frac{\varepsilon}{(1-\varepsilon) m(x)} \int_A m(x \mid \psi) q(\psi) v_\Psi(d\psi)}{1 + \frac{\varepsilon}{(1-\varepsilon) m(x)} \{ \int_A m(x \mid \psi) q(\psi) v_\Psi(d\psi) + \int_{A^c} m(x \mid \psi) q(\psi) v_\Psi(d\psi) \}} \\
&\leq \frac{\Pi_\Psi(A \mid x) + \frac{\varepsilon}{(1-\varepsilon) m(x)} \int_A m(x \mid \psi) q(\psi) v_\Psi(d\psi)}{1 + \frac{\varepsilon}{(1-\varepsilon) m(x)} \int_A m(x \mid \psi) q(\psi) v_\Psi(d\psi)}.
\end{aligned}
$$

Result (i) then follows since $(p+y)/(1+y) = 1 - (1-p)/(1+y)$ is increasing in $y \geq 0$ when $1 - p > 0$ and clearly the maximum value of $\int_A m(x \mid \psi) q(\psi) v_\Psi(d\psi)$ is attained when q places all of its mass on the set $\{\psi : m(x \mid \psi) = \sup_{\psi \in A} m(x \mid \psi)\}$, in which case $\int_A m(x \mid \psi) q(\psi) v_\Psi(d\psi) = \sup_{\psi \in A} m(x \mid \psi)$.

For result (ii),

$$
\begin{aligned}
\Pi_\Psi^\varepsilon(A \mid x) &\geq \frac{\Pi_\Psi(A \mid x) + \frac{\varepsilon}{(1-\varepsilon) m(x)} \int_A m(x \mid \psi) q(\psi) v_\Psi(d\psi)}{1 + \frac{\varepsilon}{(1-\varepsilon) m(x)} \{ \int_A m(x \mid \psi) q(\psi) v_\Psi(d\psi) + \sup_{\psi \in A^c} m(x \mid \psi) \}} \\
&\geq \frac{\Pi_\Psi(A \mid x)}{1 + \frac{\varepsilon}{(1-\varepsilon) m(x)} \sup_{\psi \in A^c} m(x \mid \psi)}
\end{aligned}
$$

where the first inequality is obvious and the second follows since $(p+y)/(1+y+b) = 1 - (1-p+b)/(1+y+b)$ is increasing in $y \geq 0$ when $1-p+b > 0$ and so the minimum is attained at $y = 0$. It is immediate that the above lower bound for $\Pi_\Psi^\varepsilon(A \mid x)$ is attained by the Q specified in (ii).

For (iii), a direct calculation yields

$$
\begin{aligned}
\delta(A) &= \Pi_\Psi^{upper}(A \mid x) - \Pi_\Psi^{lower}(A \mid x) \\
&= \frac{\Pi_\Psi(A \mid x)(1 + \varepsilon^* r(A^c)) + \varepsilon^* r(A)(1 + \varepsilon^* r(A^c)) - \Pi_\Psi(A \mid x)(1 + \varepsilon^* r(A))}{(1 + \varepsilon^* r(A))(1 + \varepsilon^* r(A^c))} \\
&= \frac{\Pi_\Psi(A \mid x) \varepsilon^* (r(A^c) - r(A))}{(1 + \varepsilon^* r(A))(1 + \varepsilon^* r(A^c))} + \frac{\varepsilon^* r(A)}{1 + \varepsilon^* r(A)}.
\end{aligned}
$$

For (iv),

$$\delta(A^c) = \sup_Q (1 - \Pi_\Psi^\varepsilon(A \,|\, x)) - \inf_Q (1 - \Pi_\Psi^\varepsilon(A \,|\, x))$$

$$= \sup_Q (-\Pi_\Psi^\varepsilon(A \,|\, x)) - \inf_Q (-\Pi_\Psi^\varepsilon(A \,|\, x)) = \delta(A).$$

4.9.1.9 Proof of Proposition 4.7.19

(i) For any set A with $r(A) = r(\Psi)$, then $r(A^c) - r(A) = r(A^c) - r(\Psi) \le 0$. Therefore,

$$\frac{\Pi_\Psi(A \,|\, x)\,\varepsilon^*(r(A^c) - r(A))}{(1 + \varepsilon^* r(A))(1 + \varepsilon^* r(A^c))} + \frac{\varepsilon^* r(A)}{1 + \varepsilon^* r(A)}$$

$$= \frac{\Pi_\Psi(A \,|\, x)\,\varepsilon^*(r(A^c) - r(\Psi))}{(1 + \varepsilon^* r(\Psi))(1 + \varepsilon^* r(A^c))} + \frac{\varepsilon^* r(\Psi)}{1 + \varepsilon^* r(\Psi)}$$

$$\ge \frac{\Pi_\Psi(C_{\Psi,\gamma}(x) \,|\, x)\,\varepsilon^*(r(A^c) - r(\Psi))}{(1 + \varepsilon^* r(\Psi))(1 + \varepsilon^* r(A^c))} + \frac{\varepsilon^* r(\Psi)}{1 + \varepsilon^* r(\Psi)}.$$

Now

$$\frac{r(A^c) - r(\Psi)}{1 + \varepsilon^* r(A^c)} \tag{4.38}$$

is increasing in $r(A^c)$, so we need to show that $A = C_{\Psi,\gamma}(x)$ minimizes $r(A^c)$ among all A satisfying $\Pi_\Psi(A \,|\, x) \le \Pi_\Psi(C_{\Psi,\gamma}(x) \,|\, x)$ and $r(A) = r(\Psi)$. Suppose that $r(A^c) < r(C_{\Psi,\gamma}^c(x))$ and let $B = \{\psi : RB_\Psi(\psi \,|\, x) > r(A^c)\}$. Note that $r(C_{\Psi,\gamma}^c(x)) \le \inf_{\psi \in C_{\Psi,\gamma}(x)} RB_\Psi(\psi \,|\, x)$ and so $C_{\Psi,\gamma}(x) \subset B$, which implies $\Pi_\Psi(B \,|\, x) > \Pi_\Psi(C_{\Psi,\gamma}(x) \,|\, x)$ but also $B \subset A$, which contradicts $\Pi_\Psi(A \,|\, x) \le \Pi_\Psi(C_{\Psi,\gamma}(x) \,|\, x)$ and therefore $r(A^c) \ge r(C_{\Psi,\gamma}^c(x))$. This establishes that (4.38) is minimized by $A = C_{\Psi,\gamma}(x)$.

(ii) Now consider all the sets A with $r(A^c) = r(\Psi)$. Since $\delta(A) = \delta(A^c)$, it is equivalent to minimize $\delta(A^c)$ among all sets A^c satisfying $\Pi_\Psi(A^c \,|\, x) \le \Pi_\Psi\left(C_{\Psi,\gamma}^c(x) \,|\, x\right) = 1 - \gamma^*(x)$ and $r(A^c) = r(\Psi)$. By part (i), this is minimized by taking $A^c = C_{\Psi,1-\gamma^*(x)}(x)$ and the result is proved.

(iii) The optimal solutions to the optimization problems in parts (i) and (ii), namely, $C_{\Psi,\gamma}(x)$ and $C_{\Psi,1-\gamma}^c(x)$, respectively, both have posterior content equal to γ. As such, one of these sets is the solution to the optimization problem stated in (iii). Then,

$$\delta(C_{\Psi,\gamma}(x)) - \delta(C_{\Psi,1-\gamma}^c(x)) = \delta(C_{\Psi,\gamma}(x)) - \delta(C_{\Psi,1-\gamma}(x))$$

$$= \frac{\gamma\varepsilon^*(r(C_{\Psi,\gamma}^c(x)) - r(\Psi))}{(1 + \varepsilon^* r(\Psi))(1 + \varepsilon r(C_{\Psi,\gamma}^c(x)))} - \frac{\gamma\varepsilon^*(r(C_{\Psi,1-\gamma}^c(x)) - r(\Psi))}{(1 + \varepsilon^* r(\Psi))(1 + \varepsilon r(C_{\Psi,1-\gamma}^c(x)))}$$

$$= \frac{\gamma\varepsilon^*}{(1 + \varepsilon^* r(\Psi))} \left\{ \frac{r(C_{\Psi,\gamma}^c(x)) - r(\Psi)}{1 + \varepsilon r(C_{\Psi,\gamma}^c(x))} - \frac{r(C_{\Psi,1-\gamma}^c(x)) - r(\Psi)}{1 + \varepsilon r(C_{\Psi,1-\gamma}^c(x))} \right\}.$$

The result follows from this because $C_{\Psi,\gamma}^c(x) \subset C_{\Psi,1-\gamma}^c(x)$, so $r(C_{\Psi,\gamma}^c(x)) \le r(C_{\Psi,1-\gamma}^c(x))$, and (4.38) is increasing in $r(A^c)$.

Chapter 5

Choosing and Checking the Model and Prior

5.1 Introduction

As has been repeatedly emphasized in this book, *all* statistical analyses depend on choices made by the statistician. At a minimum, a model $\{f_\theta : \theta \in \Theta\}$ is chosen. It has been argued that more than this is needed to be unambiguous about how to measure statistical evidence concerning various questions of interest. In particular, a proper prior π on Θ needs to be chosen. A natural question is then: *how* are these ingredients to be selected? Certainly this shouldn't be done arbitrarily, as that would undermine the validity of the subsequent analysis. So part of this chapter is concerned with this aspect of a statistical analysis.

Another concern with the selection of the ingredients is whether or not they are in reality "good" choices. There are several aspects to this. For example, an assessment of whether or not the choices bias the results in some fashion seems necessary. This was the topic of Section 4.6. Also, there is the possibility that the subjective choices made are contradicted, in some sense, by the data x. After all, the data *should* be objective, and will be if it is collected properly, and thus represents a relevant standard to judge the subjective choices against. So for the model and prior, methods are needed for checking these against the data to satisfy the principle of empirical criticism. It is reasonable to say that a theory of inference is not valid for scientific applications if this principle is not applicable to all the ingredients required by the theory.

Considering bias and checking the ingredients against the data is how statistics can deal with natural concerns about the effects of subjectivity that arise in scientific contexts. The role of these aspects of a statistical analysis is to provide assurances that the conclusions drawn are not being distorted by subjective beliefs that reflect personal biases or that are strongly contradicted by the data. Carrying out these steps, however, does not guarantee that the choices made are correct. While the inference step proceeds as if the choices made are correct, models and priors are devices employed so that we can reason logically and purposefully in statistical contexts rather than believing in their absolute reality. The validity of any statistical analysis is dependent on the ingredients, but once these are chosen, inference should proceed unambiguously and without arbitrariness. Such an approach to inference was presented in Chapter 4.

5.2   Choosing the Model

This section is embarrassingly short. It is short because there seems to be very little in the literature concerning how to go about choosing a model and it is embarrassing because one could consider this, after the proper collection of the data, the most important aspect of a statistical analysis. Perhaps a simple example provides some insight into this.

Example 5.2.1 *Choosing a Model.*

Suppose a real-valued response y is measured in some units. Given that a measurement process is being conducted as part of the data collection, something must be known about what kinds of values can be expected. This suggests there is an interval of values $(a,b) \subset R^1$ where it is *known* that y will lie. This interval could be selected quite conservatively. Also, any measurement is always measured to finite accuracy, so in reality (a,b) needs to be divided into subintervals $(a,a+\delta], (a+\delta, a+2\delta), \ldots$, where δ reflects the accuracy and each midpoint of an interval gives a possible measurement value. As such, the statistical model should only include distributions where the values taken are given by these midpoints.

If the observed values are exactly measured, such as counts, then a statistical model composed entirely of discrete distributions may be necessary. Otherwise, an approximating model consisting of continuous distributions might be appropriate and even preferable. The primary reason for this approximation is to simplify the analysis. If the approximation is inadequate, the model checking phase of the analysis will be relied on to alert us to this.

Once it is settled on whether to use a discrete or a continuous model, it is necessary to consider what distributions to put into the model. In either case, putting all possible distributions into the model seems inappropriate. For it is known that, without conducting a census, it is not possible to identify all characteristics of the distribution in question so, in the end, there are only a few characteristics that will be of interest. If the model is taken to be too big, then the meaning and relevance of these characteristics can become suspect. For example, suppose interest is in the mean of the distribution. If the distribution is highly skewed or multimodal, then the relevance of this quantity is unclear, and a similar comment applies to other measures of location. Accordingly, some restrictions on the family of distributions in the model are required so that the characteristics of interest are meaningful. Again the model checking phase will be used to establish whether or not the restrictions imposed are reasonable.

Accordingly, a collection of characteristics of the distribution of y that are of interest, as well as a family of distributions for which these characteristics are meaningful, need to be identified. To provide a simple example, suppose that interest is in the location and scale of a continuous approximating distribution and that these are characterized, respectively, by the mean μ and the standard deviation σ of y. It would then seem sensible that a family of unimodal, symmetric distributions, each centered at some μ and with standard deviation given by some σ, be selected for the model. Of course, this assumption may be wrong but, if it is reasonable, then some simplicity in the model is achieved. Again the model checking phase is relied upon to catch any inadequacies.

There is an immediate problem, however, as it is not the case that all choices of (μ, σ) are compatible with what is known about y, namely, $y \in (a, b)$. There are several possible approaches to resolving this problem. One strategy is to determine a set of possibilities for (μ, σ) that agree with $y \in (a, b)$. While laudable, it seems extremely difficult to implement this in practice. An easier approach is to, at least formally, abandon the restriction $y \in (a, b)$ and allow for distributions in the model that place mass outside (a, b). For example, consider a model which allows y to be any value in R^1 even though it is known that some of these values are impossible, e.g., negative heights. A reasonable approach then is to use a prior on (μ, σ) to impose a soft constraint by effectively making the amount of mass off (a, b) negligible for those values of (μ, σ) that are believed a priori to be relevant; see Example 5.3.1. The choice of the prior may, however, not be appropriate for the application and determining this is the role of the prior checking phase of the analysis.

Given that our interest is in (μ, σ), a simple form for the model is $\{f_{(\mu, \sigma)} : (\mu, \sigma) \in R^1 \times (0, \infty)\}$ where $f_{(\mu, \sigma)}(y) = \sigma^{-1} f_{(0,1)}((y - \mu)/\sigma)$ and $f_{(0,1)}$ is a symmetric, unimodal density with mean 0 and variance 1. Of course a more complicated model could be chosen, where several distributions in the model have the same value of (μ, σ), but this entails adding additional parameters. Let us focus, however, on the simplest situation, so we are left with choosing $f_{(0,1)}$.

A common choice for $f_{(0,1)}$ is the $N(0, 1)$ density. There are a number of natural reasons to do this; for example, many empirical measurements are known to be approximately normally distributed based on previous studies. An alternative, if longer tails are considered appropriate, would be to take $f_{(0,1)}$ to be a t_λ density, for some $\lambda > 2$, standardized to have unit variance. Of course, there are many other possible choices but in the end a rationale is necessary for whatever choice is made. ∎

The process described in Example 5.2.1 leads us to a location-scale model for a response y. The intention is not to say that such a model is always appropriate, as each situation will have unique characteristics that need to be taken into account. For example, suppose that the distribution is possibly bimodal or skewed, with the consequence that more parameters are needed in the model to reflect this greater generality. For example, if it is desirable to allow the tail length to be uncertain, then the model could be enlarged to

$$\{f_{(\mu, \sigma, \lambda)} : (\mu, \sigma) \in R^1 \times (0, \infty), \lambda \in (2, \infty]\}, \tag{5.1}$$

where $f_{(\mu, \sigma, \lambda)}(y) = \sigma^{-1} f_{(0,1,\lambda)}((y - \mu)/\sigma)$ and $f_{(0,1,\lambda)}$ is a standardized t_λ density. So now there are three parameters in the model.

While it might seem like a good idea to use models that allow for a greater variety of distributions, there are costs associated with this. One cost is that the higher the dimension of the parameter space, the more difficult inference becomes. For example, typically larger sample sizes are necessary to get the same accuracies for inferences as the number of parameters grows.

A more serious issue is that, as more parameters are added, it is necessary to really understand the role the parameters are playing. Otherwise, selecting an appropriate prior is problematical and the interpretation of inferences can become ambiguous. For example, using (5.1), for fixed λ we have a clear idea of how changing μ and σ will affect the response but for fixed (μ, σ), the role of changes in λ seems less clear. Certainly as λ decreases it is expected that more outliers will occur in a sample but how to relate this to λ is not obvious. This is reflected in the fact that it is not clear how to go about eliciting a prior on λ. For such parameters, it seems obvious that inference is more difficult.

The situation of Example 5.3.1 is relatively straightforward and much more complicated models than this are necessary for applications. For example, predictor variables and multidimensional response variables are quite common in statistical applications. Our main point here is that, in devising such models, it is necessary to be careful to ensure that the model is amenable to inference. Simply writing down a complicated model and then attempting to fit this to data to make inferences does not seem likely to lead to real understanding about questions of interest. There are different viewpoints on this issue, however, with an extreme contrary position being that the role of models is simply to produce predictions and, as such, arbitrary complexity can be permitted with the only caution being that overfitting be avoided. While there is a place for this, our concern here is with the gold standard for statistical inference and, for us, the latter approach falls well short of that.

5.3   Choosing the Prior

Throughout this book the discussion has been restricted to the use of proper priors. For us, this restriction seems necessary because it puts the development of a theory of inference squarely within the domain of probability theory. It is the case, however, that improper priors are commonly used in practice. The reasons for this are many, but perhaps the main point is that with complicated models it can be difficult to come up with a suitable elicitation procedure for a proper prior. This does not avoid, however, the need to develop the theory of inference based upon the elicitation of proper priors. Current attempts at developing a theory based upon improper priors have close connections with frequentist ideas but, as already discussed, there are issues with the principle of frequentism itself that remain unresolved.

5.3.1   Eliciting Proper Priors

One commonly heard complaint about requiring a prior as part of the ingredients to a statistical analysis is that it is often difficult to come up with something suitable. This suggests, however, that there is some deficiency concerning our understanding of the role of the parameter θ in the model $\{f_\theta : \theta \in \Theta\}$. For example, as illustrated in Example 5.3.1, when the parameters can be related to things known about response variables, then it seems that a reasonable prior can be obtained via elicitation. So at least part of the reason for this complaint is that the model may be too complicated or not well enough understood for the application at hand. As such, this cannot be viewed as a criticism of the need to specify a prior.

For the gold standard of inference, the solution of a statistical problem requires the prescription of an elicitation process for the prior. As discussed in Section 5.2, there are typically a finite number of characteristics of the distribution under study that are of interest and, as such, it seems pointless to consider all possible distributions and the model is restricted accordingly. Similarly, when considering the choice of prior, it is not the case that our knowledge is so fine that a prior can be chosen of arbitrary complexity. So attention is restricted to some family of priors that can adequately express the information available about the response. It is well to remember too that models and priors are devices for reasoning and are not to be construed as being absolutely "true." As with the model, various checks against the data are relied on to determine whether or not a particular prior is reasonable; see Section 5.6.

Consider now a commonly used statistical model to illustrate the process of elicitation. While each problem is different, this example demonstrates some general principles.

Example 5.3.1 *Priors for Location-Scale Models.*

Suppose that the model is given by $\{f_{(\mu,\sigma)} : (\mu,\sigma) \in R^1 \times (0,\infty)\}$ where $f_{(\mu,\sigma)}(y) = \sigma^{-1}f_{(0,1)}((y-\mu)/\sigma)$ and $f_{(0,1)}$ is a symmetric, unimodal density with mean 0 and variance 1. There is no need to take $f_{(0,1)}$ to be an $N(0,1)$ density but this is done here for simplicity purposes. For other choices, simply replace certain quantiles for the $N(0,1)$ distribution with the same quantiles for the distribution given by $f_{(0,1)}$ to prescribe the elicitation process.

The prior is specified hierarchically as

$$1/\sigma^2 \sim \text{gamma}_{rate}(\alpha_0,\beta_0),$$
$$\mu\,|\,\sigma^2 \sim N(\mu_0,\tau_0^2\sigma^2). \tag{5.2}$$

Here μ_0, τ_0^2 and (α_0,β_0) are hyperparameters to be specified via elicitation. This prior has the virtue that, with a normal model, the posterior of (μ,σ^2) is also of the form given in (5.2) but with different constants, which makes computations with the posterior somewhat easier. A prior with the property that the posterior is in the same family as the prior is known as a *conjugate prior*. Certainly there is no reason to restrict to a family of conjugate priors. The only real question, however, is whether or not a choice from the family can adequately reflect the prior information possessed about (μ,σ^2). After making a choice of the prior and observing the data, a check for prior-data conflict should be made to assess whether or not our choice was reasonable.

For the elicitation of the hyperparameters, consider a method developed in Cao, Evans and Guttman (2014) for a more general problem. Suppose that it is known with virtual certainty, based on knowledge of the basic measurements being taken, that μ will lie in the interval $(-m_0 + \mu_0, \mu_0 + m_0)$ for some $m_0 > 0$. The form of the interval reflects the fact that a natural choice for μ_0 is the mid-point of whatever interval is specified for μ. With this choice one hyperparameter has been specified. The phrase *virtual certainty* is interpreted here as probability greater than or equal to

0.99, although other choices could be made. This implies that $0.99 \leq 2\Phi(m_0/\tau_0\sigma) - 1$ and, letting $z_{0.995}$ denote the 0.995 quantile of the $N(0,1)$ distribution, then

$$\sigma \leq m_0/\tau_0 z_{0.995}. \tag{5.3}$$

Note that (5.3) only uses prior information about the location.

An interval that will contain an observation x with virtual certainty is given by $\mu \pm \sigma z_{0.995}$. Let s_1 and s_2 be lower and upper bounds, respectively, on the half-length of this interval based upon what is known about the response. Therefore, $s_1 \leq \sigma z_{0.995} \leq s_2$ or, equivalently,

$$s_1/z_{0.995} \leq \sigma \leq s_2/z_{0.995} \tag{5.4}$$

holds with virtual certainty. Combining (5.4) with (5.3) implies that $\tau_0 = m_0/s_2$ and note that τ_0 can be made bigger by choosing m_0 bigger.

To obtain the relevant values of α_0 and β_0, let $G(\alpha, \beta, \cdot)$ denote the cdf of the $\text{gamma}_{rate}(\alpha, \beta)$ distribution and note that $G(\alpha, \beta, x) = G(\alpha, 1, \beta x)$. Therefore, the interval for $1/\sigma^2$ implied by (5.4) contains $1/\sigma^2$ with virtual certainty, when α_0, β_0 satisfy

$$G^{-1}(\alpha_0, \beta_0, 0.995) = s_1^{-2} z_{0.995}^2,$$
$$G^{-1}(\alpha_0, \beta, 0.005) = s_2^{-2} z_{0.995}^2,$$

or equivalently

$$G(\alpha_0, 1, \beta_0 s_1^{-2} z_{0.995}^2) = 0.995, \tag{5.5}$$
$$G(\alpha_0, 1, \beta_0 s_2^{-2} z_{0.995}^2) = 0.005. \tag{5.6}$$

It is a simple matter to solve these equations for (α_0, β_0). For this choose an initial value for α_0 and, using (5.5), find w such that $G(\alpha_0, 1, w) = 0.995$, which implies $\beta_0 = w/s_1^{-2} z_{0.995}^2$. If the left side of (5.6) is less (greater) than 0.005, then decrease (increase) the value of α_0 and repeat step 1. Continue iterating this process until satisfactory convergence is attained. ∎

The example demonstrates that knowledge about realistically possible response values leads directly to a prior for (μ, σ^2). Frankly, it is difficult to conceive of a problem where a person with application-dependent knowledge wouldn't have this kind of information about a real-valued, measured response. There isn't any need to pretend to only have vague knowledge about possible values for μ and σ^2. While there might seem to be some merit in being vague, the Jeffreys–Lindley paradox teaches that, in general, this is not a good idea.

Example 5.3.1 is substantially generalized in Cao, Evans and Guttman (2014) to provide an elicitation procedure for a multivariate normal prior.

5.3.2   *Improper Priors*

Perhaps the most compelling argument for the consideration of improper priors arises as an extension of the principle of insufficient reason, discussed in Section 2.2, when

Θ is not compact. So improper priors are commonly chosen to reflect a lack of information or ignorance about the true distribution in the model and this often reflects the complexity of the model being used. It is our position, however, that at its highest level, an application of statistical reasoning demands the elicitation of a proper prior.

In any case, the difficulties that arise with elicitation with complex models lead to a desire to choose priors based on some kind of default mechanism that reflects a lack of information about suitable values of θ. These default priors are often *improper*, in the sense that the integral of the prior over Θ is infinite. Such a prior is supposed to represent ignorance and, for this reason, is often referred to as being a *noninformative prior*. When using an improper prior $\pi(\theta)$, an analysis is implemented just as when π is a proper probability density. So, after observing x, replace initial "beliefs" about θ as given by π with the "posterior" given by $\pi(\theta | x) = \pi(\theta) f_\theta(x) / m(x)$ where $m(x) = \int_\Theta \pi(\theta) f_\theta(x) \mu(d\theta)$.

Consider a simple example.

Example 5.3.2 *Location Normal.*

Suppose $x = (x_1, \ldots, x_n)$ is a sample from an $N(\mu, 1)$ distribution where $\mu \in R^1$ is unknown. Complete ignorance about μ is often represented by taking $\pi(\mu) \equiv c$, for some $c > 0$, so that every value of μ receives the same weight. Then

$$m(x) = c \int_{-\infty}^{\infty} (2\pi)^{-\frac{n}{2}} \exp\left\{ -\frac{n}{2}(\bar{x} - \mu)^2 \right\} \exp\left\{ -\frac{n}{2} \sum_{i=1}^{n} (x_i - \bar{x})^2 \right\} d\mu$$

$$= c(2\pi)^{-\frac{n-1}{2}} n^{-\frac{1}{2}} \exp\left\{ -\frac{n}{2} \sum_{i=1}^{n} (x_i - \bar{x})^2 \right\}$$

and so

$$\pi(\mu | x) = (2\pi)^{-\frac{1}{2}} n^{\frac{1}{2}} \exp\left\{ -\frac{n}{2}(\mu - \bar{x})^2 \right\},$$

which is the density of the $N(\mu, 1/n)$ distribution. Note that $\pi(\cdot | x)$ does not depend on c. ∎

More generally, when making inference about a location parameter $\mu \in R^1$ and a scaling parameter $\sigma > 0$, it is common to take the prior on μ to be proportional to 1 and the prior on σ^2 to be proportional to $1/\sigma^2$. As such, whenever the set of possible values for μ or σ is unbounded, the prior is improper. As discussed in Example 5.3.1, however, using such a prior in this context seems wholly unnecessary, given what must be known about the response.

There are a variety of problems associated with improper priors. Perhaps the most obvious problem is that there is no guarantee that $m(x) < \infty$ and, when it isn't, $\pi(\cdot | x)$ is not defined; see Example 5.3.3. Therefore, in a given context it is necessary to restrict to those π that give a finite value for $m(x)$ and often this class of priors is data dependent.

A deeper issue concerns why π is replaced by $\pi(\cdot | x)$ to specify beliefs. As has been emphasized, when π is proper, this step is justified by the principle of conditional probability and there appears to be no such justification in the improper case.

It is perhaps inappropriate then to refer to $\pi(\cdot\,|x)$ as a posterior, as it is not a conditional distribution. Also, when π is improper, it cannot be the case that m represents a probability distribution and so all applications of the prior predictive that rely on this are lost.

When there is a parameter of interest $\psi = \Psi(\theta)$, the marginal prior π_Ψ is generally not defined, as it is possible that $\int_{\Psi^{-1}\{\psi\}} \pi(\theta)J_\Psi(\theta)\,\nu(d\theta) = \infty$ for every ψ and, even if finite, it is not clear that this is the right definition for $\pi_\Psi(\psi)$. For example, consider Example 5.3.2 extended to the case where there is a sample from a bivariate normal with unknown mean $\mu = (\mu_1, \mu_2)'$, variance I and with prior $\pi(\mu) \equiv c$. It is not obvious then how to reason to the marginal priors on μ_1 and μ_2 being proportional to a constant, although this seems quite natural.

A consequence of the ambiguity concerning marginals is that it is no longer clear how to define a measure of evidence as a change of belief for a general $\psi = \Psi(\theta)$. Even in Example 5.3.2, a difficulty arises, as the *formal* relative belief ratio (and also Bayes factor) is given by

$$RB(\mu\,|x) = c^{-1}(2\pi)^{-\frac{1}{2}}n^{\frac{1}{2}}\exp\left\{-\frac{n}{2}(\mu - \bar{x})^2\right\} \tag{5.7}$$

and this depends on the arbitrary constant c. There doesn't appear to be any obvious way to pick c and so this cannot be taken as a measure of evidence. Moreover, (5.7) is not a relative belief ratio, as the prior is not an expression of beliefs via a proper probability measure.

A more subtle problem also arises with improper priors. For this suppose that for a parameter of interest $\psi = \Psi(\theta)$, the posterior $\pi_\Psi(\cdot\,|x)$ satisfies $\pi_\Psi(\psi\,|x) = g(\psi, Z(x))$ for some function Z of the data x. The function Z has marginal statistical model

$$f_{Z,\theta}(z) = \int_{Z^{-1}\{z\}} f_\theta(x)J_Z(x)\,\mu_{Z^{-1}\{z\}}(dx).$$

Suppose another statistician only observes $Z(x)$ but has the same prior and so obtains posterior $\pi_\Psi(\psi\,|Z(x))$. It is natural to suppose that $\pi_\Psi(\cdot\,|x) = \pi_\Psi(\cdot\,|Z(x))$ but Dawid, Stone and Zidek (1973) show that this is not necessarily the case when the prior is improper. This is known as the *marginalization paradox*. With proper priors, however, no such paradox can arise. This follows from

$$\pi_\Psi(\psi\,|x) = \frac{\pi_\Psi(\psi)}{m(x)}\int_{\Psi^{-1}\{\psi\}} f_\theta(x)\pi(\theta\,|\,\psi)\,\upsilon_{\Psi^{-1}\{\psi\}}(d\theta),$$

which implies $\int_{\Psi^{-1}\{\psi\}} f_\theta(x)\pi(\theta\,|\,\psi)\,\upsilon_{\Psi^{-1}\{\psi\}}(d\theta) = m(x)g(\psi, Z(x))/\pi_\Psi(\psi)$ and, noting that $m_Z(Z(x)) = \int_{Z^{-1}\{Z(x)\}} m(w)J_Z(w)\,\mu_{Z^{-1}\{Z(x)\}}(dw)$, then

$$\pi_\Psi(\psi\,|Z(x)) = \frac{\pi_\Psi(\psi)}{m_Z(Z(x))}\int_{\Psi^{-1}\{\psi\}} f_{Z,\theta}(Z(x))\pi(\theta\,|\,\psi)\,\upsilon_{\Psi^{-1}\{\psi\}}(d\theta)$$

$$= \frac{\pi_\Psi(\psi)}{m_Z(Z(x))}\int_{\Psi^{-1}\{\psi\}}\left\{\int_{Z^{-1}\{Z(x)\}} f_\theta(w)J_Z(w)\,\mu_{Z^{-1}\{Z(x)\}}(dw)\right\}\times$$
$$\pi(\theta\,|\,\psi)\,\upsilon_{\Psi^{-1}\{\psi\}}(d\theta)$$

$$
= \frac{\pi_\Psi(\psi)}{m_Z(Z(x))} \int_{Z^{-1}\{Z(x)\}} \left\{ \int_{\Psi^{-1}\{\psi\}} f_\theta(w) \pi(\theta \mid \psi) \, \upsilon_{\Psi^{-1}\{\psi\}}(d\theta) \right\} \times
$$
$$
J_Z(w) \, \mu_{Z^{-1}\{Z(x)\}}(dw)
$$
$$
= \frac{g(\psi, Z(x))}{m_Z(Z(x))} \int_{Z^{-1}\{Z(x)\}} m(w) J_Z(w) \, \mu_{Z^{-1}\{Z(x)\}}(dw) = g(\psi, Z(x)).
$$

Irrespective of whether a prior is improper or not, the basic idea of noninformativity is ambiguous, as the following example illustrates.

Example 5.3.3 *Binomial.*

Suppose that $x \sim \text{binomial}(n, \theta)$ for some unknown $\theta \in [0, 1]$. In many ways a $U(0, 1)$ prior seems natural here to represent noninformativity (recall the discussion in Example 1.4.2). For a number of reasons, *Haldane's prior*, given by $\pi(\theta) = c\theta^{-1}(1 - \theta)^{-1}$, is recommended as being an appropriate choice and note that this is improper. Perhaps the most common recommendation is *Jeffreys' prior*, which, in this case, is proportional to a $\text{beta}(1/2, 1/2)$ density. For the three priors, the posterior is $\text{beta}(x + 1, n - x + 1)$, $\text{beta}(x, n - x)$ and $\text{beta}(x + 1/2, n - x + 1/2)$, respectively. So, whenever $x = 0$ or $x = n$, the posterior induced by Haldane's prior is improper. Arguments can be advanced for all three priors as being appropriate characterizations of noninformativity. ■

A general motivation for the use of improper priors is that the posterior that results can arise as the limit of increasingly less informative priors. An example illustrates this.

Example 5.3.4 *Location Normal (continued).*

Suppose the model is as in Example 5.3.2 but now use the proper prior $\mu \sim N(\mu_0, \tau_0^2)$. The posterior is then $\mu \mid x \sim N((n + 1/\tau_0^2)^{-1}(\mu_0/\tau_0^2 + n\bar{x}), (n + 1/\tau_0^2)^{-1})$. It would seem reasonable in this context to interpret larger values of τ_0^2 as being less informative about the true value than smaller values. As $\tau_0^2 \to \infty$, the posterior converges to the posterior obtained using a flat prior. While this is interesting in terms of measuring beliefs, more relevant is what happens to a relative belief ratio $RB(\mu \mid x)$ as $\tau_0^2 \to \infty$ as a measurement of evidence. The calculations follow just as in Example 4.4.2 and $RB(\mu \mid x) \to \infty$ as $\tau_0^2 \to \infty$. Example 4.6.1 shows that this phenomenon is an expression of the fact that, as the prior becomes more diffuse, bias is being introduced into the problem in favor of μ being the true value.

An alternative way to obtain a noninformative limit is to place a uniform prior on $(-c, c)$ and look at the limit as $c \to \infty$. Clearly, for $c > 0$ large enough, the posterior has density

$$
\pi_c(\mu \mid x) = \frac{(2\pi)^{-\frac{1}{2}} n^{\frac{1}{2}} \exp\left\{ -\frac{n}{2}(\mu - \bar{x})^2 \right\}}{2(1 - \Phi(\sqrt{n}(c - \bar{x})))}
$$

for $-c < \mu < c$ and this converges to the posterior based on a flat prior as $c \to \infty$. The relative belief ratio for given c equals $2c\pi_c(\mu \mid x)$ and this also converges to ∞ as $c \to \infty$. ■

In recent years a concerted effort has been made to deal with the problems associated with noninformative priors. Most notable is the idea of a *reference prior* due to Bernardo (1979) and further developed in Berger and Bernardo (1992), Berger, Bernardo and Sun (2009) and many other papers. The basic approach is to apply restrictions to the class of improper priors considered to be reasonable, based on information-theoretic ideas. Often this leads to *Jeffreys' prior*, which, for $\theta \in R^k$, is defined by

$$\pi(\theta) = \left\{ -E_\theta \left(\det \left(\frac{\partial^2 \log(f_\theta(X))}{\partial \theta_i \partial \theta_j} \right) \right) \right\}^{1/2} \qquad (5.8)$$

when all the second order partial derivatives and their expectations exist for all $\theta \in \Theta$. The quantity inside the curly brackets is known as the *Fisher information function*. With the location-normal model of Example 5.3.2, Jeffreys' prior equals $\pi(\theta) = n^{1/2}$, namely, it is a flat prior. With the binomial model of Example 5.3.3, Jeffreys' prior equals $\pi(\theta) = n^{1/2}\theta^{-1/2}(1-\theta)^{-1/2}$, which is proportional to a beta$(1/2, 1/2)$ prior.

This text has focused on the development of statistical theory within the context of proper priors. Perhaps our primary reason for doing this is that it allows for a precise measure of evidence based on the intuitively satisfying idea of change in belief from a priori to a posteriori. Whether or not it is reasonable to describe a priori beliefs by an improper prior is an open question and, even if we accept this, it is not clear how to measure evidence in such a context. Further research could possibly resolve these issues. Recall that the aim of the theory of inference developed in Chapter 4 is to represent a gold standard. In contexts where this cannot be attained, due to problems eliciting priors, the noninformative approach seems essential.

5.4   Checking the Ingredients

Once a model and prior have been chosen and the data collected, it might seem that the next step is to carry out the inference step described in Chapter 4. But what if the choices made are grossly in error? In such a case it seems that inferences made based on these ingredients cannot be viewed as valid. On the other hand, there does not seem to be any way to guarantee that the ingredients are correct and so, to some extent, doubts about the validity of the inferences will always be with us.

The solution to this problem is to weaken somewhat our criteria for when inferences are valid. Whatever inferences are made, these are dependent on the choices made for the model and prior. These inferences are in a sense necessary, as they follow from our definition of how to measure statistical evidence based on the ingredients. So it is only really the ingredients that might lead to doubts concerning the inferences. Rather than ask for an absolute guarantee that the ingredients are correct, it is reasonable to ask only whether or not the subjective ingredients are in any meaningful sense contradicted by the objective data. Furthermore, invoking the principle of empirical criticism, no use should be made in the analysis of any ingredients that cannot be checked against the data.

So methodology is required for checking whether or not the model $\{f_\theta : \theta \in \Theta\}$ and the prior π are contradicted by the data x. Our approach to checking the ingredients will be to prescribe appropriate p-values for the task. Note that, while

p-values have been criticized for inference, their role here is more suitable, as there are no alternatives and no assertions that the model or prior is correct. In essence, it is only required to determine whether or not the data is surprising given the model and prior, and the p-value seems well suited for this task.

If the data is surprising for the model, then there is no need to check the prior, as the prior depends on the model. It is possible, however, that the data is not surprising for the model but is surprising for the prior. As such, it is appropriate to check the model first and then possibly check the prior, depending on the outcome. Certainly, proceeding in this way leads to a more precise understanding of where a failure occurs, when this is the case.

There is a basic factorization of the joint probability model for (θ, x) that corresponds to the various aspects of a statistical analysis, namely, checking a model, checking a prior and inference. For any minimal sufficient statistic T for the model, the joint probability measure for (θ, x) can be factored as

$$\Pi \times P_\theta = \Pi(\cdot \,|\, x) \times M = \Pi(\cdot \,|\, x) \times M_T \times P(\cdot \,|\, T(x)) \qquad (5.9)$$

where $\Pi(\cdot \,|\, x) = \Pi(\cdot \,|\, T(x))$ is the posterior probability measure for θ, M_T is the prior predictive probability measure for T and $P(\cdot \,|\, T(x))$ is the conditional probability measure for x given the value $T(x)$.

Each of the measures in (5.9) plays a different role in the analysis. Broadly speaking, this proceeds as follows.

(i) First assess whether or not the data x is surprising when compared to $P(\cdot \,|\, T(x))$ as a check on the model $\{f_\theta : \theta \in \Theta\}$. This comparison is discussed in Section 5.5.

(ii) If the model passes the checks in (i), then check the prior by seeing if the observed value $T(x)$ is a surprising value from M_T. This comparison is discussed in Section 5.6.

(iii) If the model and prior pass their checks, then use $\Pi(\cdot \,|\, x)$ to make probability statements about the unknown θ as part of inference, as discussed in Chapter 4.

The process outlined in (i), (ii) and (iii) gives the simplest formulation of a statistical analysis. As noted, there is a natural order to this, as the model is checked *before* checking the prior, etc.

This process can be generalized in a number of ways. Suppose that statistic $A(T)$ is a nontrivial maximal ancillary for the model. Then (5.9) can be modified to

$$\Pi \times P_\theta = \Pi(\cdot \,|\, x) \times M_T(\cdot \,|\, A(T(x))) \times P_{A(T)} \times P(\cdot \,|\, T(x)) \qquad (5.10)$$

where M_T has been factored as $M_T(\cdot \,|\, A(T(x))) \times P_{A(T)}$. It is relevant then to use $M_T(\cdot \,|\, A(T(x)))$ in the prior checking phase instead of M_T, since the variation due to $A(T)$ has nothing to do with θ, and thus nothing to do with the prior. As such, the effect of this variation is removed by conditioning and basing this on a maximal ancillary removes the effect of the maximum amount of irrelevant variation. The statistic $A(T)$ is also available to use in the model checking phase by comparing the observed value $A(T(x))$ to $P_{A(T)}$. Based on the recommendation in Durbin (1970), only ancillaries that are functions of the minimal sufficient statistic are considered, as this avoids the contradiction that can arise between sufficiency and ancillarity discussed in Section 3.3.3. Subsequent sections will develop checks for the model and prior.

Some authors, such as Guttman (1967), Rubin (1984), Meng (1994), Gelman, Meng and Stern (1996), Bayarri and Berger (2000) and Bayarri and Castellanos (2007) have developed posterior checks of the model and prior together. The reason that these checks are a posteriori is not clear, however, as this does not arise from an application of the principle of conditional probability. While not ruling out posterior checks, our development here follows the structure laid out in (5.9) and (5.10). Box (1980) proposed a method for simultaneously checking both the model and prior based on the prior predictive M. This is closest to what is proposed here but, as we will see, not confounding the model and prior in a single check leads to improvements. In particular, more accurate information is obtained about the locus of failures when they arise.

A more general principle can be cited as support for splitting a statistical analysis up into component parts according to the factorizations. This is associated with avoiding so-called *double use of the data*. This somewhat vague phrase has often been cited as a criticism of some inference methodology without being precise about its meaning. It is clear, however, that it is entirely appropriate to make multiple uses of the data as part of a statistical analysis. For example, stating an estimate using the data and then quoting an assessment of its accuracy based on the same data seems not only appropriate but necessary. Similarly, measuring the evidence concerning the truth of a hypothesis and then a separate assessment of its strength uses the same data for both roles.

In Evans (2007) it is argued that the basic idea of avoiding "double use of the data" can be given precision by saying that each component of, for example, (5.9) is available for one and only one purpose in a statistical analysis. For example, the sufficiency principle effectively states that $P(\cdot \mid T(x))$ is not to be used for inferences about θ, as it is considered irrelevant for this purpose and, if it were used in this way, then the inferences are possibly misleading. Similarly, if $\Pi(\cdot \mid x)$ is used in model checking, then this could be misleading and this surely underlies the points made in Robins, van der Vaart and Ventura (2000) concerning posterior model checks and overly conservative p-values. So for our developments here, avoiding double use of the data simply means using the components of various factorizations for the different purposes as described.

5.5   Checking the Model

Model checking is commonly recommended and many methods have been proposed for this. There is no attempt made here at an in-depth examination of this topic. Furthermore, at this time, there don't appear to be many basic principles guiding this. One plausible principle is to obey the factorization (5.10) and base our assessment of the model on $P_{A(T)} \times P(\cdot \mid T)$ where T is a minimal sufficient statistic and $A(T)$ is a maximal ancillary. This is motivated by a desire not to confound model checking with either checking the prior or inference. Much of what follows in this section can be found in Evans and Jang (2010).

5.5.1 Checking a Single Distribution

As a first step, consider the problem of checking whether or not observed data $x_0 \in \mathscr{X}$ could have come from some assumed probability measure P with density f with respect to volume measure on $\mathscr{X}$. The comments here will be seen to apply to the more general situations discussed subsequently.

Suppose that there is a *discrepancy statistic* $D : \mathscr{X} \to \mathscr{D}$. It is natural to ask how D is determined. At this point, there doesn't seem to be a good answer to this other than to say that, when x_0 is observed, then $D(x_0)$ should measure, in some sense, deviations from the model. There are clearly bad choices for D, such as a constant function, but there doesn't appear to be a general theory that prescribes what good functions are. Typically this is determined by the type of deviation from model correctness that might be expected in an application. In any case, the fundamental step in checking the model is the comparison of $D(x_0)$ to the distribution P_D of D to determine if this is a reasonable value.

If f_D is the density of D with respect to volume measure on $\mathscr{D}$, then a natural measure of the location of $D(x_0)$ in the distribution P_D, is the p-value

$$P_D(f_D(d) \leq f_D(D(x_0))), \tag{5.11}$$

where $d \sim P_D$. When (5.11) is small, $D(x_0)$ lies in a region of low probability, like a tail or antimode, and this indicates that there is a potential problem with the model, as a surprising value of the discrepancy statistic has been observed.

Note that (5.11) is treated here as a *measure of surprise* and is not connected with measuring evidence. The value of (5.11) simply serves to locate the value $D(x_0)$ within its distribution and, as a probability, it is a measure of belief and not evidence. A large value of (5.11) is not evidence in favor of the model. To see this, note that a data value z ostensibly from f, that lies in a region where f is relatively high, does not indicate that f is correct, as there are many densities for which this holds. Rather, all that can be said is that the value z does not contradict f. Our earlier discussions about evidence make it clear that it is necessary to have alternatives, one of which is known to be true, before it can be stated that evidence in favor of a particular value has been obtained. Also note that the commonly used p-value $P_D(d \geq D(x_0))$ does not locate $D(x_0)$ in its distribution unless $\mathscr{D} = [0, \infty)$ and f_D is monotone decreasing, when it is equivalent to (5.11).

In an application, suppose there are two discrepancy statistics D_1 and D_2 that seem appropriate. Notice that in (5.11) there is no requirement that D be real valued. In fact, D can be multidimensional and the components are not even required to be real valued. Furthermore, when there is more than one discrepancy statistic of interest, it seems logical to combine these, as in $D = (D_1, D_2)$, and then assess the model via (5.11). Even if D_1 and D_2 were independent, it is not at all clear that it makes sense to do the assessments separately and, in fact, doing so would not be equivalent to the use of (5.11).

Another natural question then is, how many discrepancy statistics D_i should be used when forming the overall discrepancy statistic $D = (D_1, \ldots, D_k)$? To a great extent this issue is not a real problem because inevitably adding new discrepancies seems redundant and will not change the value of (5.11). Still, the answer to this question is not entirely clear and our only suggestion is that good judgment be employed. The point of view taken here is that models are just devices used for statistical reasoning and are not to be regarded as being true. It is gross violations that may destroy the validity of inferences and it doesn't seem sensible to worry about any deviation whatsoever.

One significant problem with (5.11) arises in the continuous case, as it is not invariant to change of variable. To see this, suppose that $D(x) = x$ so that $P(f(x) \leq f(x_0))$ is used to assess the reasonableness of f after observing x_0. Suppose instead that $D : \mathscr{X} \to \mathscr{X}$ is any smooth, 1 to 1, onto map with density $f_D(d) = f(D^{-1}(d))J_D(D^{-1}(d))$. If the assessment is based on the observed value $d_0 = D(x_0)$, then the relevant p-value is $P_D(f_D(d) \leq f_D(d_0)) = P(f(x)J_D(x) \leq f(x_0)J_D(x_0))$. The two p-values are generally not equal unless $J_D(x)$ is constant in x. Recall, however, the discussion in Section 1.4 and note that the essential finiteness of an application entails deciding on an appropriate discretization. If this is based on the response x, then it does not make sense to allow volume distortions induced by transformations to influence the assessment of the assumption and so, rather than use $P_D(f_D(d) \leq f_D(d_0))$, it is more reasonable to use

$$P_D(f_D(d)/J_D(D^{-1}(d)) \leq f_D(d_0)/J_D(D^{-1}(d_0))) = P(f(x) \leq f(x_0)) \qquad (5.12)$$

and the p-value is invariant. More details on this argument are provided in Evans and Jang (2010).

In general, however, it will be desirable to use $D : \mathscr{X} \to \mathscr{D}$ where D is onto and smooth, but not necessarily 1 to 1, so

$$f_D(d) = \int_{D^{-1}\{d\}} f(x)J_D(x)\,\mu_{D^{-1}\{d\}}(dx).$$

An argument in Evans and Jang (2010), again based on the underlying discretization of $\mathscr{X}$, leads to using the p-value

$$P_D(f_D^*(d) \leq f_D^*(D(x_0))), \qquad (5.13)$$

where

$$f_D^*(d) = \int_{D^{-1}\{d\}} f(x)\,\mu_{D^{-1}\{d\}}(dx)$$

is assumed finite for every d. Note that f_D^* is the density of P_D with respect to the measure $(f_D(d)/f_D^*(d))\mu_{\mathscr{D}}(dd)$ and the ratio $f_D(d)/f_D^*(d)$ measures the effect of the volume distortion induced by D on the density f_D. The following result establishes an invariant p-value.

Proposition 5.5.1 *The p-value $P_D(f_D^*(u) \leq f_D^*(D(x_0)))$ is invariant under 1 to 1, onto, smooth transformations of $\mathscr{D}$.*

Proof. Let $V : \mathscr{D} \to \mathscr{D}$ be 1 to 1, onto and smooth. The density of V is $f_V(v) = \int_{D^{-1}V^{-1}\{v\}} f(x) J_{V(D)}(x) \mu_{D^{-1}V^{-1}\{v\}}(dx)$ and $V^{-1}\{v\} = V^{-1}(v)$ since V is 1 to 1, so $f_V^*(v) = \int_{D^{-1}V^{-1}\{v\}} f(x) \mu_{D^{-1}V^{-1}\{v\}}(dx) = f_D^*(V^{-1}(v))$, which implies $P_V(f_V^*(v) \le f_V^*(V(D(x_0)))) = P_D(f_D^*(d) \le f_D^*(D(x_0)))$. $\blacksquare$

Notice that the volume distortions caused by the transformation D have been removed from (5.13). In cases where $f_D^*(d)$ is not always finite, then limiting p-values under increasing truncations of $\mathscr{D}$ need to be considered as well.

It is worth noting several simulation approaches for evaluating $f_D^*(u)$. Putting $g(x) = J_D^{-1}(x) I_B(D(x))$ for $B \subset \mathscr{D}$ gives

$$
\begin{aligned}
E_P(g(x)) &= \int_{\mathscr{X}} g(x) f(x) \mu(dx) \\
&= \int_{\mathscr{D}} \left\{ \int_{D^{-1}\{d\}} g(x) f(x|d) \mu_{D^{-1}\{d\}}(dx) \right\} f_D(d) \mu_{\mathscr{D}}(dd) \\
&= \int_B E_{P(\cdot|d)}(J_D^{-1}(x)) f_D(d) \mu_{\mathscr{D}}(dd) \\
&= \int_B \left\{ \int_{D^{-1}\{d\}} J_D^{-1}(x) \frac{f(x) J_D(x)}{f_D(d)} \mu_{D^{-1}\{d\}}(dx) \right\} f_D(d) \mu_{\mathscr{D}}(dd) \\
&= \int_B \left\{ \int_{D^{-1}\{d\}} f(x) \mu_{D^{-1}\{d\}}(dx) \right\} \mu_{\mathscr{D}}(dd) = \int_B f_D^*(d) \mu_{\mathscr{D}}(dd).
\end{aligned}
$$

Therefore,

$$
f_D^*(d) = f_D(d) E_{P(\cdot|D)(d)}(J_D^{-1}(x))
$$

and $E_{P(\cdot|D)(d)}(J_D^{-1}(x))$ can be estimated by simulation from $P(\cdot|d)$. Alternatively, if $x_1, \ldots, x_n$ is a sample from f, then kernel density estimator $\hat{f}_D(d) = n^{-1} \sum_{i=1}^{n} K_h(D(x_i) - d)$, with $K_h(d) = I_{(-1,1)}(d/h)/2h$, can be used to approximate $f_D(d)$ when $\mathscr{D} \subset R$. For small $h > 0$ and n large, then

$$
f_D^*(d) = f_D(d) E_{P(\cdot|d)}(J_D^{-1}(x)) \approx n^{-1} \sum_{i=1}^{n} J_D^{-1}(x_i) K_h(D(x_i) - d).
$$

Also,

$$
E_{P(\cdot|d)}(J_D^{-1}(x)) \approx \sum_{i=1}^{n} J_D^{-1}(x_i) K_h(D(x_i) - d) / \sum_{i=1}^{n} K_h(D(x_i) - d),
$$

the Nadaraya–Watson estimator. The approximation is carried out at some of the $D(x_i)$ values.

5.5.2   Checks Based on Minimal Sufficiency

Now consider the role of $P(\cdot|T(x_0))$ in model checking. Again suppose there is a discrepancy statistic $D : \mathscr{X} \to \mathscr{D}$. The previous comments made about D still apply. Now, however, the comparison of the observed value $D(x_0)$ is with the conditional distribution $P_D(\cdot|T(x_0))$. If $f_D(\cdot|T(x_0))$ denotes the conditional density of D

with respect to volume measure on $\mathscr{D}$, then the location of $D(x_0)$ with respect to $P_D(\cdot\,|\,T(x_0))$ can be measured by the p-value

$$P_D(f_D(d\,|\,T(x_0)) \le f_D(D(x_0)\,|\,T(x_0))\,|\,T(x_0)). \qquad (5.14)$$

If (5.14) is small, then $D(x_0)$ lies in a region of low density and thus typically in a region of low probability, such as a tail or antimode, and this can be taken as an indication of model inadequacy.

There is again the problem of the lack of invariance of (5.14) but now this arises in two distinct ways. There is, of course, the change of variable issue associated with D. Suppose, however, that $T : \mathscr{X} \to \mathscr{T}$ is smooth and onto, so the conditional density of x given $T(x)$ equals, for any $\theta \in \Theta$,

$$f(x\,|\,T(x)) = f_\theta(x)J_T(x)/f_{\theta,T}(T(x)) \qquad (5.15)$$

with respect to volume measure $\mu_{T^{-1}\{T(x)\}}$ on $T^{-1}\{T(x)\}$, where

$$f_{\theta,T}(t) = \int_{T^{-1}\{t\}} f_\theta(z)J_T(z)\,\mu_{T^{-1}\{t\}}(dz) \qquad (5.16)$$

is the marginal density of T with respect to volume measure $\mu_{\mathscr{T}}$ on $\mathscr{T}$. The factorization theorem gives that $f_\theta(x) = g_\theta(T(x))h(x)$ for some g_θ and h. Applying this to (5.15) and (5.16) leads to

$$f(x\,|\,T(x)) = h(x)J_T(x)/\int_{T^{-1}\{t\}} h(z)J_T(z)\,\mu_{T^{-1}\{t\}}(dz).$$

Therefore, a volume distortion through $J_T(x)$ is a fundamental aspect of the conditional density and this must be taken into account when considering model checking, as it depends on which T, among the 1 to 1 equivalent versions of a minimal sufficient statistic, is used.

For general discrepancy statistic D, the joint density of (D,T) is given by

$$f_{\theta,D,T}(d,t) = \int_{D^{-1}\{d\}\cap T^{-1}\{t\}} f_\theta(z)J_{D,T}(z)\,\mu_{D^{-1}\{d\}\cap T^{-1}\{t\}}(dz).$$

Therefore, the conditional density of D given $T(x) = t$, with respect to volume measure on $D(D^{-1}\{d\}\cap T^{-1}\{t\})$, is

$$f_D(d\,|\,t) = \frac{f_{\theta,D,T}(d,t)}{f_{\theta,T}(t)} = \frac{\int_{D^{-1}\{d\}\cap T^{-1}\{t\}} f_\theta(z)J_{D,T}(z)\,\mu_{D^{-1}\{d\}\cap T^{-1}\{t\}}(dz)}{\int_{T^{-1}\{t\}} f_\theta(z)J_T(z)\,\mu_{T^{-1}\{t\}}(dz)}$$

$$= \frac{\int_{D^{-1}\{d\}\cap T^{-1}\{t\}} h(z)J_{D,T}(z)\,\mu_{D^{-1}\{d\}\cap T^{-1}\{t\}}(dz)}{\int_{T^{-1}\{t\}} h(z)J_T(z)\,\mu_{T^{-1}\{t\}}(dz)}.$$

From this it follows that (5.14) equals

$$P_D\left(\frac{\int_{D^{-1}\{d\}\cap T^{-1}\{T(x_0)\}} h(z)J_{D,T}(z)\,\mu_{D^{-1}\{d\}\cap T^{-1}\{T(x_0)\}}(dz) \le}{\int_{D^{-1}\{D(x_0)\}\cap T^{-1}\{T(x_0)\}} h(z)J_{D,T}(z)\,\mu_{D^{-1}\{D(x_0)\}\cap T^{-1}\{T(x_0)\}}(dz)} \,\middle|\, T(x_0) \right)$$

and this does not depend on the denominator $\int_{T^{-1}\{t\}} h(z) J_T(z) \mu_{T^{-1}\{t\}}(dz)$. It is immediate then that the invariant p-value that is not affected by volume distortions, either through T or D, is given by

$$P_D \left(\frac{\int_{D^{-1}\{d\} \cap T^{-1}\{T(x_0)\}} h(z) \mu_{D^{-1}\{d\} \cap T^{-1}\{T(x_0)\}}(dz) \leq}{\int_{D^{-1}\{D(x_0)\} \cap T^{-1}\{T(x_0)\}} h(z) \mu_{D^{-1}\{D(x_0)\} \cap T^{-1}\{T(x_0)\}}(dz)} \,\middle|\, T(x_0) \right). \quad (5.17)$$

Consider an example of implementing (5.17).

Example 5.5.1 *Model Checking for the Location-Scale Normal Model.*

Suppose that $x = (x_1, \ldots, x_n)$ is a sample of n from the $N(\mu, \sigma^2)$ distribution with $\mu \in R^1$ and $\sigma^2 > 0$ unknown. Then $T(x) = (\bar{x}, ||x - \bar{x}1_n||)$ is minimal sufficient. Putting $u = U(x) = (x - \bar{x}1_n)/||x - \bar{x}1_n||$, then $x = \bar{x}1_n + ||x - \bar{x}1_n||u$ and note that $\bar{x}, ||x - \bar{x}1_n||$ and u are statistically independent with u uniformly distributed on $S^{n-1} \cap L^{\perp}\{1_n\}$ where S^{n-1} is the unit sphere in R^n and $L\{1_n\}$ is the linear span of $\{1_n\}$. In this case h is constant (so we can take it to be 1) and $T^{-1}\{(\bar{x}_0, ||x_0 - \bar{x}_0 1_n||)\}$ is the $(n-2)$-dimensional sphere $\bar{x}_0 1_n + ||x_0 - \bar{x}_0 1_n||(S^{n-1} \cap L^{\perp}\{1_n\})$.

It is natural here to consider functions of $U(x)$ as discrepancy statistics for checking the model. For example, the family $T_q(u) = \sum_{i=1}^{d} u_i^q$ is of some interest, as this gives effectively the skewness and kurtosis statistics when $q = 3$ and 4, respectively. In this case, $\int_{D^{-1}\{d\} \cap T^{-1}\{T(x_0)\}} h(z) \mu_{D^{-1}\{d\} \cap T^{-1}\{T(x_0)\}}(dz)$ is the volume of the $(n-3)$-dimensional submanifold of $\bar{x}_0 1_n + ||x_0 - \bar{x}_0 1_n||(S^{n-1} \cap L^{\perp}\{1_n\})$ given by $(T_q \circ U)^{-1}\{d\} \cap T^{-1}\{(\bar{x}_0, ||x_0 - \bar{x}_0 1_n||)\}$.

Alternatively, the invariant p-value can be computed more directly here by assuming $(\mu, \sigma) = (0, 1)$, letting f denote the density of a sample of n from the $N(0, 1)$ distribution and computing

$$P_{(0,1)}(f^*_{T_q \circ U}(T_q(U(x))) \leq f^*_{T_q \circ U}(T_q(U(x_0))) \mid T(x_0))$$
$$= P_U(f^*_{T_q \circ U}(T_q(u)) \leq f^*_{T_q \circ U}(T_q(u_0)))$$

where $f^*_{T_q \circ U}(d) = \int_{(T_q \circ U)^{-1}\{d\}} f(x) \mu_{(T_q \circ U)^{-1}\{d\}}(dx)$ and u is uniformly distributed on $S^{n-1} \cap L^{\perp}\{1_n\}$. The volume distortion induced by $T_q \circ U$ can be computed explicitly as $J_{T_q \circ U}(x) = p(T_{2q-2}(u(x)) - T_{q-1}^2(u(x))/n - T_q^2(u(x)))^{-1/2}/||x - \bar{x}1_n||$. We see that this is not a function of $T_q \circ U$ and also $J_{T_q \circ U}(x) = J_{T_q \circ U}(-x)$. Since $f(x) = f(-x)$ and $(T_q \circ U)^{-1}\{-d\} = (-1)^p (T_q \circ U)^{-1}\{d\}$ when q is a nonnegative integer, then $f^*_{T_q \circ U}(d)$ and the density

$$f_{T_q \circ U}(d) = \int_{(T_q \circ U)^{-1}\{d\}} f(x) J_{T_q \circ U}(x) \mu_{(T_q \circ U)^{-1}\{d\}}(dx)$$

are symmetric about 0 when q is odd. If both $f^*_{T_q \circ U}$ and $f_{T_q \circ U}$ are unimodal, then this implies that the p-values based on the densities, tail probabilities of $|T_q \circ U|$ and the invariant p-values are the same when q is odd.

Figure 5.1 is a plot of the densities and invariant p-values for tests of skewness for several sample sizes n. The p-values are two sided. The invariant p-values are the

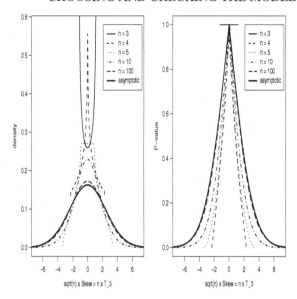

Figure 5.1 *Densities and invariant p-values for test of skewness for various sample sizes n when sampling from normal.*

same as those based on the density of $T_3 \circ U$ and tail probabilities of $|T_3 \circ U|$, for all cases except $n = 3$. Note that the asymptotic approximation to the exact p-value can be poor but is quite good for $n = 100$. When $n = 3$, the invariant p-value is identically equal to 1, i.e., there is never evidence against the model. In this case, it can be shown that $J_{T_3}^{-2}(x) = 1/6 - T_3^2(x)$ and the density of T_3 at t is proportional to $(1 - 6t^2)^{-1/2}$ for $-1/\sqrt{6} < t < 1/\sqrt{6}$, namely, all the density is due to the volume distortion caused by T_3. Notice too that the density is U-shaped with infinite singularities at the end-points. Accordingly, if the density were used for the p-value, then we would reject the model for values of T_3 near 0 and this doesn't make sense. It wouldn't seem to make sense to reject for large values of $|T_3|$ either, at least based on the shape of the distribution. The invariant p-value is telling us that there is no test for skewness based on T_3 when $n = 3$. Intuitively this seems reasonable, as we need two degrees of freedom for location and scale and, to check for skewness, at least two more are needed to see if there is asymmetry about the center.

Figure 5.2 is a plot of the densities and invariant p-values for tests of kurtosis for several sample sizes n. The densities are quite irregular for small sample sizes and skewed. The invariant p-values, those based on the density and tail probabilities, are all different in this case. The p-values based on the densities and asymptotics are quite similar for $n = 100$ while this is not the case for the invariant p-values. This indicates that the volume distortion is still having an appreciable effect when $n = 100$. ■

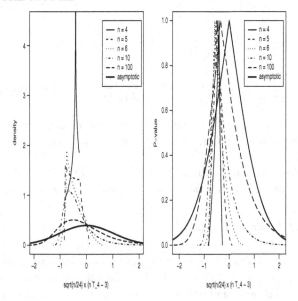

Figure 5.2 *Densities and invariant p-values for test of kurtosis for various sample sizes n when sampling from normal.*

5.5.3 Checks Based on Ancillaries

Consider now a check on the model based on an ancillary $U = A \circ T$, where $U : \mathscr{X} \to \mathscr{U}$ is onto and smooth. It may be that $U = A(T)$ is a suitable choice of discrepancy and then the observed value $U(x_0)$ can be compared with P_U but, in general, discrepancies $D \circ U$ that are functions of U need to be considered.

Naturally, we will want any p-value based on $D \circ U$ to be invariant. Note that, for any $\theta \in \Theta$,

$$f_D(d) = \int_{U^{-1}D^{-1}\{d\}} f_\theta(x) J_{D \circ U}(x) \mu_{U^{-1}D^{-1}\{d\}}(dx)$$

$$= \int_{D^{-1}\{d\}} f_U(u) J_D(u) \mu_{\mathscr{U}}(du)$$

$$= \int_{D^{-1}\{d\}} J_D(u) \int_{U^{-1}\{u\}} f_\theta(x) J_U(x) \mu_{U^{-1}\{u\}}(dx) \mu_{D^{-1}\{d\}}(du)$$

so there are two sources of volume distortion through U and D. Notice, however, that there is no reason to suppose that $\int_{U^{-1}D^{-1}\{d\}} f_\theta(x) \mu_{U^{-1}D^{-1}\{d\}}(dx)$ or $\int_{U^{-1}\{u\}} f_\theta(x) \mu_{U^{-1}\{u\}}(dx)$ is independent of θ and so no obvious way to define an invariant p-value.

It is the case, however, that in many examples an invariant p-value that doesn't depend on θ can be defined via the same route, namely,

$$P_D(f^*_{\theta,D}(d) \le f^*_{\theta,D}(d(x_0))),\tag{5.18}$$

with

$$f_{\theta,D}^*(d) = \int_{U^{-1}D^{-1}\{d\}} f_\theta(x)\,\mu_{U^{-1}D^{-1}\{d\}}(dx)$$

$$= \int_{D^{-1}\{d\}} J_D(u) \int_{U^{-1}\{u\}} f_\theta(x) \frac{J_U(x)}{J_{D\circ U}(x)}\,\mu_{U^{-1}\{u\}}(dx)\,\mu_{D^{-1}\{d\}}(du).$$

For example, in discrete problems all the volume distortion factors $J_D(u), J_U(x)$ and $J_{D\circ U}(x)$ are constantly 1 and the independence of $f_{\theta,D}^*$ from θ follows from the ancillarity of U.

With continuous models it can also happen that (5.18) does not depend on θ. The following example serves as an archetype for a common situation where this is the case.

Example 5.5.2 *Location-Scale Models.*

Suppose $x \in R^n$ and the model is $x = \mu 1_n + \sigma z$ where z is distributed with density f with respect to volume measure on R^n, and $\mu \in R^1, \sigma > 0$ are unknown. Then x has density $f_{\mu,\sigma}(x) = \sigma^{-n} f((x - \mu 1_n)/\sigma)$. Take the parameter space to be $\Theta = \{(\mu,\sigma) : \mu \in R^1, \sigma > 0\}$ and note that there is a group product defined on Θ via $(\mu_1,\sigma_1)(\mu_2,\sigma_2) = (\mu_1 + \sigma_1\mu_2, \sigma_1\sigma_2)$. This group acts on R^n via $(\mu,\sigma)x = \mu 1_n + \sigma x$. A maximal invariant is then given by $U(x) = (x - \bar{x}1_n)/\|x - \bar{x}1_n\|$ and this is ancillary. Note that $U^{-1}\{u\} = \{x : x = a1_n + cu, \text{ for some } (a,c) \in \Theta\} = \Theta u$, i.e., $U^{-1}\{u\}$ is an orbit of the group action. Clearly, this orbit is half of a two-dimensional plane in R^n and so volume measure on this set is just area.

To base the checking on U itself requires evaluating

$$f_{\mu,\sigma,U}^*(u) = \int_{U^{-1}\{u\}} f_{\mu,\sigma}(x)\,\mu_{U^{-1}\{u\}}(dx) = \int_0^\infty \int_{-\infty}^\infty f_{\mu,\sigma}(a1_n + cu)\,\sqrt{n}\,da\,dc$$

$$= \int_0^\infty \int_{-\infty}^\infty \sigma^{-n} f\left(\frac{a-\mu}{\sigma}1_n + \frac{c}{\sigma}u\right)\sqrt{n}\,da\,dc$$

$$= \sigma^{-(n-2)}\int_0^\infty \int_{-\infty}^\infty f(a1_n + cu)\,\sqrt{n}\,da\,dc.$$

Accordingly, the p-value for model checking is given by

$$P_U(f_{\mu,\sigma,U}^*(u) \le f_{\mu,\sigma,U}^*(u_0))$$

$$= P_U\left(\int_0^\infty \int_{-\infty}^\infty f(a1_n + cu)\,da\,dc \le \int_0^\infty \int_{-\infty}^\infty f(a1_n + cu_0)\,da\,dc\right)$$

and this is independent of the model parameter and so is a valid p-value for checking the model. If a function $D \circ U$ is used instead, then a similar argument shows that the p-value is independent of (μ,σ), as the volume distortion factor due to D does not depend on the parameter. Note that when f is the $N(0,1)$ density, then basing model checking on the ancillary U or the conditional distribution of the data given a minimal sufficient statistic produces the same results. ∎

More generally, suppose $\{f_g : g \in G\}$ is a group model, namely, G is a group, with a smooth product, acting freely and smoothly on $\mathscr{X}$ and $f_g(x) = f(g^{-1}x)J_g(g^{-1}x)$ for some fixed density f. Now suppose that $[\cdot] : \mathscr{X} \to G$ is smooth and satisfies $[gx] = g[x]$, so $A(x) = [x]^{-1}x$ is a maximal invariant and is thus ancillary. So $a = A(x) \in \mathscr{X}, x = [x]A(x)$ and $A^{-1}\{a\}$ is the orbit $\{ga : g \in G\}$. If v_G^* denotes volume measure on G, then $\mu_{A^{-1}\{a\}} = K(a)v_G^*$ for some positive function K. Let $z = g^{-1}x$ so that $[z] = g^{-1}[x]$ and let $J_g^*([z])$ denote the Jacobian of the transformation $[z] \to [x]$. Therefore,

$$f_{g,A}^*(a) = \int_{A^{-1}\{a\}} f_g(x)\,\mu_{A^{-1}\{a\}}(dx) = \int_{\{ga : g \in G\}} f_g([x]a)\,\mu_{A^{-1}\{a\}}(dx)\mu_{A^{-1}\{a\}}(dx)$$

$$= K(a)\int_G f(g^{-1}[x]u)J_g(g^{-1}[x]a)\,v_G^*(d[x])$$

$$= K(a)\int_G f([z]a)J_g(u)J_g^*([z])\,v_G^*(d[z]).$$

If $J_g(u)J_g^*([z]) = L(a)m(g)$ for some positive functions L and m, then the invariant p-value $P_A(f_{g,A}^*(a) \le f_{g,A}^*(a_0))$ is indeed independent of g. In Example 5.5.2, $J_{(\mu,\sigma)}(a) = \sigma^{-n}$ and, with $[x] = [\bar{x}, ||x - \bar{x}1_n||]$, then $J_g^*([z]) = \sigma^2$ and this condition is satisfied. More generally, this condition is satisfied in a wide range of group models, such as those discussed in Fraser (1979).

Of course, there can be several distinct maximal ancillaries. But this doesn't cause a problem for model checking, as if at least one maximal ancillary indicates model inadequacy, then that does not contradict the fact that checks based on other maximal ancillaries may not.

What does one do if the model is checked and found wanting? A common approach is to try enlarging the model to include more distributions. For example, in Example 5.2.1, if normality were assumed, then the model could be enlarged to include longer tailed distributions as well, at least if tail length were considered to be the source of the failure. Also, transforming the data in some fashion and then checking if the model holds for the transformed data is a common approach. Stating general principles that prescribe exactly how one should proceed in such a situation is certainly worth further investigation and development.

5.6   Checking the Prior

The prior π may well be considered the most controversial ingredient of the theory of inference proposed in Chapter 4. In part, this is because a prior is seen as a subjective aspect of a statistical analysis but, as discussed in Section 1.5, this is also true of the model $\{f_\theta : \theta \in \Theta\}$. It seems that subjective choices are part of any statistical analysis and this isn't limited to the choice of the prior. Of course, if a prior is not needed, why add it? Our answer is that a prior is needed if there is to be a clear definition of how to measure statistical evidence and that this is the single most important issue any statistical theory needs to address. In addition, a prior should reflect information possessed by the analyst and, as such, will lead to more informative statistical analyses. Accordingly, a prior should never be chosen in an arbitrary fashion but rather reflect what is known about the response being measured through an elicitation process.

Still, just as with a model, a chosen prior may be incorrect in the sense that it misleads the analysis either by introducing strong bias or by being strongly contradicted by the data. The issue of bias has been addressed in Section 4.6 and so a possible contradiction between the data and the prior is considered here. Such a contradiction is referred to as a *prior-data conflict*.

Some may argue that the sanctity of the prior is paramount, as it represents beliefs and any assessment of the suitability of the prior or, even worse, attempts to modify the prior based on the results of such an assessment, is incoherent. But the point of view taken here is that only the data, when properly collected, has an unassailable role in a statistical analysis. The other ingredients are choices and so subject to all the frailties that this implies. Beyond the logical reasons for checking a prior, when there is no prior-data conflict, the inferences can be expected to be more robust to the choice of the prior than otherwise; see Section 4.7.6.

This is not to suggest, however, that an analyst is free to make whatever adjustments they like to the model and prior after seeing the data. For this would almost certainly lead to poor results through overfitting. Undoubtedly, the ideal is to select a model and prior, accept these choices after checking that they are not contradicted by the data, and then carry out the inference step based on these ingredients. But if our approach refused to make adjustments after detecting problems, or worse, refused to check for problems, then it would be useless for practical inference. So a methodology is needed that allows for initial choices to be modified in a logical way when problems are detected.

Logically, checking the model comes first and, if the model is deemed acceptable, an assessment is then made of the prior. In Sections 5.6.1, 5.6.2, 5.6.3, 5.6.4 and 5.6.5 an approach to checking for prior-data conflict is presented. If it has been determined that a prior-data conflict exists, then some action may be necessary. It is well known that with a sufficient amount of data the effect of the prior on the inferences is minimal. Some diagnostics to assess this are discussed in Section 5.6.6. Sometimes, however, the inadequacy of the prior will have to be addressed. In Section 5.7 the issue of how one would go about modifying a prior in a logical way, when a conflict is detected, is considered.

The contents of this section are contained in the papers Evans and Moshonov (2006), Evans and Moshonov (2007) and Evans and Jang (2011b). These references contain additional technical details and examples.

5.6.1   *Prior-Data Conflict*

Suppose that the model $\{f_\theta : \theta \in \Theta\}$ has passed its checks. The implication drawn from this is that there is at least one value of θ such that the data x could plausibly have been generated from the distribution f_θ. But suppose that all the θ values for which this is true lie outside the effective support of the prior π, namely, these values all lie in regions of relatively low prior probability. In such a case, there is a prior-data conflict. It will be shown in Section 5.6.3 that this basic idea can be considerably refined so that individual aspects of a prior can be checked instead of just checking the whole prior.

The question now arises as to how should one check for the existence of a prior-data conflict. As with model checking, this is somewhat unclear but, however this is carried out, it is desirable that it be consistent in some well-defined sense. Also, as discussed in Section 5.4, any such check should be as separate as possible from the process of model checking and inference. The factorization (5.9) leads to using the p-value

$$M_T(m_T(t) \leq m_T(T(x_0)))$$ (5.19)

where m_T is the prior-predictive density of a minimal sufficient statistic T for the model. Note that $m_T(T(x_0)) = \int_\Theta f_{\theta,T}(T(x_0))\pi(\theta)\nu(d\theta)$ can be considered as the average value of the likelihood where the averaging is carried out using the prior. Then (5.19) is the prior probability that the average likelihood that could have been obtained from a set of data is no greater than that obtained for the observed data. Therefore, a small value of (5.19) indicates that the observed data is unusual for this prior.

Consider the following simple example.

Example 5.6.1 *Location Normal Model.*

Suppose $x = (x_1, ..., x_n)$ is a sample from an $N(\mu, 1)$ distribution with $\mu \in R^1$ and $\mu \sim N(\mu_0, \sigma_0^2)$. A minimal sufficient statistic is $T(x) = \bar{x} \sim N(\theta, 1/n)$. For the prior predictive distribution of $\bar{x}$ we can write $\bar{x} = \mu + Z$ where $Z \sim N(0, 1/n)$ independent of μ. So the prior predictive distribution is $N(\mu_0, \sigma_0^2 + 1/n)$. Accordingly, the relevant p-value is

$$M_T(m_T(\bar{x}) \leq m_T(\bar{x}_0)) = 2(1 - \Phi(|\bar{x}_0 - \mu_0|/(\sigma_0^2 + 1/n)^{1/2})).$$ (5.20)

This shows that when $\bar{x}_0$ lies in the tails of its prior distribution, then there is an indication of a prior-data conflict existing. Note that using the prior-predictive distribution of $\bar{x}$ to assess this, instead of the prior for μ, results in the standardization by $(\sigma_0^2 + 1/n)^{1/2}$ rather than σ_0. So $\bar{x}_0$ has to be somewhat further out in the tail for a prior-data conflict to exist than would be the case in a comparison of $\bar{x}_0$ to the prior of μ.

Now suppose that the true value of μ is μ_{true}. Then as $n \to \infty$ we have that (5.20) converges almost surely to $2(1 - \Phi(|\mu_{true} - \mu_0|/\sigma_0))$. Therefore, if μ_{true} lies in the tails of the prior, then asymptotically (5.20) will detect this and lead to evidence that a prior-data conflict exists.

Also note that when $\sigma_0 \to \infty$, then (5.20) converges to 1 and so no indication will be found of a prior-data conflict. This is not surprising, as a limiting diffuse prior simply indicates that all values are equally likely, so no conflict should be found. Furthermore, as $\sigma_0 \to 0$, so we have a very precise prior, then (5.20) will definitely find evidence of a prior-data conflict for large n, unless μ_0 is indeed the true value. ■

The value of (5.20) illustrates an important general property of the p-value given by (5.19). In particular, it is established in Evans and Jang (2011b) that, under quite general conditions,

$$M_T(m_T(t) \leq m_T(T(x_0))) \to \Pi(\pi(\theta) \leq \pi(\theta_{true}))$$

as the amount of data increases. This consistency result establishes the correctness of (5.19) for the purposes of checking the prior, as clearly this shows that the p-value is assessing whether or not θ_{true} lies in a region of low probability for the prior.

The following example illustrates a connection between checking for prior-data conflict and noninformativity.

Example 5.6.2 *Bernoulli Model.*

Suppose $x = (x_1, ..., x_n)$ is a sample from a Bernoulli(θ) model with $\theta \sim$ beta(α, β). A minimal sufficient statistic is $T(x) = \sum x_i \sim$ binomial(n, θ). The prior-predictive probability function of $T(x)$ is then given by

$$m_T(t) = \binom{n}{t} \frac{\Gamma(\alpha + \beta)}{\Gamma(\alpha)\Gamma(\beta)} \frac{\Gamma(t + \alpha)\Gamma(n - t + \beta)}{\Gamma(n + \alpha + \beta)}.$$

To assess whether or not $T(x) = t_0$ is surprising requires comparing t_0 to m_T via the tail probability $M_T(m_T(t) \leq m_T(t_0))$ which, in this case, must be done numerically. In essence, this is the prior probability of obtaining a value of the minimal sufficient statistic with probability of occurrence no greater than that for the observed value. Because of cancellations, the p-value can be written as

$$M_T\left(\frac{\Gamma(t + \alpha)\Gamma(n - t + \beta)}{\Gamma(t + 1)\Gamma(n - t + 1)} \leq \frac{\Gamma(t_0 + \alpha)\Gamma(n - t_0 + \beta)}{\Gamma(t_0 + 1)\Gamma(n - t_0 + 1)}\right). \tag{5.21}$$

Notice that when $\alpha = \beta = 1$, then (5.21) equals 1 and there is never any indication of prior-data conflict. This makes sense, as when the sampling model is correct, there can be no conflict between the data and a noninformative prior. ∎

It is interesting to note that there is a subtle difference between binomial and negative-binomial sampling when it comes to prior-data conflict.

Example 5.6.3 *Negative-Binomial Sampling.*

Suppose that $t \sim$ negative-binomial(k, θ) and t_0 is observed where $\theta \in (0, 1]$. Then t is minimal sufficient. Note that the likelihood function is a positive multiple of the likelihood function in Example 5.6.2 when we obtain t_0 zeros in a sample of size $n = t_0 + k$. It is then clear that posterior inferences about θ are the same from the two models. The checks for prior-data conflict, however, are somewhat different.

Suppose that the prior $\theta \sim U(0, 1)$. In Example 5.6.2, the prior predictive for the minimal sufficient statistic is uniform and there is never an indication of a prior-data conflict existing. The prior predictive under negative binomial sampling, however, is given by

$$m(t) = \binom{t + k - 1}{k - 1} \int_0^1 \theta^k (1 - \theta)^t \, d\theta = \frac{k}{(t + k)(t + k + 1)} \tag{5.22}$$

and this is not uniform. Indeed, it cannot be uniform because the support for this distribution is the nonnegative integers. Since (5.22) is decreasing in t,

$$M(m(t) \leq m(t_0)) = \sum_{t = t_0}^{\infty} \frac{k}{(t + k)(t + k + 1)} = \frac{k}{t_0 + k}. \tag{5.23}$$

This p-value is small only when t_0 is large compared to k, namely, only when it requires many tosses to get the k-th head, which suggests that θ is small.

While this p-value may seem unusual when compared to the binomial case, there is a subtle difference between the two situations. In particular, with the binomial(n, θ) it is permissible to have $\theta = 0$ but this is not possible with the negative-binomial(k, θ), as there is no such thing as the negative-binomial$(k, 0)$ distribution.

Recall that the uniform prior on $[0, 1]$ is to be considered as an approximation to a prior placed on a finite grid of equispaced points and one of these can be $\theta = 0$ for binomial sampling. But this approximation is not sensible with the negative binomial. Consider instead, for the negative-binomial, placing a uniform prior on $[\varepsilon, 1]$ with $\varepsilon > 0$ and let m_ε denote the prior predictive. So we are requiring that θ be bounded away from 0. Then the dominated convergence theorem establishes that $\lim_{\varepsilon \to 0} M_\varepsilon (m_\varepsilon(t) \leq m_\varepsilon(t_0))$ equals (5.23). Accordingly, (5.23) can be viewed as an approximation to the p-value obtained based on a prior that is 0 in a small neighborhood of $\theta = 0$ and uniform otherwise. As such, a small value of (5.23) is indicating that $\theta_{true} < \varepsilon$ for a fixed ε relevant to the application and this contradicts the prior. ■

The following is an important example that can be easily generalized to many of the linear models used in statistical analyses.

Example 5.6.4 *Location-Scale Normal Model.*

Suppose that $x = (x_1, ..., x_n)$ is a sample from an $N(\mu, \sigma^2)$ distribution where $\mu \in R^1$ and $\sigma > 0$ are unknown. With $s^2 = (n-1)^{-1} \sum (x_i - \bar{x})^2$, then $T(x) = (\bar{x}, s^2)$ is a minimal sufficient statistic for this model. Then

$$\bar{x} | \mu, \sigma^2 \sim N(\mu, \sigma^2/n)$$
$$s^2 | \mu, \sigma^2 \sim \text{gamma}_{rate}((n-1)/2, (n-1)/2\sigma^2)$$

with $\bar{x}$ and s^2 conditionally independent given (μ, σ^2). Suppose the prior on (μ, σ^2) is given by

$$\mu | \sigma^2 \sim N(\mu_0, \tau_0^2 \sigma^2), \sigma^{-2} \sim \text{gamma}_{rate}(\alpha_0, \beta_0). \tag{5.24}$$

From this it follows that the joint conditional prior predictive of $(\bar{x}, s^2)$ given σ^2 is

$$\bar{x} | \sigma^2 \sim N(\mu_0, \sigma^2(\tau_0^2 + 1/n))$$
$$s^2 | \sigma^2 \sim \text{gamma}_{rate}((n-1)/2, (n-1)/2\sigma^2)$$

with $\bar{x}$ and s^2 conditionally independent given σ^2. Multiplying the joint conditional density given σ^2 by the prior for σ^2 and integrating out σ^2 gives

$$m_T(\bar{x}, s^2) \propto (s^2)^{(n-1)/2-1} \beta_x^{-n/2-\alpha_0} \tag{5.25}$$

where

$$\beta_x = \beta_0 + (n-1)s^2/2 + (\bar{x} - \mu_0)^2/2(\tau_0^2 + 1/n).$$

The p-value $M_T (m_T(\bar{x}, s^2) \leq m_T(\bar{x}_0, s_0^2))$ can then be computed, for specified values of the hyperparameters α_0, β_0, μ_0 and τ_0^2, via simulation. For this, repeat the following steps many times: generate (μ, σ^2) using (5.24), generate $(\bar{x}, s^2)$ from their joint distribution given (μ, σ^2) and record whether or not (5.25) is less than or equal to the observed value of this quantity. The proportion of times the inequality holds gives the estimate of (5.19). Note that the norming constant of m_T is not needed.

For example, suppose that $(\mu, \sigma)_{true} = (0, 1)$ and that for a sample of size $n = 20$ from this distribution we obtained $\bar{x}_0 = 0.0358324$ and $s_0^2 = 0.836563$. Then for the prior specified by $\tau_0^2 = 1$, $\mu_0 = 50$, $\alpha_0 = 1$ and $\beta_0 = 5$, the above simulation process, based on a Monte Carlo sample of size $N = 10^3$, gives the p-value 0.0019 and so a clear indication of a prior-data conflict is obtained.

Rather than computing a p-value based on the full prior predictive distribution of the minimal sufficient statistic, it is also possible to use the prior predictive distribution of some function of the minimal sufficient statistic. For example, we could instead use the marginal prior predictive distributions of $\bar{x}$ and s^2.

Now note that $\bar{x} = \mu + (\sigma^2/n)z$ where $z \sim N(0, 1)$ and when $\mu \,|\, \sigma^2 \sim N(\mu_0, \tau_0^2 \sigma^2)$, then $\bar{x} \,|\, \sigma^2 \sim N(\mu_0, \sigma^2(\tau_0^2 + 1/n))$. Therefore, after multiplying the $N(\mu_0, \sigma^2(\tau_0^2 + 1/n))$ density by the marginal prior for $1/\sigma^2$ and integrating out σ^2, the marginal prior predictive distribution of $\bar{x}$ is $t_{2\alpha_0}(\mu_0, (\beta_0 (\tau_0^2 + 1/n) \alpha_0^{-1})^{1/2})$ where $t_\lambda (0, 1)$ denotes the t distribution with λ degrees of freedom and $t_\lambda (\mu, \sigma) = \mu + \sigma t_\lambda (0, 1)$. Therefore, the p-value based on $\bar{x}$ alone is $2(1 - G_{2\alpha_0}(|\bar{x}_0 - \mu_0|/(\beta_0 (\tau_0^2 + 1/n) \alpha_0^{-1})^{1/2}))$ where $G_{2\alpha_0}$ is the cdf of the $t_{2\alpha_0}(0, 1)$ distribution. For the above generated data this gives the p-value 0.0021, which also indicates the existence of a prior-data conflict.

Similarly, $s^2 \sim (\beta_0/\alpha_0)F_{(n-1, 2\alpha_0)}$ is the marginal prior predictive of s^2. For the above example then compare $s_0^2/5 = 0.1673126$ with the $F(19, 2)$ distribution. The p-value obtained by computing the probability of obtaining a value from the $F(19, 2)$ distribution with density smaller than that obtained at the observed value equals

$$P(F(19, 2) \leq 0.1673126) + \Pr(F(19, 2) > 1.5295) = 0.47832,$$

which does not indicate any prior-data conflict. The value 1.5295 for the right tail was determined numerically. These two tests seem to indicate that the conflict arises from the location of the prior and not its spread (see Example 5.6.7).

As we will subsequently discuss, the checking of the prior can be decomposed into checking the individual components of (5.24). When a prior-data conflict exists, this allows for a somewhat clearer detection of the origin of the conflict when only one component fails. ∎

5.6.2 Prior-Data Conflict and Ancillaries

When an ancillary $A(T)$ exists, the p-value (5.19) needs to be modified. This is because the value of (5.19) will be affected by the variation exhibited by $A(T)$ and this variation has nothing to do with the true value of θ and thus can tell us nothing about prior-data conflict. The appropriate way to get rid of the effect of this variation is to condition on the observed value $A(T(x_0))$, replacing (5.19) by

$$M_T(m_T(t \,|\, A(T(x_0))) \leq m_T(T(x_0) \,|\, A(T(x_0))) \,|\, A(T(x_0))). \qquad (5.26)$$

Of course, $A(T)$ is taken to be a maximal ancillary so that the conditioning is on the maximum amount of ancillary information. As previously noted, the lack of a unique

maximal ancillary causes difficulties for inference but not in this context. For, if one
of the values (5.26) is small for some maximal ancillary, then this is an indication of
a prior-data conflict and this does not contradict the fact that p-values based on other
maximal ancillaries are not small. Note that this is based on a further factorization of
M_T as $M_T = M_{A(T)} \times M_T(\cdot | A(T))$.

The necessity of conditioning on an ancillary is avoided whenever there is a complete minimal sufficient statistic T. In this situation Basu's theorem implies that every
ancillary is independent of T, so $A(T)$ is constant, and the prior predictive distribution of T does not exhibit variation that can be ascribed to $A(T)$. Therefore, for the
examples discussed in Section 5.6.1, there is no need to condition on ancillaries.

Consider now a simple example where conditioning is clearly necessary.

Example 5.6.5 *Two Measuring Instruments.*

Consider once again Example 3.3.1. It is clear that $T(i,x) = (i,x)$ is minimal sufficient and so $A(i,x) = i$ is a function of T. Suppose an $N(\theta_0, 1)$ prior is placed on θ.
Therefore, when $i = 1$ the conditional prior predictive of T is $x \sim N(\theta_0, \sigma_1^2 + 1)$, and
when $i = 2$ the conditional prior predictive of T is $x \sim N(\theta_0, \sigma_2^2 + 1)$. The unconditional prior predictive is the mixture of these two distributions and it seems obvious
that the check for prior-data conflict should depend on which measuring instrument
is being used. Since $\sigma_1^2 \gg \sigma_2^2$, when p_1 is not close to 1, it is much less likely that
prior-data conflict will occur with measuring instrument 1, if this is assessed using
the $N(\theta_0, \sigma_1^2 + 1)$ distribution rather than using the mixture distribution which has
variance $p_1\sigma_1^2 + p_2\sigma_2^2$. ∎

A more substantive example of conditioning arises in the following example.

Example 5.6.6 *Location Cauchy.*

Suppose $x = (x_1, \ldots, x_n)$ is a sample from a distribution with density proportional
to $1/(1 + (x - \theta)^2)$ where $\theta \in R^1$. The order statistic $T(x) = (x_{(1)}, \ldots, x_{(n)})$ is a
minimal sufficient statistic. Furthermore, $A(T(x)) = (x_{(2)} - x_{(1)}, \ldots, x_{(n)} - x_{(1)}) = (a_1, \ldots, a_{n-1})$ is ancillary. Clearly the conditional distribution of T given $A(T)$ can be
expressed as the conditional distribution of $x_{(1)}$ given $A(T)$ and this has conditional
density proportional to

$$1/\{(1 + (x_{(1)} - \theta)^2)\Pi_{i=1}^{n-1}(1 + (x_{(1)} + a_i - \theta)^2)\}.$$

Note that the normalizing constant will not involve θ and so the conditional prior
predictive density of $x_{(1)}$ is

$$m_T(x_{(1)} | a_1, \ldots, a_{n-1})$$

$$\propto \int_{-\infty}^{\infty} \frac{1}{\left(1 + (x_{(1)} - \theta)^2\right)} \prod_{i=1}^{n-1} \frac{1}{\left(1 + (x_{(1)} + a_i - \theta)^2\right)} \pi(\theta) \, d\theta$$

$$= \int_{-\infty}^{\infty} \frac{1}{(1 + v^2)} \prod_{i=1}^{n-1} \frac{1}{\left(1 + (v + a_i)^2\right)} \pi(x_{(1)} - v) \, dv.$$

Integrating $x_{(1)}$ out of the above expression shows that the normalizing constant is
given by $\int_{-\infty}^{\infty} \{(1 + v^2)\Pi_{i=1}^{n-1}(1 + (v + a_i)^2)\}^{-1} dv$. Checking for any prior-data con-

flict is then carried out by comparing the observed value of $x_{(1)}$ with the distribution given by $m_T(\cdot \mid a_1, \ldots, a_{n-1})$. Evaluating the p-value associated with this conditional prior predictive requires numerical integration. ∎

Example 5.6.6 can be generalized to a wide class of statistical models based on group transformations, as discussed in Fraser (1979).

5.6.3   Hierarchical Priors

In Box (1980) a method was suggested for jointly checking the model and the prior based upon the full marginal prior predictive M for the data. As already noted, M factors as $M = M_T \times P(\cdot \mid T)$ for any minimal sufficient statistic T and so Box's check confounds a check on the sampling model with a check on the prior. Separate checks are advocated here, as more is learned with this approach. Indeed, this can be further refined to check on individual aspects of the prior.

Suppose $\theta = (\theta_1, \theta_2)$ and the prior is specified as

$$\pi(\theta_1, \theta_2) = \pi_1(\theta_1)\pi_2(\theta_2 \mid \theta_1)$$

where π_1 is the marginal prior for θ_1 and $\pi_2(\cdot \mid \theta_1)$ is the conditional prior for θ_2 given θ_1. Of course any multidimensional prior can be written this way, but here it is meant that the prior π is constructed by specifying π_1 and π_2. It is natural then to see if the methods for checking the full prior π can be generalized to checking the components π_1 and π_2 separately to help identify the source of any conflict more precisely. A prior may be specified with more than two components, but the discussion is restricted here to the two component case, as the generalization is straightforward.

As in the previous section, suppose that T is a minimal sufficient statistic for the model and $A(T)$ is maximal ancillary for θ. Consider first checking π_2 for conflict with the data. Suppose there is a statistic $V(T)$ that is *ancillary for* θ_2 in the sense that the sampling distribution of $V(T)$ is independent of θ_2 and depends on θ_1. If it is concluded that the observed value $T(x_0)$ is a surprising value from $M_T(\cdot \mid A(T(x_0)))$, this could arise because $V(T(x_0))$ is a surprising value from $M_{V(T)}(\cdot \mid A(T(x_0)))$, the conditional prior predictive of $V(T)$ given $A(T)$. The following result has implications for checking the prior on θ_1.

Proposition 5.6.1 *If $V(T)$ is ancillary for θ_2, then $M_{V(T)}(\cdot \mid A(T))$ does not depend on π_2.*

Proof. Let $\Theta_2(\theta_1) = \{\theta_2 : (\theta_1, \theta_2) \in \Theta\}$; then for $B \subset \mathscr{V}$,

$$M_{V(T)}(B \mid A(T)) = \int_{\Theta} P_{V,\theta}(B \mid A(T)) \Pi(d\theta)$$

$$= \int_{\Theta_1} \int_{\Theta_2(\theta_1)} P_{V,\theta_1}(B \mid A(T)) \Pi_2(d\theta_2 \mid \theta_1) \Pi_1(d\theta_1)$$

$$= \int_{\Omega_1} P_{V,\theta_1}(B \mid A(T)) \left\{ \int_{\Theta_2(\theta_1)} \Pi_2(d\theta_2 \mid \theta_1) \right\} \Pi_1(d\theta_1)$$

$$= \int_{\Omega_1} P_{V,\theta_1}(B \mid A(T)) \Pi_1(d\theta_1),$$

which establishes the result. ∎

Therefore, a surprising value of $V(T(x_0))$ cannot be due to a conflict with π_2. Now argue just as in the previous section for ancillary statistics. To assess whether or not $T(x_0)$ is conflicting with π_2, the sensible thing to do is to remove the variation in $M_T(\cdot\,|A(T))$ due to $V(T)$ when making the comparison and this is done by conditioning on $V(T)$, namely, compare $T(x_0)$ to $M_T(\cdot\,|A(T(x_0)),V(T(x_0)))$. Of course we want to remove the maximum effect of the ancillary variation and so we take $V(T)$ to be a maximal ancillary for θ_2. Again there could be a number of distinct maximal ancillaries for θ_2 and these all provide valid checks.

The marginal distribution of $V(T)$ depends on θ_1 and so $V(T(x_0))$ can be compared to $M_{V(T)}(\cdot\,|A(T(x_0)))$ to assess whether or not there is any conflict with π_1. Notice that Proposition 5.6.1 implies that this comparison does not depend in any way upon π_2. In contrast, $M_T(\cdot\,|A(T(x_0)),V(T(x_0)))$ will generally depend upon π_1 as well as π_2. This suggests that we first check for prior-data conflict with π_1 by comparing $V(T(x_0))$ to $M_{V(T)}(\cdot\,|A(T(x_0)))$ and, if no conflict is found, then proceed to check for prior-data conflict with π_2 by comparing $T(x_0)$ to $M_T(\cdot\,|A(T(x_0)),V(T(x_0)))$. This is analogous to the proviso that it doesn't make sense to check for prior-data conflict unless it is first agreed that the sampling model is reasonable.

This approach is seen to be based on a further factorization of the conditional prior predictive as

$$M_T(\cdot\,|A(T)) = M_{V(T)}(\cdot\,|A(T)) \times M_T(\cdot\,|A(T),V(T)).$$

The first factor is concerned with checking π_1 and the second factor is concerned with checking π_2. Consider an example.

Example 5.6.7 *Location-Scale Normal.*

Suppose the situation is as described in Example 5.6.4 and recall that $T(x) = (\bar{x}, s^2)$. Here π_1 is the prior on $\theta_1 = \sigma^2$ and $\pi_2(\cdot\,|\theta_1)$ is the conditional prior on $\theta_2 = \mu$ given σ^2. It is immediate that s^2 is ancillary for μ. The marginal prior predictive of s^2 is given by $s^2 \sim (\beta_0/\alpha_0)F_{(n-1,2\alpha_0)}$ and the conditional prior predictive of $\bar{x}$ given s^2 is distributed as $t_{n+2\alpha_0-1}(\mu_0,\tilde{\sigma})$ where

$$\tilde{\sigma}^2 = \left(\tau_0^2 + 1/n\right)\left(2\beta_0 + (n-1)s^2\right)/(n+2\alpha_0-1).$$

Now consider the numerical values of the hyperparameters and data specified in Example 5.6.4. To assess if there is any conflict with π_1, we compare $s_0^2/5 = 0.1673126$ with the $F(19,2)$ distribution. In Example 5.6.4 the relevant p-value was computed to be 0.47832 and this doesn't indicate any prior-data conflict. To assess if there is any conflict with π_2, compare $(\bar{x}-\mu_0)/\tilde{\sigma} = -31.81617$ to the $t_{21}(0,1)$ distribution. This is clearly a very extreme value and in fact the p-value is 0 to 7 decimals. This check has appropriately detected the discrepancy between the prior and the location of the data. Example 5.6.4 also considered using the marginal prior predictive distribution of $\bar{x}$ to assess prior-data conflict. This resulted in a p-value of 0.0021 and so implies that a prior-data conflict exists. Intuitively, this seemed to indicate a problem with the specification of the part of the prior for μ, but there was no clear rationale for this. The hierarchical approach, however, indicates a very clear problem with the specification of π_2 and does so much more dramatically. ∎

The need to specify an order for the checking and the need for a statistic $V(T)$ ancillary for θ_2 indicates that checking for individual components may not be available generally. There is, however, a fairly wide class of models and decompositions where such a $V(T)$ exists. For example, an explicit construction can be given for $V(T)$ when the basic statistical model corresponds to a group model with a specific structure. In particular, when the parameter space is a group that can be written as $\Theta = \Theta_2\Theta_1$ with $\theta_1 \in \Theta_1, \theta_2 \in \Theta_2$ and Θ_1, Θ_2 subgroups with the product being a semidirect product, then $V(T)$ is easily obtained. Example 5.6.7 exhibits this structure and the construction procedure leads to $V(\bar{x}, s^2) = s$. In fact, $V(\bar{x}, s^2) = s$ works for any location-scale model with the same prior decomposition.

As documented in Fraser (1979), there are many models in statistics that have this structure. For example, consider a regression model where a basic observation $y \in R^1$ has $E(y|x) = x^t\theta_2, Var(y|x) = \theta_1$ with $\theta_2 \in R^p$ unknown, $\theta_1 > 0$ unknown and the conditional distributions otherwise fully specified, e.g., normal. Then taking the group product to be $(\theta_1, \theta_2)(\theta_1', \theta_2') = (\theta_1\theta_1', \theta_2 + \theta_1\theta_2')$, we have that this structure obtains. This can also be generalized to multivariate regression with any error distribution.

5.6.4   Hierarchical Models

There is a another context where priors are specified hierarchically, namely, hierarchical modeling. A slight change of notation is made here so that $\theta_2 \in \Theta_2$ denotes the sampling model parameter and $\theta_1 \in \Theta_1$ denotes a hyperparameter that specifies the prior for θ_2 as $\pi_2(\cdot|\theta_1)$. Further, suppose there is prior π_1 on θ_1. This induces the prior measure $\Pi_2^*(A) = \int_{\Omega_1} \Pi_2(A|\theta_1)\Pi_1(d\theta_1)$ for $A \subset \Theta_2$ and all the methods discussed so far, based on the minimal statistic T for the model $\{P_{\theta_2} : \theta_2 \in \Theta_2\}$, are available to check whether or not Π_2^* conflicts with the data. Again we might like to check for conflicts with the individual components of the prior, but the situation is different from the previous problem because θ_1 is not part of the model parameter. Therefore, the methods just discussed for hierarchical priors are not available for this problem and a different approach is needed.

The joint distribution of (θ_1, θ_2, x) can be factored as

$$\Pi_1 \times \Pi_2(\cdot|\theta_1) \times P_{\theta_2, T} \times P(\cdot|T)$$

and recall that $P(\cdot|T)$ is used to check the sampling model $\{P_{\theta_2} : \theta_2 \in \Omega_2\}$. Now observe that there is another way to at least formally generate a model for x from the joint distribution. With T a minimal sufficient statistic for $\{P_{\theta_2} : \theta_2 \in \Theta_2\}$, put

$$M(dx|\theta_1) = \int_{\Theta_2} P_{\theta_2}(dx)\Pi_2(d\theta_2|\theta_1) = P(dx|T)(t)\int_{\Theta_2} P_{\theta_2, T}(dt)\Pi_2(d\theta_2|\theta_1)$$

$$= P(dx|T)(t) \times M_T(dt|\theta_1).$$

The model $\{M(\cdot|\theta_1) : \theta_1 \in \Theta_1\}$ is only formal, as, strictly speaking, when the model $\{P_{\theta_2} : \theta_2 \in \Theta_2\}$ is correct, it is not the case that $x \sim M(\cdot|\theta_1)$ for some $\theta_1 \in \Omega_1$, except in certain special circumstances. Note that $M(\cdot|\theta_1)$ is the conditional prior predictive distribution for x given θ_1 and $M_T(\cdot \,|\, \theta_1)$ is the conditional prior predictive distribution for T given θ_1.

Now consider the model for T given by $\{M_T(\cdot \,|\, \theta_1) : \theta_1 \in \Theta_1\}$ and let $V(T)$ be a minimal sufficient statistic for this model. Then

$$M_T(\cdot \,|\, \theta_1) = M_{V(T)}(\cdot \,|\, \theta_1) \times M_T(\cdot \,|\, V(T)),$$

where $M_T(\cdot \,|\, V(T))$ is the conditional prior predictive distribution of T given $V(T)$, and $M_{V(T)}(\cdot \,|\, \theta_1)$ is the conditional prior predictive distribution of $V(T)$ given θ_1. Then the joint distribution of (θ_1, x) can be factored as

$$\Pi_1(\cdot \,|\, V(T)) \times M_V \times M_T(\cdot \,|\, V(T)) \times P(\cdot \,|\, T) \qquad (5.27)$$

where M_V is the prior predictive distribution of V, and $\Pi_1(\cdot \,|\, V)$ is the posterior distribution of θ_1. Note that a simple argument establishes that the factors in (5.27) are the same, whether determined from the joint distribution of (θ_1, θ_2, x) or the joint distribution of (θ_1, x). In particular, the posterior distribution of θ_1 only depends on the data through $V(T(x_0))$.

Now, using the same arguments as in Section 5.6.3, consider how each of the factors in (5.27) is to be used. First $P(\cdot \,|\, T)$ is available for checking the basic sampling model $\{P_{\theta_2} : \theta_2 \in \Omega_2\}$. If no indication is found against $\{P_{\theta_2} : \theta_2 \in \Omega_2\}$, proceed to check the model $\{M_T(\cdot \,|\, \theta_1) : \theta_1 \in \Theta_1\}$ for T using $M_T(\cdot \,|\, V)$ and note that this does not depend on Π_1. If an indication is found against $\{M_T(\cdot \,|\, \theta_1) : \theta_1 \in \Theta_1\}$, then, because the sampling model $\{P_{\theta_2} : \theta_2 \in \Omega_2\}$ has been accepted, and so consequently the model $\{P_{\theta_2, T} : \theta_2 \in \Omega_2\}$ for T, this must occur because of a conflict between the observed value $T(x_0)$ and Π_2. If no indication is found against $\{M_T(\cdot \,|\, \theta_1) : \theta_1 \in \Theta_1\}$, then we can check for a conflict with Π_1 using M_V. Finally, if there is no conflict with Π_1, then $\Pi_1(\cdot \,|\, V)$ is available for inference about θ_1. Of course, if there is no conflict with Π_1 and Π_2, then we can also make inference about the parameter of interest θ_2. Consider an example of this.

Example 5.6.8 *Random Effects.*

Suppose that $x = (x_1, \ldots, x_n)$ is a sample from the $N_p(\mu, I)$ distribution where $\mu \sim N_p(\mu_0, \sigma^2 I)$ with $\mu_0 \in R^p$ fixed and $\sigma^{-2} \sim$ chi-squared(α_0) with α_0 fixed. So here $\theta_2 = \mu$ and $\theta_1 = \sigma^2$. Then $T = \bar{x}, M_T(\cdot \,|\, \theta_1)$ is the $N_p(\mu_0, (\theta_1 + 1/n)I)$ distribution and $V = (\bar{x} - \mu_0)^t (\bar{x} - \mu_0)$.

The $N_p(\mu, I)$ sampling model is checked using the conditional distribution of x given $\bar{x}$. This can involve checks for nonnormality as in Example 5.5.1.

Given $V = v$, the conditional distribution of $\bar{x}$ is uniform on the sphere of radius $v^{1/2}$ centered at μ_0. So in this case $M(\cdot \,|\, V)$ would seem to imply that there will never be an indication against the $N_p(\mu_0, \sigma^2 I)$ factor in the prior, at least when computing p-values as we have prescribed. At first this seems anomalous, but consider that this factor allows for any value for σ^2 and, by an appropriate choice of σ^2, any prior-data conflict can be avoided, in the sense that the likelihood and prior support for μ will overlap. Whether or not appropriate values for σ^2 are allowed depends only on the prior Π_1 for σ^2.

Contrast this with Example 5.6.4, where the sampling model depends on both θ_1 and θ_2, while the sampling model only depends on θ_2 here. In Example 5.6.4 there will also be values of θ_1 such that the prior for θ_2 will not conflict with the data, but these values of θ_1 may not be realistic in light of the likelihood. So in Example 5.6.4 a conflict may exist with Π_2 irrespective of the prior placed on θ_1.

The sampling distribution of $(\sigma^2 + 1/n)^{-1}(\bar{x} - \mu_0)^t(\bar{x} - \mu_0)$ is chi-squared(p). So M_V has density m_V given by

$$\frac{\Gamma(p/2 + \alpha_0/2)}{\Gamma(p/2)\Gamma(\alpha_0/2)} v^{p/2-1} \int_0^\infty \left(\frac{1}{1+u/n}\right)^{p/2} \exp\left\{-\frac{1}{2}\frac{v}{1+u/n}\right\} g(u)\, du$$

where g is the chi-squared $(p + \alpha_0)$ density. Although nonstandard, this is easily numerically evaluated on a grid of values and the relevant p-value computed for checking the prior distribution of σ^2. ∎

5.6.5 Invariant p-Values for Checking for Prior-Data Conflict

Any 1 to 1 function of a minimal sufficient statistic is minimal sufficient. In the continuous case the p-value $M_T(m_T(t) \leq m_T(T(x_0)))$ will typically suffer from a lack of invariance under smooth, 1 to 1 transformations of the minimal sufficient statistic T. This comment will also apply to many of the other p-values developed here for the purpose of checking prior-data conflict. The solution to this problem is just as in Section 5.5, where the relevant density is modified by an appropriate factor.

As an example of this, consider first the situation where T is a complete minimal sufficient statistic, so ancillaries can be ignored, and the check is based on the value $T(x_0)$. The relevant invariant p-value for checking for prior-data conflict is then

$$M_T(m_T^*(t) \leq m_T^*(t_0)) \tag{5.28}$$

where

$$m_T^*(t) = \int_{T^{-1}\{t\}} m(x)\, \mu_{T^{-1}\{t\}}(dx) = m_T(t) E_{P(\cdot\,|T)(t)}(J_T^{-1}(x))$$

$$= \int_\Theta f_{\theta,T}^*(x)\, \Pi(d\theta).$$

Notice that, whenever $J_T(x)$ is constant, then no correction needs to be applied for the p-value. This situation arises whenever $T(x)$ has a discrete distribution or is an affine function of x. For example, $T(x) = \bar{x}$ is an affine function of x.

Now suppose that $A(T)$ is maximal ancillary. The following result is then relevant.

Proposition 5.6.2 *Suppose that T is minimal sufficient and $A(T)$ is ancillary. The invariant p-value based on the conditional prior predictive of T given $A(t_0) = a_0$ equals*

$$M_T(m_T^*(t) \leq m_T^*(t_0)\,|A(t_0)). \tag{5.29}$$

Proof. Let $t_0 = T(x_0)$. For $t \in A^{-1}\{a_0\}$, then $m_T(t\,|\,a_0) = \int_\Theta f_{\theta,T}(t\,|\,a_0)\,\Pi(d\theta)$. Note that $T^{-1}\{t\} \cap T^{-1}A^{-1}\{a_0\} = T^{-1}\{t\}$ when $t \in A^{-1}\{a_0\}$ and is the empty set otherwise. If $t \in A^{-1}\{a_0\}$, then

$$f_{\theta,T}(t\,|\,a_0) = \frac{f_{\theta,T}(t)J_A(t)}{f_{A(T)}(a_0)} = \frac{J_A(t)}{f_{A(T)}(a_0)} \int_{T^{-1}\{t\}} f_\theta(x)J_T(x)\,\mu_{T^{-1}\{t\}}(dx).$$

Therefore, removing the volume distortions due to T and $A(T)$, we have that $m_T^*(t\,|\,a_0) = \int_\Theta \int_{T^{-1}\{t\}} (f_\theta(x)/f_{A(T)}(a_0))\,\mu_{T^{-1}\{t\}}(dx)\,\Pi(d\theta) = m_T^*(t)/f_{A(T)}(a_0)$ and the result follows. ∎

So (5.29) is obtained by averaging, with respect to the prior, the relevant functions for checking each P_θ measure and then comparing the observed value of this function with its distribution under the prior predictive given the ancillary $A(T)$.

The following illustrates the necessary correction.

Example 5.6.9 *Location-Scale Normal.*

For a sample x of size n from the $N(\mu,\sigma^2)$ model, $T(x) = (\bar{x}, (n-1)^{-1}||x - \bar{x}1_n||^2) = (\bar{x}, s^2)$ is minimal sufficient. Then for a prior π on (μ, σ^2)

$$m_T(\bar{x}, s^2) = \int_0^\infty \int_{-\infty}^\infty \int_{T^{-1}\{(\bar{x},s^2)\}} f(x\,|\,\mu, \sigma^2) J_T(x)\, v_{T^{-1}\{(\bar{x},s^2)\}}(dx)\,\pi(\mu, \sigma^2)\,d\mu\,d\sigma^2$$

where $f(\cdot\,|\,\mu,\sigma^2)$ is the joint density of the sample, $v_{T^{-1}\{(\bar{x},s^2)\}}$ is surface area measure on the $(n-2)$-dimensional sphere $\bar{x}1_n + ||x - \bar{x}1_n||(S^{n-1} \cap L^\perp\{1_n\})$, and $J_T(x) = n^{1/2}(n-1)^{1/2}/2s$. So $m_T^*(\bar{x}, s^2) = 2sm_T(\bar{x}, s^2)/n^{1/2}(n-1)^{1/2}$ and the p-values based on m_T and m_T^* differ by very little. In fact this difference disappears as n grows. The arbitrariness of the p-value based on the density is demonstrated by the fact that, if we had instead chosen the minimal sufficient statistic to be $T(x) = (n\bar{x}, ||x - \bar{x}1_n||)$, then the p-value based on m_T equals the invariant p-value. ∎

The change in the p-value from (5.19) to (5.28) is typically very small. In Evans and Jang (2011b) a convergence result is established for $M_T(m_T^*(t) \leq m_T^*(t_0))$ as the amount of data increases.

5.6.6 *Diagnostics for the Effect of Prior-Data Conflict*

Suppose we have concluded that there is prior-data conflict. The question then remains as to what to do next. There are circumstances, however, where the answer can be as simple as collecting more data. For it is known that, with sufficient data, the effect of the prior on the inferences is small. If collecting more data is not possible, however, and the prior does have a material effect on the inferences, then modifying the prior seems necessary and this is the subject of Section 5.7.

So, when a prior-data conflict is detected, a quantitative assessment is needed of whether or not there is enough data to ignore this and proceed with inference. In some circumstances there is a natural candidate for a noninformative prior expressed in terms of prior-data conflict. For example, in Example 5.6.2 it is seen that the

$U(0, 1)$ prior never produces a conflict and so could be considered as noninformative. Accordingly, we could then compare the inferences obtained from a particular prior with those obtained from the uniform prior to see whether or not the prior is having an appreciable effect. The significance of the size of the differences obtained depends on the particular application, as a small difference in one context may be considered quite large in another. Consider the following example.

Example 5.6.10 *Bernoulli Model.*

Suppose the situation is as described in Example 5.6.2 with a beta(α, β) prior distribution for θ. The posterior distribution is beta$(\alpha + n\bar{x}, \beta + n(1 - \bar{x}))$. The relative belief estimate of θ is always $\bar{x}$, independent of the prior, but the assessment of the accuracy of this estimate is dependent on the prior through the posterior. So it is natural to compare the 0.95 relative belief intervals for θ under the two priors. If there is little difference, then it makes sense to say that the effect of the beta(α, β) prior can be ignored even though it does conflict. Similarly, if the inference of interest is to assess the hypothesis $H_0 = \{\theta_0\}$, then comparing the relative belief ratios and the strength of the evidence under the two priors seems like a sensible way to assess whether or not the conflict induced by the prior needs to be addressed. ∎

Example 5.6.10 has a unique feature in that a natural candidate for a noninformative prior seems to exist. More generally, this is not the case, although limits of increasingly diffuse priors often have the property of avoiding all prior-data conflicts in the limit; see Evans and Moshonov (2006). In such a context inferences based on a particular prior can be compared with limiting inferences to see if the prior-data conflict can be ignored. Diagnosing when a prior-data conflict can be ignored is a topic worthy of further study and it is acknowledged that the developments here are quite limited.

5.7 Modifying the Prior

Suppose that a prior has been found to have a serious conflict with the data. It is natural then to consider modifying the prior so that the conflict is avoided. A logical approach is needed, however, as doing this in an arbitrary, ad hoc fashion is not convincing. The methodology presented here is based on defining what it means for one prior to be more informative than another and this is done in terms of prior-data conflict. A full description of the theory can be found in Evans and Jang (2011a) and is outlined here. This in turn is based on ideas found in Gelman (2006) and Gelman, Jakulin, Pittau and Su (2008) concerning weakly informative priors.

Suppose there are two proper priors Π_1 and Π_2 on a parameter space Θ for a statistical model $\{f_\theta : \theta \in \Theta\}$. A natural question to ask is: how do we compare the amount of information each of these priors puts into the problem? While there are natural intuitive ways in which this can be expressed, such as prior variances, it seems difficult to characterize this precisely in general, e.g., a prior need not have a variance.

Consider the following context. An analyst has a prior Π_1 that represents the information at hand concerning θ but prefers to use a prior that is somewhat conservative, with respect to the amount of information put into the analysis, when compared to Π_1. So Π_1 is considered as a base prior and all other priors are to be compared to it with respect to information content. Actually, the necessity of making such a substitution for Π_1 could be seen as arising whenever Π_1 produces prior-data conflict. In fact, our methodology for comparing priors is based on comparing their a priori behavior with respect to producing prior-data conflicts.

First consider the situation where checking for prior-data conflict is effected through the p-value $M_{1,T}(m_{1,T}^*(t) \leq m_{1,T}^*(t_0))$ where $M_{i,T}$ is the prior-predictive distribution of the minimal sufficient statistic T under the prior Π_i and $m_{1,T}^*$ is as described in Section 5.6.5. Before observing the data, however, there is no way of knowing if there will be a prior-data conflict. Accordingly, since the analyst has determined that Π_1 best reflects the available information, it is reasonable to consider the prior distribution of $M_{1,T}(m_{1,T}^*(t) \leq m_{1,T}^*(t_0))$ when $t_0 \sim M_{1,T}$. Of course, this is effectively uniformly distributed (exactly so if $m_{1,T}^*(t)$ has a continuous distribution when $t_0 \sim M_{1,T}$) and this expresses the fact that all the information about assessing whether or not a prior-data conflict exists is contained in the p-value, with no need to compare the p-value to its distribution.

Consider now, however, the distribution of $M_{2,T}(m_{2,T}^*(t) \leq m_{2,T}^*(t_0))$, which is the p-value for checking Π_2. Given that a priori the appropriate distribution of t_0 is $M_{1,T}$, at least for inferences about an unobserved value of t_0, then $M_{2,T}(m_{2,T}^*(t) \leq m_{2,T}^*(t_0))$ is not uniformly distributed. In fact, from the distribution of $M_{2,T}(m_{2,T}^*(t) \leq m_{2,T}^*(t_0))$ an intuitively reasonable idea of what it means for Π_2 to be weakly informative with respect to Π_1 can be obtained. For suppose that the prior distribution of $M_{2,T}(m_{2,T}^*(t) \leq m_{2,T}^*(t_0))$ concentrates around 1. This implies that, if Π_2 is used as the prior when Π_1 is appropriate, then there is a small prior probability of a prior-data conflict. Similarly, if the prior distribution of $M_{2,T}(m_{2,T}^*(t) \leq m_{2,T}^*(t_0))$ concentrates around 0, then there is a large prior probability of a prior-data conflict. If one prior distribution results in a larger prior probability of there being a prior-data conflict than another, then it seems reasonable to say that the first prior is more informative than the second. In fact, a completely noninformative prior should never produce prior-data conflicts.

So it is natural to compare the distribution of $P_2(t_0) = M_{2,T}(m_{2,T}^*(t) \leq m_{2,T}^*(t_0))$ when $t_0 \sim M_{1,T}$ to the distribution of $P_1(t_0) = M_{1,T}(m_{1,T}^*(t) \leq m_{1,T}^*(t_0))$ when $t_0 \sim M_{1,T}$, and do this in a way that is relevant to the prior probability of obtaining a prior-data conflict. One such approach is to select a γ-quantile $x_\gamma \in [0,1]$ of the distribution of $P_1(t_0)$, and then compute the prior probability

$$M_{1,T}(P_2(t_0) \leq x_\gamma). \tag{5.30}$$

The value γ is a cut-off, dependent on the application, such that an indication of a prior-data conflict with Π_1 exists whenever $P_1(t_0) \leq \gamma$. If $m_{1,T}^*(t)$ has a continuous distribution when $t_0 \sim M_{1,T}$, then $x_\gamma = \gamma$. This leads to the following definitions.

Definition 5.7.1 *If $M_{1,T}(P_2(t_0) \leq x_\gamma) \leq x_\gamma$, then the prior Π_2 is weakly informative relative to Π_1 at level γ. If Π_2 is weakly informative relative to Π_1 at level γ for every $\gamma \leq \gamma_0$, then Π_2 is uniformly weakly informative relative to Π_1 at level γ_0. If Π_2 is weakly informative relative to Π_1 at level γ for every $\gamma \in [0,1]$, then Π_2 is uniformly weakly informative relative to Π_1.*

Clearly, when Π_2 is weakly informative relative to Π_1 at level γ, then this is an indication that the prior distribution of $M_{2T}(m_{2T}^*(t) \leq m_{2T}^*(t_0))$ is more concentrated around 1 than the prior distribution of $M_{1T}(m_{1T}^*(t) \leq m_{1T}^*(t_0))$. Choosing a prior Π_2 that is uniformly weakly informative relative to Π_1 seems appealing, but there is still the necessity of choosing a value γ in an application to determine whether or not there is an indication of a prior-data conflict. As such, the criterion of uniformly weakly informative at level γ_0 seems more generally applicable.

It also makes sense to consider Π_2 being *asymptotically weakly informative at level* γ with respect to Π_1 whenever (5.30) is bounded above by γ in the limit as the amount of data increases. This can simplify matters considerably as an asymptotically weakly informative prior may be easy to find and still weakly informative for finite amounts of data.

Once γ has been selected, the degree of weak informativity of a prior Π_2 relative to Π_1 can be assessed by comparing $M_{1,T}(P_2(t_0) \leq x_\gamma)$ to x_γ via the ratio

$$1 - M_{1,T}(P_2(t_0) \leq x_\gamma)/x_\gamma. \tag{5.31}$$

If Π_2 is weakly informative relative to Π_1 at level γ, then (5.31) gives the proportion of fewer prior-data conflicts to be expected a priori when using Π_2 rather than Π_1. Thus (5.31) provides a measure of how much less informative Π_2 is than Π_1. For example, we might ask for a prior Π_2 such that (5.31) equals 50%.

Using (5.30) directly is difficult but the following result gives a simpler expression to work with.

Lemma 5.7.1 *Suppose $P_i(t)$ has a continuous distribution under $M_{i,T}$ for $i = 1, 2$. Then there exists r_γ such that $M_{1,T}(P_2(t) \leq \gamma) = M_{1,T}(m_{2,T}^*(t) \leq r_\gamma)$, and Π_2 is weakly informative at level γ with respect to Π_1 whenever $M_{1,T}(m_{2,T}^*(t) \leq r_\gamma) \leq \gamma$. Furthermore, Π_2 is uniformly weakly informative with respect to Π_1 if and only if $M_{1,T}(m_{2,T}^*(t) \leq m_{2,T}^*(t_0)) \leq M_{2,T}(m_{2,T}^*(t) \leq m_{2,T}^*(t_0))$ for every t_0.*

Proof. Necessarily $x_\gamma = \gamma$ since $P_1(t)$ has a continuous distribution under $M_{1,T}$. Suppose $m_{i,T}^*(t)$ has a point mass at r_0 when $t \sim M_{i,T}$. The assumption $M_{i,T}(m_{i,T}^*(t) = r_0) > 0$ implies $(m_{i,T}^*)^{-1}\{r_0\} \neq \emptyset$. Then, pick $t_{r_0} \in (m_{i,T}^*)^{-1}\{r_0\}$ so that $m_{i,T}^*(t_{r_0}) = r_0$ and let $\eta_i = P_i(t_{r_0})$. Then, $P_i(t)$ has point mass at η_i because $M_{i,T}(P_i(t) = \eta_i) \geq M_{i,T}(m_{i,T}^*(t) = m_{i,T}^*(t_{r_0})) = M_{i,T}(m_{i,T}^*(t) = r_0) > 0$. This is a contradiction and so $m_{i,T}^*(t)$ has a continuous distribution when $t \sim M_{i,T}$. Let $r_\gamma = \sup\{r \in \mathscr{R} : M_{2,T}(m_{2,T}^*(t) \leq r) \leq \gamma\}$ where $\mathscr{R} = \overline{\{m_{2,T}^*(t) : t \in \mathscr{T}\}}$ and $\mathscr{T}$ is the range space of T. Then, $M_{2,T}(m_{2,T}^*(t) \leq r_\gamma) = \gamma$ and $M_{2,T}(m_{2,T}^*(t) \leq r_\gamma + \varepsilon) > \gamma$ for all $\varepsilon > 0$. Thus, we have that $\{t : P_2(t) \leq \gamma\} = \{t : m_{2,T}^*(t) \leq r_\gamma\}$, $M_{1,T}(P_2(t) \leq \gamma) = M_{1,T}(m_{2,T}^*(t) \leq r_\gamma)$, and Π_2 is weakly informative at level γ relative to Π_1 if and only if $M_{1,T}(m_{2,T}^*(t) \leq r_\gamma) \leq \gamma$. The fact that $\{r_\gamma : \gamma \in [0,1]\} \subset \mathscr{R}$ implies the last statement. ∎

It can be shown that $P_i(t)$ has a continuous distribution under $M_{i,T}$ if and only if $m_{i,T}^*(t)$ has a continuous distribution under $M_{i,T}$.

The following example shows that (5.30) behaves as it should in a simple context.

Example 5.7.1 *Comparing Normal Priors.*

Suppose $x = (x_1, \ldots, x_n)$ is a sample from an $N(\mu, 1)$ distribution where μ is unknown. Then $t = T(x) = \bar{x} \sim N(\mu, 1/n)$ is minimal sufficient and, since T is linear, there is constant volume distortion and so this can be ignored. Let the prior Π_1 on μ be an $N(\mu_0, \sigma_1^2)$ distribution with μ_0 and σ_1^2 known. Then $M_{1,T}$ is the $N(\mu_0, 1/n + \sigma_1^2)$ distribution. Now suppose that Π_2 is an $N(\mu_0, \sigma_2^2)$ distribution with σ_2^2 known so $M_{2,T}$ is the $N(\mu_0, 1/n + \sigma_2^2)$ distribution. It is supposed that the location μ_0 of the base prior is reasonable, although in general this may not be the case. Therefore,

$$P_2(t_0) = M_{2,T}(m_{2,T}^*(t) \le m_{2,T}^*(t_0)) = M_{2,T}(m_{2,T}(t) \le m_{2,T}(t_0))$$
$$= M_{2,T}((t - \mu_0)^2 \ge (t_0 - \mu_0)^2) = 1 - G_1((t_0 - \mu_0)^2/(1/n + \sigma_2^2)),$$

where G_k denotes the chi-squared(k) distribution function. Under $M_{1,T}$, $(t_0 - \mu_0)^2/(1/n + \sigma_1^2) \sim$ chi-squared(1). Therefore,

$$M_{1,T}(P_2(t_0) \le \gamma) = M_{1,T}(1 - G_1((t_0 - \mu_0)^2/(1/n + \sigma_2^2)) \le \gamma)$$
$$= M_{1,T}\left(\frac{(t_0 - \mu_0)^2}{1/n + \sigma_1^2} \ge \frac{1/n + \sigma_2^2}{1/n + \sigma_1^2} G_1^{-1}(1 - \gamma)\right)$$
$$= 1 - G_1\left(\frac{1/n + \sigma_2^2}{1/n + \sigma_1^2} G_1^{-1}(1 - \gamma)\right). \quad (5.32)$$

It is immediate that (5.32) is less than γ if and only if $\sigma_2 > \sigma_1$. In other words, Π_2 will be uniformly weakly informative relative to Π_1 if and only if Π_2 is more diffuse than Π_1. Note that $M_{1,T}(P_2(t_0) \le \gamma)$ converges to 0 as $\sigma_2^2 \to \infty$ to reflect noninformativity. Also, as $n \to \infty$, then (5.32) increases to $1 - G_1((\sigma_2^2/\sigma_1^2)G_1^{-1}(1 - \gamma))$ and so n can be ignored and σ_2^2 chosen conservatively based on this limit to obtain an asymptotically uniformly weakly informative prior and this value of σ_2^2 will also determine a weakly informative prior for finite n.

If it is specified that (5.31) equal $p \in [0, 1]$, then (5.32) implies that $\sigma_2^2 = (1/n + \sigma_1^2)(G_1^{-1}(1 - \gamma + p\gamma)/G_1^{-1}(1 - \gamma)) - 1/n$. Such a choice will give a proportion p fewer prior-data conflicts at level γ than the base prior. This decreases to $\sigma_1^2 G_1^{-1}(1 - \gamma + p\gamma)/G_1^{-1}(1 - \gamma)$ as $n \to \infty$ and so the more data the less extra variance is needed for Π_2 to be weakly informative.

Now suppose that $t \sim N_k(\mu, n^{-1}I)$ with Π_1 given by $\mu \sim N_k(\mu_0, \Sigma_1)$. Note that $M_{i,T}$ is the $N_k(\mu_0, n^{-1}I + \Sigma_i)$ distribution. It is then easy to see that $P_2(t_0) = 1 - G_k((t_0 - \mu_0)'(n^{-1}I + \Sigma_2)^{-1}(t_0 - \mu_0))$ and

$$M_{1,T}(P_2(t_0) \le \gamma) = M_{1,T}((t_0 - \mu_0)'(n^{-1}I + \Sigma_2)^{-1}(t_0 - \mu_0) \ge G_k^{-1}(1 - \gamma)). \quad (5.33)$$

Note that (5.33) increases to the probability that $(t_0 - \mu_0)'\Sigma_2^{-1}(t_0 - \mu_0) \ge G_k^{-1}(1 - \gamma)$, when $t_0 \sim N_k(\mu_0, \Sigma_1)$, as $n \to \infty$. This probability can be easily computed via simulation.

The following result is proved in Evans and Jang (2011a).

Proposition 5.7.1 *For a sample of n from the statistical model* $\{N_k(\mu,I) : \mu \in R^k\}$, *an* $N_k(\mu_0,\Sigma_2)$ *prior is uniformly weakly informative with respect to an* $N_k(\mu_0,\Sigma_1)$ *prior if and only if* $\Sigma_2 - \Sigma_1$ *is positive semidefinite.*

The general necessary part of this result is much more difficult to establish than the $k = 1$ case and shows that an $N_k(\mu_0,\Sigma_2)$ prior cannot be uniformly weakly informative with respect to an $N_k(\mu_0,\Sigma_1)$ prior unless $\Sigma_2 \geq \Sigma_1$. If Σ_1 and Σ_2 are arbitrary $k \times k$ positive definite matrices, then $r\Sigma_2 \geq \Sigma_1$ whenever $r \geq \lambda_k(\Sigma_1)/\lambda_1(\Sigma_2)$ where $\lambda_i(\Sigma)$ denotes the i-th ordered eigenvalue of Σ. Also, if $\Sigma_i = QD_iQ'$ is the spectral decomposition of Σ_i, then $\Sigma_2 \geq \Sigma_1$ whenever $\lambda_i(\Sigma_2) \geq \lambda_i(\Sigma_1)$ for $i = 1,\ldots,k$. These facts can be used in the selection of Σ_2. ∎

It is not uncommon to find t priors being substituted for normal priors on location parameters based on the idea that the longer tails of the t prior mean less information being placed into the problem. The following example shows that this is not necessarily the case.

Example 5.7.2 *Comparing a t Prior with a Normal Prior.*

Suppose $x = (x_1,\ldots,x_n)$ is a sample from an $N(\mu,1)$ distribution where μ is unknown. Take Π_1 to be an $N(\mu_0,\sigma_1^2)$ distribution and Π_2 to be a $t_1(\mu_0,\sigma_2,\lambda)$ distribution, namely, $t_1(\mu_0,\sigma_2^2,\lambda)$ denotes the distribution of $\mu_0 + \sigma_2 z$ with z distributed as a one-dimensional t distribution with λ degrees of freedom. We want to determine σ_2 and λ so that the $t_1(\mu_0,\sigma_2^2,\lambda)$ prior is weakly informative with respect to the normal prior.

Consider first the limiting case as $n \to \infty$. The limiting prior predictive distribution of the minimal sufficient statistic $T(x) = \bar{x}$ is $N(\mu_0,\sigma_1^2)$ while $P_2(t_0)$ converges in distribution to $1 - H_{1,\lambda}((t_0 - \mu_0)^2/\sigma_2^2)$ where $H_{1,\lambda}$ is the distribution function of an $F_{1,\lambda}$ distribution. This implies that (5.30) converges to $1 - G_1((\sigma_2^2/\sigma_1^2)H_{1,\lambda}^{-1}(1-\gamma))$ and this is less than or equal to γ if and only if $\sigma_2^2/\sigma_1^2 \geq G_1^{-1}(1-\gamma)/H_{1,\lambda}^{-1}(1-\gamma)$. So to have that Π_2 is asymptotically weakly informative with respect to Π_1 at level γ, then σ_2 must be large enough. Clearly, Π_2 is asymptotically uniformly weakly informative with respect to Π_1 if and only if

$$\sigma_2^2/\sigma_1^2 \geq \sup_{\gamma \in [0,1]} G_1^{-1}(1-\gamma)/H_{1,\lambda}^{-1}(1-\gamma).$$

For example, consider the following table.

λ	0.5	1	3	100
$\sup_{\gamma \in [0,1]} G_1^{-1}(1-\gamma)/H_{1,\lambda}^{-1}(1-\gamma)$	0.4569	0.6366	0.8488	0.9950

So for a t_1 (Cauchy) prior it is necessary that $\sigma_2^2 \geq \sigma_1^2(0.6366)$ for this prior to be uniformly weakly informative with respect to an $N(\mu_0,\sigma_1^2)$ prior. A $t_1(\mu_0,\sigma_2^2,3)$ prior has variance $3\sigma_2^2$ and, if σ_2^2 is chosen so that this prior also has variance σ_1^2, then $\sigma_2^2/\sigma_1^2 = 1/3$ and this is less than 0.8488 and so it is not uniformly weakly informative. So a $t_1(\mu_0,\sigma_2^2,3)$ prior has to have variance at least equal to $(2.5464)\sigma_1^2$ if it is to be uniformly weakly informative with respect to an $N(\mu_0,\sigma_1^2)$ prior. This is somewhat surprising and undoubtedly is caused by the peakedness of the t distribution. Note that $\sup_{\gamma \in [0,1]} G_1^{-1}(1-\gamma)/H_{1,\lambda}^{-1}(1-\gamma)) \to 1$ as $\lambda \to \infty$, so this increase in variance for the t prior over the normal prior decreases as we increase the degrees of freedom.

The situation for finite n is covered by the following result proved in Evans and Jang (2011a).

Proposition 5.7.2 *For a sample of n from the statistical model $\{N(\mu,1) : \mu \in R^1\}$, a $t_1(\mu_0, \sigma_2^2, \lambda)$ prior is uniformly weakly informative with respect to an $N_1(\mu_0, \sigma_1^2)$ prior whenever $\sigma_2^2 \geq \sigma_{0n}^2$, where σ_{0n}^2 satisfies $(1/n + \sigma_1^2)^{-1/2} = \int_0^\infty (1/n + \sigma_{0n}^2/u)^{-1/2} k_\lambda(u)\, du$ with k_λ the $\mathrm{gamma_{rate}}(\lambda/2, \lambda/2)$ density. Furthermore, σ_{0n}^2/σ_1^2 increases to*

$$\sup_{\gamma \in [0,1]} \frac{G_1^{-1}(1-\gamma)}{H_{1,\lambda}^{-1}(1-\gamma)} = \frac{2}{\lambda} \frac{\Gamma^2((\lambda+1)/2)}{\Gamma^2(\lambda/2)} \qquad (5.34)$$

as $n \to \infty$ and so a $t_1(\mu_0, \sigma_2^2, \lambda)$ prior is asymptotically uniformly weakly informative if and only if σ_2^2/σ_1^2 is greater than or equal to (5.34).

Proposition 5.7.2 establishes that (5.34) can be used to select a uniformly weakly informative t prior.

Figure 5.3 is a plot of the values of (5.30) that arise with various $t_1(0, \sigma_2^2, 3)$ priors where σ_2^2 is chosen in a variety of ways, together with the 45-degree line. A uniformly weakly informative prior will have (5.30) always below the 45-degree line, while a uniformly weakly informative prior at level γ_0 will have (5.30) below the 45-degree line to the left of γ_0 and possibly above to the right of γ_0. For example, when $\sigma_2^2 = 1/\sqrt{3}$, then the $t_1(0, \sigma_2^2, 3)$ prior and the $N(0,1)$ prior have the same variance. It is seen that this prior is only uniformly weakly informative at level $\gamma_0 = 0.0357$ and it is not uniformly weakly informative.

Note that (5.31) converges to $1 - G_1((\sigma_2^2/\sigma_1^2)H_{1,\lambda}^{-1}(1-\gamma))/\gamma$ as $n \to \infty$, and setting this equal to p implies that $\sigma_2^2 = \sigma_1^2 G_1^{-1}(1-\gamma+\gamma p)/H_{1,\lambda}^{-1}(1-\gamma)$, which converges to the result obtained in Example 5.7.1 as $\lambda \to \infty$. So when $\lambda = 3, \gamma = 0.05$ and $p = 0.5$, then $\sigma_2^2/\sigma_1^2 = 5.0239/10.1280 = 0.49604$.

This analysis indicates that care has to be exercised when considering the scaling of the t prior if it is required that the t prior be less informative than a normal prior, at least when uniform weak informativity is the goal. This is undoubtedly due to the peakedness of the t prior, as we are putting more prior probability near the center than with a similarly scaled normal prior.

Consider now comparing a multivariate t prior to a multivariate normal prior. Let $t_k(\mu_0, \Sigma_2, \lambda)$ denote the k-dimensional t distribution given by $\mu_0 + \Sigma_2^{1/2} z$, where $\Sigma_2^{1/2}$ is a square root of the positive definite matrix Σ_2 and z has a k-dimensional t distribution with λ degrees of freedom. This is somewhat more complicated than the normal case but the following result, proved in Evans and Jang (2011a), provides sufficient conditions for asymptotic uniform weak informativity.

Proposition 5.7.3 *When sampling from the statistical model $\{N_k(\mu,I) : \mu \in R^k\}$, a $t_k(\mu_0, \Sigma_2, \lambda)$ prior is asymptotically uniformly weakly informative relative to*

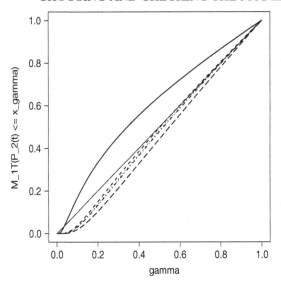

Figure 5.3 *Plot of (5.30) versus γ for $t_1(0,\sigma_2^2,3)$ priors relative to an $N(0,1)$ prior when $n = 20$ where σ_2^2 is chosen to match variances (thick solid line), match the MAD (dashed line), just achieve uniform weak informativity (dotted line), just achieve asymptotic uniform weak informativity (dash-dot line), and equal to 1 (long-dashed line).*

an $N_k(\mu_0,\Sigma_1)$ prior whenever $\Sigma_2 - \tau_\lambda^2\Sigma_1$ is positive semidefinite, where $\tau_\lambda^2 = (2/\lambda)\Gamma^{2/k}((k+\lambda)/2)/\Gamma^{2/k}(\lambda/2)$.

For the choice of Σ_2 note that, if Σ_1 and Σ_2 are arbitrary $k \times k$ positive definite matrices, then $r\Sigma_2 \geq \tau_\lambda^2\Sigma_1$ whenever $r \geq \tau_\lambda^2\lambda_k(\Sigma_1)/\lambda_1(\Sigma_2)$ where $\lambda_i(\Sigma)$ denotes the i-th ordered eigenvalue of Σ. Also, if $\Sigma_i = QD_iQ'$ is the spectral decomposition of Σ_i, then $\Sigma_2 \geq \Sigma_1$ whenever $\lambda_i(\Sigma_2) \geq \tau_\lambda^2\lambda_i(\Sigma_1)$ for $i = 1,\ldots,k$. ∎

Now consider priors for scale parameters.

Example 5.7.3 *Comparing Inverse Gamma Priors.*

Suppose that $x = (x_1,\ldots,x_n)$ is a sample from an $N(0,\sigma^2)$ distribution where σ^2 is unknown. Then $t = T(x) = (x_1^2 + \cdots + x_n^2)/n$ is minimal sufficient and $T \sim \text{gamma}_{\text{rate}}(n/2,n/2\sigma^2)$. Let Π_i be an inverse gamma prior on σ^2, namely, $\sigma^{-2} \sim \text{gamma}_{\text{rate}}(\alpha_i,\beta_i)$. This implies that $\alpha_i T/\beta_i \sim F(n,2\alpha_i)$ and, since $J_T(x) = (4x'x/n)^{-1/2} = (4t/n)^{-1/2}, m_{i,T,n}^*(t) = m_{i,T,n}(t)(4t/n)^{1/2} \propto t^{(n-1)/2}(1 + nt/2\beta_i)^{-n/2-\alpha_i}$, then

$$P_{i,n}(t_0) = M_{i,T,n}(t^{(n-1)/2}(1+nt/2\beta_i)^{-n/2-\alpha_i} \leq t_0^{(n-1)/2}(1+nt_0/2\beta_i)^{-n/2-\alpha_i}).$$

Now consider the weak informativity of a $\text{gamma}_{\text{rate}}(\alpha_2,\beta_2)$ prior relative to the $\text{gamma}_{\text{rate}}(\alpha_1,\beta_1)$ prior. For finite n this is a difficult problem so consider the asymptotic case. When the prior is Π_i and $n \to \infty$, then $m_{i,T,n}(t) \to m_{i,T}(t) = (\beta_i^{\alpha_i}/\Gamma(\alpha_i))t^{-\alpha_i-1}e^{-\beta_i/t}$, namely, $1/t \sim \text{gamma}_{\text{rate}}(\alpha_i,\beta_i)$ in the limit. Therefore,

$P_{2,n}(t_0) \to P_2(t_0) = \Pi_2(t^{-\alpha_2 - 1/2}e^{-\beta_2/t} \leq t_0^{-\alpha_2 - 1/2}e^{-\beta_2/t_0})$ and conditions on (α_2, β_2) are required so that $\Pi_1(P_2(t) \leq \gamma) \leq \gamma$.

This problem is greatly simplified if a natural restriction is imposed on (α_2, β_2), namely, restrict the location of the bulk of the mass for Π_2 to be roughly in the same place as the bulk of the mass for Π_1. Accordingly, we could require the priors to have the same means or modes but, as it turns out, the constraint that requires the modes of the $m_{i,T}^*$ functions to be the same is easier to work with. Actually, $m_{i,T,n}^*(t)$ converges to 0 but there is cancellation in the inequalities defining $P_{i,n}(t_0)$ and so define

$$m_{i,T,n}^*(t) = t^{-\alpha_i - 1/2}e^{-\beta_i/t},$$

which has its mode at $t = \beta_i/(\alpha_i + 1/2)$. This restriction implies that $\beta_2/(\alpha_2 + 1/2) = \beta_1/(\alpha_1 + 1/2)$ so (α_2, β_2) lies on the line through the points $(0, \beta_1/2(\alpha_1 + 1/2))$ and (α_1, β_1). The following result is established in Evans and Jang (2011a).

Proposition 5.7.4 *Suppose a gamma$_{rate}(\alpha_1, \beta_1)$ prior is placed on $1/\sigma^2$ when sampling from the statistical model $\{N(0, \sigma^2) : \sigma^2 > 0\}$. Then a gamma$_{rate}(\alpha_2, \beta_2)$ prior on $1/\sigma^2$, with $\beta_2/(\alpha_2 + 1/2) = \beta_1/(\alpha_1 + 1/2)$, is asymptotically weakly informative with respect to the gamma$_{rate}(\alpha_1, \beta_1)$ prior whenever $\alpha_2 \leq \alpha_1$ and $\beta_2 = \beta_1(\alpha_2 + 1/2)/(\alpha_1 + 1/2)$ or, equivalently, whenever $\beta_1/2(\alpha_1 + 1/2) \leq \beta_2 \leq \beta_1$ and $\alpha_2 = (\alpha_1 + 1/2)\beta_2/\beta_1 - 1/2$.*

Of particular interest here is that the rate parameter β_2 cannot be arbitrarily close to 0 and still be guaranteed asymptotic weak informativity. ∎

Now suppose that there exists a meaningful ancillary $A(T)$. Therefore, $M_{i,T}(m_{i,T}^*(t) \leq m_{i,T}^*(t_0))$ is replaced in all the definitions by $M_{i,T}(m_{i,T}^*(t) \leq m_{i,T}^*(t_0)|A(T))$ and (5.30) is replaced by

$$M_{1,T}(P_2(t_0|A(T)) \leq x_\gamma|A(T)). \tag{5.35}$$

Of course, $A(T)$ should be a maximal ancillary and there may be several maximal ancillaries. With several maximal ancillaries it is necessary to look at the effect of each maximal ancillary on the analysis and make our assessment about Π_2 based on this. For example, the maximum value of (5.35) over all maximal ancillaries can be used to assess whether or not Π_2 is weakly informative relative to Π_1. When this is small, then there is a small prior probability of finding evidence against the null hypothesis of no prior-data conflict when using Π_2. Consider an example where there are several maximal ancillaries.

Example 5.7.4 *Nonunique Maximal Ancillaries.*

Suppose x is a sample of n from the

$$\text{Multinomial}(1, (1 - \theta)/6, (1 + \theta)/6, (2 - \theta)/6, (2 + \theta)/6)$$

distribution where $\theta \in [-1, 1]$ is unknown. Then the counts (f_1, f_2, f_3, f_4) constitute a minimal sufficient statistic and $A_1 = (f_1 + f_2, f_3 + f_4) = (a_{11}, a_{12})$ is ancillary as is $A_2 = (f_1 + f_4, f_2 + f_3) = (a_{21}, a_{22})$.

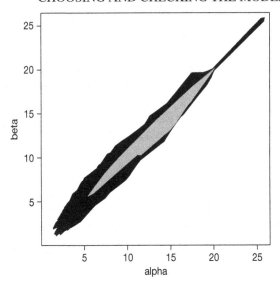

Figure 5.4 *Plot of all* (α, β) *corresponding to* $beta(\alpha, \beta)$ *priors that are weakly informative at level* $\gamma = 0.05$ *(light and dark shading) and uniformly weakly informative (light shading) in Example 5.7.4.*

Then $T = (f_1, f_2, f_3, f_4) \mid A_1$ is given by

$$f_1 \mid A_1 \sim \text{binomial} (a_{11}, (1 - \theta)/2)$$

independent of

$$f_3 \mid A_1 \sim \text{binomial} (a_{12}, (2 - \theta)/4)$$

giving

$$m_T(f_1, f_3 \mid A_1) = \binom{a_{11}}{f_1} \binom{a_{12}}{f_3} \times$$

$$\int_{-1}^{1} \left(\frac{1 - \theta}{2} \right)^{f_1} \left(\frac{1 + \theta}{2} \right)^{a_{11} - f_1} \left(\frac{2 - \theta}{4} \right)^{f_3} \left(\frac{2 + \theta}{4} \right)^{a_{12} - f_3} \pi(\theta) \, d\theta.$$

A similar result holds for the conditional distribution given A_2.

For example, suppose π is a beta(20, 20) distribution on $[-1, 1]$, so the prior concentrates about 0, and for a sample of $n = 18$, $A_1 = f_1 + f_2 = 10$ and $A_2 = f_1 + f_4 = 8$. Figure 5.4 is a plot of all the values of (α, β) that correspond to a beta(α, β) prior that is weakly informative relative to the beta(20, 20) prior at level $\gamma = 0.05$. So for each such (α, β) it is the case that (5.35) is less than or equal to 0.05 for both A_1 and A_2. ■

The paper by Evans and Jang (2011a) contains further discussion and examples on weak informativity. In particular, it is shown that in logistic regression, an arbitrarily diffuse prior on the regression coefficients is not weakly informative with respect to a given normal prior on these quantities. Placing a weakly informative prior on these quantities is more complicated than this and requires the analysis provided in the reference.

The key development in this section is that there is a logical way to modify a prior when prior-data conflict is encountered. This may mean only modifying a particular component of a hierarchically specified prior when only this component fails. The analysis of this situation is exactly as described here for the full model parameter. For example, Evans and Jang (2011a) provide details on how this is to be carried out for a normal linear regression model when the prior is prescribed by specifying a prior on the variance and then a conditional prior on the regression coefficients given the variance, just as in Example 5.6.4.

5.8   Concluding Comments

This chapter has been concerned with the ingredients in a statistical analysis necessary for an application of the definition of the measure of statistical evidence via relative belief. The necessary ingredients are a model $\{f_\theta : \theta \in \Theta\}$ and a prior π.

The first problem to be addressed is how these ingredients are to be chosen. At this point in time there doesn't seem to be much to say about the choice of the model, at least in terms of stating general principles to follow. One reasonable principle, however, is that it doesn't make sense to believe that by choosing a very complicated model then inferences will be better. In fact, the opposite seems to be true, as the additional complexity simply adds to difficulties in implementation and interpretation.

The choice of the prior is dependent on the model. The right way to choose a prior is surely based on elicitation, which means having an understanding of what values of θ are reasonable. This in turn depends on understanding what response values are reasonable. So one consequence of selecting a complicated model is that it becomes harder to see what the relationship is between parameters and responses, the more parameters there are in the model. In the end, for both the model and the prior, we are primarily reliant on the good judgment of the practitioner, but at least the choices made can be checked for their reasonableness.

No matter how the ingredients are chosen, they may be *wrong* in the sense that they are unreasonable in light of the data obtained. So, as part of any statistical analysis, it is necessary to check the ingredients chosen against the data. If the ingredients are determined to be flawed, then any inferences based on these choices are undermined with respect to their validity. Also, checking the model and the prior against the data is part of how statistics can deal with the inherent subjectivity in any statistical analysis. There should be an explicit recognition that subjectivity is *always* part of a statistical analysis. A positive consequence of accepting this is that it is now possible to address an important problem, namely, the necessity of assessing and controlling the effects of subjectivity. This seems like a more appropriate role for statistics in science as opposed to arguing for a mythical objectivity or for the virtues of subjectivity based on some kind of coherency.

Chapter 6

Conclusions

The primary purpose of this book has been to develop an approach to statistical analyses that is based on being explicit about measuring statistical evidence and having the measure of evidence determine all inferences. One could argue about the appropriateness of the specific measure of evidence chosen, namely, the relative belief ratio, but if there is a better measure, then by all means it should be used and the corresponding theory developed. For the author the relative belief ratio is justified as the measure of evidence by the many nice properties possessed by the associated theory. The key point is that any theory of inference has to be based on a well-defined measure of statistical evidence.

Whatever measure of evidence is employed, there will be a need to calibrate this quantity. The proposal here has been to compare the observed value of the measure of evidence against its posterior distribution to assess beliefs as to whether weak, moderate or strong evidence has been obtained. The separation of the measure of the evidence from its calibration helps to resolve a number of difficulties associated with characterizing statistical evidence. This has been demonstrated in several examples and is a unique feature of relative belief theory.

A key aspect of a relative belief inference, whether hypothesis assessment or estimation, is that the inference also includes an assessment of accuracy. This seems in tune with what is required in scientific applications as opposed to contexts where a categorical decision is the goal. Still, even in decision-oriented situations, a role can be seen for a presentation of the evidence, independent of what decision is taken.

One of the strengths of the theory of inference derived here is its simplicity. Based on the ingredients of model, prior and data, only three basic principles are required, namely,

1. the *principle of conditional probability* prescribing how beliefs are updated,
2. the *principle of evidence* prescribing when there is evidence in favor, neutral or against a specified value of a quantity of interest and
3. the *relative belief preference ordering* prescribing how the evidence for different values of an unknown quantity of interest are to be compared.

Once these principles are agreed upon, then inference necessarily follows, as the statistician is not required to make arbitrary choices to produce an answer.

It is readily acknowledged that an application of relative belief theory is dependent on subjective choices made by an analyst. This includes both the sampling model for the response and the prior describing beliefs within the model. This is both a strength and a weakness. It is a strength because good judgment applied to the choices can lead to more accurate and informative inferences. It is a weakness because subjectivity leads to doubts concerning the validity of the choices made and thus the validity of the inferences drawn. Certainly in scientific contexts, this is an issue of considerable importance.

For scientific work, the goal is the highest level of objectivity attainable. But it has been argued here that absolute objectivity is impossible in statistical contexts. At best, the data, when appropriately collected, can be considered objective. Given the objectivity of the data, the appropriateness of the various subjective choices can then be assessed by model checking and checking for prior-data conflict. In fact, the *principle of empirical criticism* requires that no ingredients be used as part of a statistical analysis that cannot be checked against the data. Theories of statistical inference that violate this principle are, in our view, invalid for scientific applications. As discussed in Chapter 5, relative belief theory satisfies this principle.

There is another important issue associated with subjectivity and this is the concern with bias. For example, can one choose the ingredients of a statistical analysis to get the answers one wants? Surely this lies at the heart of criticisms of Bayesian inference and it is an appropriate concern for any theory of inference. But it has been shown here that, with a definition of how evidence is to be measured, it is possible to measure bias a priori. If a conclusion of strong bias is applicable, this must vitiate any inference that simply reflects the bias. For example, if there is a large prior probability that evidence in favor of a hypothesis will be obtained, and then such evidence is observed, it seems contrary to good sense to accept the inference as sound. In essence, a statistical analysis in such a situation seems like a waste of time unless the evidence is strong enough to overcome the bias, as in finding evidence against when there is strong bias in favor of a hypothesis. As such, part of relative belief theory is an assessment of bias so that an appropriate weighting can be applied to the conclusions.

The cure for bias is primarily data. The more data there is, the less our inferences are corrupted by bias and recall that the data is objective when collected properly. This is a satisfying result, as bias can be controlled through design. Certainly, in well-designed statistical investigations bias should be minimized.

The developments concerning the assessment of bias and the checking of the ingredients is certainly very close to a frequentist approach. There is nothing contradictory about this, as there is no role for the posterior in these issues. Also, the treatment does not suffer from the ambiguity and seemingly irreconcilable difficulties that frequentist theories for inference often produce. Such issues are avoided in the relative belief approach to inference because of the necessity of the inference step implied by adding the prior and the three principles. There is also much that is common with pure likelihood theory and, in particular, the emphasis on evidence. While various approaches to Bayesian theory share features with relative belief, there are also key differences, such as not being decision based and making a sharp distinction between belief and evidence.

There is no claim made here that a full development of all of statistical theory has been accomplished. There is certainly much more work to do. This text has emphasized the logic and theory of relative belief so the examples have been purposefully chosen to be simple to avoid implementation issues. Actually, the computational issues are typically no greater than those encountered in many Bayesian inference problems. In particular, there is a need to approximate integrals, for example, see Evans and Swartz (2000). Computational problems may, however, sometimes force compromises. As has been emphasized, a valid theory of statistics represents a gold standard and there is no need to recoil from compromises and approximations when these are necessary. The important point is to keep the standard clearly in mind.

Application of relative belief theory to a number of problems of practical and theoretical importance can be found in several published papers. Evans and Shakhatreh (2014) demonstrate that the relative belief approach can avoid difficulties in a problem that the other approaches to inference have failed to resolve. Baskurt and Evans (2013) present an application to ANOVA problems and Baskurt and Evans (2015) apply relative belief to the analysis of logistic regression models. Evans, Gilula, Guttman and Swartz (1997) and Evans, Gilula and Guttman (2012) present applications to inference with contingency tables. Cao, Evans and Guttman (2014) apply relative belief theory to factor analysis. Muthukumarana and Evans (2014) discuss the use of this approach when considering problems involving bioequivalence.

Application to large scale inference problems needs to be considered. This may require an explicit representation of the compromise to the ideal required due to computational complexity. Whatever the application, however, applying a precise definition of how to measure statistical evidence will help to keep statistical issues at the fore as opposed to mathematical or computational features.

There may be features of a statistical problem that lie beyond the model and the prior and that need to be taken into account when taking a decision. As such, the ultimate decision could go against what the evidence says. For example, a hypothesis may be rejected even though there is evidence that the hypothesis is true. This is not a contradiction *provided* the evidence is clearly stated and the decision is justified accordingly. Separating the presentation of the evidence from decision making helps to resolve various conflicts between statisticians of different schools of thought. Our concern has been with the presentation of the evidence. Minimizing expected loss, or maximizing expected utility, is not seen as the primary focus of the subject of statistics.

Other authors have recognized the centrality of an adequate definition of how to measure statistical evidence for a theory of inference. In fact, any author who has taken a nondecision theoretic point of view to inference, as in Fraser (1979), is essentially in this spirit. The works of Edwards and Royall, as discussed in Section 3.2, Birnbaum, as discussed in Section 3.3, Good, as discussed in Section 4.3.2, and Jeffreys (1961) through the Bayes factor, are all directly concerned with developing a theory of statistical inference based upon measuring statistical evidence. Vieland (2014) and Vieland, Das, Hodge and Seok (2013) are focused on developing a characterization of statistical evidence based on close analogies with concepts from statistical physics. As cited in Section 4.3, there is also extensive discussion of the issues surrounding measuring evidence in the philosophy of science literature. It is the thesis of this text that the relative belief ratio provides an appropriate measure of statistical evidence and a sound basis for the theory of statistical inference.

The following statement emphasizes the essential distinction between belief and evidence. From the author's experience it takes some reflection before one becomes fully comfortable with this. Hopefully this text helps in that regard.

Evidence is what causes beliefs to change and so evidence is measured by change in belief.

Appendix A

The Definition of Density

As noted in Section 1.4, models containing infinities are to be considered as approximations to situations where all of the ingredients are in reality finite. When using a discrete probability distribution with an infinite set for the domain, then this is to be considered as an approximation to a situation where the true domain is finite. There seem to be no serious problems associated with this, provided the infinite domain has no limit points, as then anomalies can arise. Continuous probability models are not to be considered as representing the truth, as measurements are always taken to some finite accuracy and so responses are discrete. Reflecting their role as approximations, densities arise as limits and freedom in their definition on sets of measure zero is not allowed. This is obviously important for the definition of the relative belief ratio for the continuous case.

So whether using probability models containing infinities for measured responses or for unobserved parameters, these are always considered as approximations to truly finite contexts. This imposes few constraints on statistical theory but helps in clearing up apparent anomalies. Perhaps the simplest of these is the need to explain why any data can arise from a continuous distribution when each possible value arises with probability zero. In effect, restrictions on the ingredients chosen for a statistical analysis are required to ensure that they are relevant for statistical contexts and are not considered as just abstract mathematical objects. This appendix provides some details concerning these restrictions.

A.1 Defining Densities

Suppose that P is a probability measure on the Borel sets of an open set $\Omega \subset R^k$ and v is volume measure on R^k. Basically we follow Rudin (1974), Chapter 8.

Definition A.1.1 *The family of Borel sets $\{N_\delta(\omega) : \delta > 0\}$ shrinks (converges) nicely to ω as $\delta \to 0$, if there exists a number $\alpha > 0$ such that for every δ there is a ball $B_{r_\delta}(\omega)$ with $N_\delta(\omega) \subset B_{r_\delta}(\omega)$ and $v(N_\delta(\omega)) \geq \alpha v(B_{r_\delta}(\omega))$ and $r_\delta \to 0$ as $\delta \to 0$.*

It is not required that $\omega \in N_\delta(\omega)$ but this will commonly be the case. The restriction on the sets in the sequence ensures that they retain real volume in the relevant space as they shrink.

Now define the density of P with respect to v.

Definition A.1.2 *If $f(\omega) = \lim_{\delta \to 0} P(N_\delta(\omega))/\nu(N_\delta(\omega))$ for every family of Borel sets $\{N_\delta(\omega) : \delta > 0\}$ converging nicely to ω, then $f(\omega)$ is the derivative of P with respect to ν at ω.*

It is proved in Rudin (1974) that $f(\omega)$ exists almost everywhere ν and, when P is absolutely continuous with respect to ν, then $P(B) = \int_B f(\omega)\,\nu(d\omega)$ for every Borel set B. So f is a density of P with respect to ν.

The following result is key for the developments in the text, as it removes any ambiguity about the definition of density.

Proposition A.1.1 *If P is absolutely continuous with respect to ν and there exists a density f of P with respect to ν that is continuous at ω_0, then $f(\omega_0)$ is the derivative of P with respect to ν at ω_0.*

Proof. Clearly the balls $B_\delta(\omega)$ shrink nicely to ω as $\delta \to 0$. Then

$$\inf_{\omega \in B_\delta(\omega_0)} f(\omega) \leq P(B_\delta(\omega_0))/\nu(B_\delta(\omega_0)) \leq \sup_{\omega \in B_\delta(\omega_0)} f(\omega)$$

and by the continuity of f at ω_0 we can make the upper and lower bounds as close to $f(\omega_0)$ as we wish by choosing δ suitably small. ∎

Therefore, when $\omega \in N_\delta(\omega)$, the $N_\delta(\omega)$ shrink converge nicely to ω and there is a version of a density of P with respect to ν that is continuous at ω, then $P(N_\delta(\omega)) \approx f(\omega)\nu(N_\delta(\omega))$ for δ small. This is what is meant by a continuous probability measure approximating a probability measure that is finite, as $N_\delta(\omega)$ can be thought of as arising from a discretization of P. Note that this approximation doesn't necessarily hold at ω if $f(\omega)$ is allowed to be defined arbitrarily on a set of ν-measure zero.

For the probability spaces considered here, the density will be defined by the limit in Definition A.1.2 and by $f(\omega) = 0$ at any ω for which the limit does not exist. This clearly handles all the standard statistical contexts and gives the usual definitions for densities.

A.2   General Spaces

It is necessary to discuss situations more general than probability measures on Euclidean spaces. For this we allow Ω to be a Riemann manifold where there is a natural analog of volume measure. General background references for the mathematics involved are Loomis and Sternberg (1968) and Tjur (1974). Some relevant details are summarized in Evans, Fraser and Monette (1985).

Suppose then that Ω is a k-dimensional Riemann manifold for some $k \in \mathbb{N}_0$ and the support measure ν on Ω is taken to be the associated volume measure. Note that, if $B \subset U$ and (ϕ, U) is a chart for the manifold, then $\nu(B) = \int_{\phi B} |\det D\phi^{-1}(u)|\,du$ where du is volume measure on R^k and $D\phi^{-1}(u)$ is the differential of ϕ^{-1} evaluated at $u \in R^k$. Note that $k = 0$ implies that Ω is a discrete set, $|\det D\phi^{-1}(u)| \equiv 1$ and ν is counting measure.

If P is a probability measure on Ω (more formally P is defined on the Borel sets of Ω) absolutely continuous with respect to ν, then the density of P with respect to ν is given by the limit

$$f(\omega) = \lim_{\delta \to 0} \frac{P(N_\delta(\omega))}{\nu(N_\delta(\omega))}$$

where $N_\delta(\omega)$ converges nicely to ω in the sense that, whenever (ϕ, U) is a chart for the manifold and $N_\delta(\omega) \subset U$, then $\phi N_\delta(\omega)$ converges nicely to $\phi(\omega)$ as in Definition A.1.1. The value $f(\omega)$ is independent of the chart (ϕ, U); see Evans, Fraser and Monette (1985).

Consider a function $\Lambda : \Omega \to \Lambda^*$, where Λ is onto, Λ^* is an l-dimensional Riemann manifold with $l \leq k$, Λ is infinitely differentiable and $D\Lambda(\omega)$ is onto for every ω. Then $\Lambda^{-1}\{\lambda\}$ is a $(k-l)$-dimensional Riemann submanifold of Ω with volume measure $\nu_{\Lambda^{-1}\{\lambda\}}$ for each $\lambda \in \Lambda$. The density of the marginal probability measure P_Λ of Λ, with respect to volume measure ν_{Λ^*} on Λ^*, is then given by

$$f_\Lambda(\lambda) = \int_{\Lambda^{-1}\{\lambda\}} f(\omega) J_\Lambda(\omega) \, \nu_{\Lambda^{-1}\{\lambda\}}(d\omega) \tag{A.1}$$

where $J_\Lambda(\omega) = \{\det D\Lambda(\omega) \circ (D\Lambda(\omega))^*\}^{-1/2}$ and A^* is the adjoint of linear operator A. The function J_Λ is referred to here as the *volume distortion factor*.

Now suppose that $f_\Lambda(\lambda_0) > 0$, $N_\delta(\lambda_0)$ converges nicely to λ_0 and $f_\Lambda(\lambda_0) = \lim_{\delta \to 0} P_\Lambda(N_\delta(\lambda_0))/\nu_{\Lambda^*}(N_\delta(\lambda_0))$ so $P_\Lambda(N_\delta(\lambda_0)) > 0$ for all δ small enough. The conditional probability measure of ω, given that $\Lambda(\omega) = \lambda_0$, is then defined by

$$
\begin{aligned}
P(A \mid \lambda_0) &= \lim_{\delta \to 0} P(A \mid N_\delta(\lambda_0)) \\
&= \lim_{\delta \to 0} \frac{P(A \cap \Lambda^{-1} N_\delta(\lambda_0))}{P_\Lambda(N_\delta(\lambda_0))} = \lim_{\delta \to 0} \frac{\int_{A \cap \Lambda^{-1} N_\delta(\lambda_0)} f(\omega) \, \nu(d\omega)}{\int_{N_\delta(\lambda_0)} f_\Lambda(\lambda) \, \nu_{\Lambda^*}(d\lambda)} \\
&= \lim_{\delta \to 0} \frac{\int_{N_\delta(\lambda_0)} \left\{ \int_{\Lambda^{-1}\{\lambda\}} I_A(\omega) f(\omega) J_\Lambda(\omega) \, \nu_{\Lambda^{-1}\{\lambda\}}(d\omega) \right\} \nu_{\Lambda^*}(d\lambda)/\nu_{\Lambda^*}(N_\delta(\lambda_0))}{\int_{N_\delta(\lambda_0)} f_\Lambda(\lambda) \, \nu_{\Lambda^*}(d\lambda)/\nu_{\Lambda^*}(N_\delta(\lambda_0))} \\
&= \frac{\int_{\Lambda^{-1}\{\lambda_0\}} I_A(\omega) f(\omega) J_\Lambda(\omega) \, \nu_{\Lambda^{-1}\{\lambda_0\}}(d\omega)}{f_\Lambda(\lambda_0)}
\end{aligned}
$$

where we have used the *measure decomposition*

$$\nu(A) = \int_\Omega I_A(\omega) \, \nu(d\omega) = \int_{\Lambda^*} \left\{ \int_{\Lambda^{-1}\{\lambda\}} I_A(\omega) J_\Lambda(\omega) \, \nu_{\Lambda^{-1}\{\lambda\}}(d\omega) \right\} \nu_{\Lambda^*}(d\lambda),$$

as discussed in Tjur (1974). From this we conclude that the conditional density given that $\Lambda(\omega) = \lambda_0$, with respect to volume measure $\nu_{\Lambda^{-1}\{\lambda_0\}}$ on $\Lambda^{-1}\{\lambda_0\}$, equals

$$f(\omega \mid \lambda_0) = \frac{f(\omega) J_\Lambda(\omega)}{f_\Lambda(\lambda_0)} \tag{A.2}$$

for $\omega \in \Lambda^{-1}\{\lambda_0\}$.

The expressions (A.1) and (A.2) give the formulas used throughout the text for marginal and conditional densities, respectively. Notice that in the discrete case $J_\Lambda(\omega) \equiv 1$ so that these formulas generalize the basic formulas in an obvious way. These formulas are useful, as there is no need in general to find a complementary transformation $\Upsilon : \Omega \to \Upsilon^*$, so that $(\Lambda, \Upsilon) : \Omega \to \Lambda^* \times \Upsilon^*$ is 1 to 1 and smooth, to obtain expressions for the marginal and conditional densities. In fact, such a complementing transformation may not exist.

One further take-away from these considerations is that, when discussing principles of statistical reasoning, we can focus on the finite context. Of course, continuous probability is important for applications as a simplifying assumption, but it is necessary to make sure that volume distortion, which can arise in such a context, is not affecting inference in some fundamental way.

References

Achinstein, P. The Book of Evidence. Oxford University Press, 2001.

Aitkin, M. Statistical Inference: An Integrated Bayesian/Likelihood Approach. Chapman and Hall/CRC, 2010.

Al-Labadi, L. and Evans, M. Optimal robustness results for some Bayesian procedures and relationship to prior-data conflict. arXiv 1504.06898, 2015.

Allais, M. Le comportement de l'homme rationnel devant le risque: critique des postulats et axiomes de l'école Américaine. Econometrica, 21, 4, 503–546, 1953.

Ash, R. Real Analysis and Probability. Academic Press, 1972.

Ayer, A. J. Probability and Evidence. MacMillan, 1972.

Barndorff-Nielsen, O. E. Diversity of evidence and Birnbaum's theorem (with discussion). Scandinavian Journal of Statistics, 22 (4), 513–522, 1995.

Baskurt, Z. and Evans, M. Hypothesis assessment and inequalities for Bayes factors and relative belief ratios. Bayesian Analysis, 8, 3, 569–590, 2013.

Baskurt, Z. and Evans, M. Goodness of fit and inference for the logistic regression model. Manuscript. 2015

Basu, D. The family of ancillary statistics. Sankhya A, 21, 247–56, 1959.

Bayarri, M. J. and Berger, J. O. P-values for composite null models (with discussion). Journal of the American Statistical Association, 95, 452, 1143–1156, 2000.

Bayarri, M. J. and Castellanos, M. E. Bayesian checking of the second levels of hierarchical models (with discussion). Statistical Science, 22, 3, 322–343, 2007.

Berger, J. O. An overview of robust Bayesian analysis (with discussion). Test, 3, 5–124, 1994.

Berger, J. O. Statistical Decision Theory and Bayesian Analysis, Second Edition. Springer, 2006.

Berger, J. O., Bayarri, M. J. and Mulder, J. The ubiquity of information inconsistency in testing and model selection. Abstract, Joint Statistical Meetings, 2014.

Berger, J. O. and Bernardo, J. M. On the development of reference priors (with discussion). Bayesian Statistics 4, edited by J. O. Berger, A. P. Dawid and A. F. M. Smith, 35–60. Oxford University Press, 1992.

Berger, J. O., Bernardo, J. M. and Sun, D. The formal definition of reference priors. The Annals of Statistics, 37, 2, 905–938, 2009.

Berger, J. O. and Berry, D. A. Statistical analysis and the illusion of objectivity. The American Scientist, 76, 159–165, 1988.

Berger, J. and Delampady, M. Testing precise hypotheses (with discussion). Statistical Science, 2, 317–352, 1987.

Berger, J. and Sellke, T. Testing a point null hypothesis: The irreconcilability of P-values and evidence (with discussion). Journal of the American Statistical Association, 82, 112–139, 1987.

Berger, J. O. and Wolpert, R. L. The Likelihood Principle (Second Edition). Institute of Mathematical Statistics, Lecture Notes-Monograph Series, 6, 1988.

Bernardo, J. M. Reference posterior distributions for Bayesian inference (with discussion). Journal of the Royal Statistical Society, B, 41, 2, 113–147, 1979.

Bernardo, J. M. Intrinsic credible regions: An objective Bayesian approach to interval estimation. Test, 14, 2, 317–384, 2005.

Bernardo, J. M. and Smith A. F. M. Bayesian Theory. John Wiley and Sons, 1993.

Birnbaum, A. On the foundations of statistical inference (with discussion). Journal of the American Statistical Association, 57, 269–332, 1962.

Bjornstad, J. F. Predictive likelihood: A review. Statistical Science, 5, 2, 242–254, 1990.

Box, G. E. P. Sampling and Bayes' inference in scientific modelling and robustness. Journal of the Royal Statistical Society, A, 143, 383–430, 1980.

Box, G. E. P. and Tiao, G. C. Bayesian Inference in Statistical Analysis, John Wiley and Sons, 1973.

Brier, G. W. Verification of forecasts expressed in terms of probability. Monthly Weather Review, 78, 1–3, 1950.

Cao, Y., Evans, M. and Guttman, I. Bayesian factor analysis via concentration. In Current Trends in Bayesian Methodology with Applications, edited by S. K. Upadhyay, U. Singh, D. K. Dey and A. Loganathan. CRC Press, Taylor & Francis Group, 2015.

Carlin, B. P. and Louis, T. A. Bayes and Empirical Bayes Methods for Data Analysis. Chapman and Hall/CRC, 1996.

Carnap, R. Logical Foundations of Probability. University of Chicago Press, 1950.

Champernowne, D. G. The construction of decimals normal in the scale of ten. Journal of the London Mathematical Society, 8, 254–260, 1933.

Christensen, D. Measuring confirmation. The Journal of Philosophy, 96, 9, 437–461, 1999.

Cornfield, J. Sequential trials, sequential analysis and the likelihood principle. The American Statistician, 29, 2, 18–23, 1966.

Cox, R. T. Probability, frequency and reasonable expectation. American Journal of Physics, 14(1), 1–13, 1946.

Cox, R. T. The Algebra of Probable Inference. The Johns Hopkins Press, 1961.

Cox, D. R. and Hinkley, D. V. Theoretical Statistics. Chapman and Hall, 1974.

Crupi, V., Tentori, K., and Gonzalez, M. On Bayesian measures of evidential support: Theoretical and empirical issues. Philosophy of Science, 74, 2, 229–252, 2007.

Dawid, A. P., Stone, M. and Zidek, J. V. Marginalization paradoxes in Bayesian and structural inferences, with discussion. Journal of the Royal Statistical Society, Series B, 2, 189–232, 1973.

de Finetti, B. Foresight: Its logical laws, and its subjective sources. Appeared originally as La Prévision: ses lois logiques, ses sources subjectives, Annales de l'Institut Henri Poincaré, 7, 1937 and translation by H. Kyburg appeared in Kyburg and Smokler (1964).

de Finetti, B. Theory of Probability, Volume 1. Wiley, 1974.

DeGroot, M. H. Optimal Statistical Decisions. McGraw-Hill, 1970.

de la Horra, J. and Fernandez, C. Bayesian analysis under ε-contaminated priors: A trade-off between robustness and precision. Journal of Statistical Planning and Inference, 38, 13–30, 1994.

Dempster, A. P. The direct use of likelihood for significance testing. *Memoirs, No. 1, Proc. of Conference on Foundational Questions in Statistical Inference*, eds. O. Barndorff-Nielsen, P. Blaesild and G. Schou, Institute of Mathematics, U. of Aarhus, 335–354 (reprinted in Statistics and Computing, 7, 247–252, 1997), 1973.

Dickey, J. M. The weighted likelihood ratio, linear hypotheses on normal location parameters. Annals of Statistics, 42, 204–223, 1971.

Dickey, J. M. and Lientz, B. P. The weighted likelihood ratio, sharp hypotheses about chances, the order of a Markov chain. Annals of Mathematical Statistics, 41, 1, 214–226, 1970.

Draper, D. The role of statistics in the discovery of the Higgs boson. Annual Review of Statistics and Its Applications, 1, 41–59, 2014.

Druilhet, P. and Marin, J.-M. Invariant HPD credible sets and MAP estimators. Bayesian Analysis, 2, 4, 681–692, 2007.

Durbin, J. On Birnbaum's theorem on the relation between sufficiency, conditionality and likelihood. Journal of the American Statistical Association, 654, 395–398, 1970.

Edwards, A. W. F. Likelihood, Expanded Edition. Johns Hopkins University Press, 1992.

Evans, M. An example concerning the likelihood function. Statistics and Probability Letters, 7, 5, 417–418, 1989.

Evans, M. Bayesian inference procedures derived via the concept of relative surprise. Communications in Statistics — Theory and Methods, 26, 5, 1125–1143, 1997.

Evans, M. Comment on "Bayesian checking of the second levels of hierarchical models" by Bayarri and Castellanos. Statistical Science, 22, 3, 344–348, 2007.

Evans, M. What does the proof of Birnbaum's theorem prove? Electronic Journal of Statistics, 7, 2645–2655, 2013.

Evans, M. Discussion of "On the Birnbaum Argument for the Strong Likelihood Principle". Statistical Science, 29, 2, 242–246, 2014.

Evans, M., Fraser, D. A. S. and Monette, G. On regularity for statistical models. Canadian Journal of Statistics, 13, 2, 137–144, 1985.

Evans, M., Fraser, D. A. S. and Monette, G. On principles and arguments to likelihood (with discussion). Canadian Journal of Statistics, 14, 3, 181–199, 1986.

Evans, M., Gilula, Z. and Guttman, I. An inferential approach to collapsing scales. Quantitative Marketing and Economics, 10, 283–304, 2012.

Evans, M., Gilula, Z., Guttman, I. and Swartz, T. Bayesian analysis of stochastically ordered distributions of categorical variables. Journal of the American Statistical Association, 92, 437, 208–214, 1997.

Evans, M., Guttman, I. and Swartz, T. Optimality and computations for relative surprise inferences. Canadian Journal of Statistics, 34, 1, 113–129, 2006.

Evans, M. and Jang, G. H. Invariant p-values for model checking. Annals of Statistics, 38, 1, 512–525, 2010.

Evans, M. and Jang, G. H. Weak informativity and the information in one prior relative to another. Statistical Science, 26, 3, 423–439, 2011.

Evans, M. and Jang, G. H. A limit result for the prior predictive. Statistics and Probability Letters, 81, 1034–1038, 2011.

Evans, M. and Jang, G. H. Inferences from prior-based loss functions. arXiv:1104.3258, 2011.

Evans, M. and Moshonov, H. Checking for prior-data conflict. Bayesian Analysis, 1, 4, 893–914, 2006.

Evans, M. and Moshonov, H. Checking for prior-data conflict with hierarchically specified priors. In Bayesian Statistics and Its Applications, jointly edited by Upadhyay, A., Singh, U., and Dey, D. Anamaya Publishers, New Delhi, 145–159, 2007.

Evans, M. and Rosenthal, J. S. Probability and Statistics: The Science of Uncertainty. Second Edition. W. H. Freeman, 2010.

Evans, M. and Shakhatreh, M. Optimal properties of some Bayesian inferences. Electronic Journal of Statistics, 2, 1268–1280, 2008.

Evans, M. and Shakhatreh, M. Consistency of Bayesian estimates for the sum of squared normal means with a normal prior. Sankhya: The Indian Journal of Statistics, 76-A, 1, 25–47, 2014.

Evans, M. and Swartz, T. Approximating Integrals via Monte Carlo and Deterministic Methods. Oxford University Press, 2000.

Evans, M. and Zou, T. Robustness of relative surprise inferences to choice of prior. Recent Advances in Statistical Methods, Proceedings of Statistics 2001 Canada: The 4th Conference in Applied Statistics Montreal, Canada 6 – 8 July 2001, edited by Y. P. Chaubey, Imperial College Press, 90–115, 2001.

Finch, H. A. Confirming power of observations metricized for decisions among hypotheses, parts I and II. Philosophy of Science, 27, 3, 293–307 and 27, 4, 391–40, 1960.

Fine, T. Theories of Probability. Academic Press, 1973.

Fishburn, P. Utility Theory for Decision Making. Wiley, 1970.

Fisher, R. A. Inverse probability. Proceedings of the Cambridge Philosophical Society, 26, 528–535, 1930.

Fraser, D. A. S. On fiducial inference. Annals of Mathematical Statistics, 32, 3, 661–676, 1961.

Fraser, D. A. S. The Structure of Inference. John Wiley and Sons, 1968.

Fraser, D. A. S. Inference and Linear Models. McGraw-Hill, 1979.

Fraser, D. A. S. Ancillaries and conditional inference. Statistical Science, 19, 2, 333–369, 2004.

Gelman, A. Prior distributions for variance parameters in hierarchical models. Bayesian Analysis 1, 3, 515–533, 2006.

Gelman, A., Carlin, J. B., Stern, H. S. and Rubin, D. B. Bayesian Data Analysis. Chapman and Hall/CRC, 2004.

Gelman, A., Jakulin, A., Pittau, M. G. and Su, Y.-S. A weakly informative default prior distribution for logistic and other regression models. Annals of Applied Statistics, 2, 4, 1360–1383, 2008.

Gelman A., Meng, X.-L. and Stern, H. Posterior predictive assessment of model fitness via realized discrepancies (with discussion). Statistica Sinica, 6, 733–807, 1996.

Ghosh, M., Reid, N. and Fraser, D. A. S. Ancillary statistics: A review. Statistica Sinica, 20, 1309–1332, 2010.

Good, I. J. Weight of evidence, corroboration, explanatory power, information and the utility of experiments. Journal of the Royal Statistical Society, B, 22, 319–331, 1960. Corrigendum. Journal of the Royal Statistical Society, B, 30, 203, 1968.

Good, I. J. The paradox of confirmation. The British Journal for the Philosophy of Science, 11, 42, 145–149, 1960.

Good, I. J. A derivation of the probabilistic explication of information. Journal of the Royal Statistical Society, B, 28, 578–581, 1966.

Good, I. J. The probabilistic explication of information, evidence, surprise, causality. Explanation and utility (with discussion). Foundations of Statistical Inference: A Symposium. Editors V. P. Godambe and D. A. Sprott. Holt, Rinehart and Winston, 108–141, 1971.

Good, I. J. Good Thinking: The Foundations of Probability and Its Applications. Dover, 1983.

Good, I. J. Weight of evidence: A brief survey (with discussion). Bayesian Statistics 2, J. M. Bernardo and A. F. M. Smith, eds., 249–269, 1985.

Guttman, I. The use of the concept of a future observation in goodness-of-fit problems. Journal of the Royal Statistical Society, B, 143, 383–430, 1967.

Halpern, J. Y. A counterexample to theorems of Cox and Fine. Journal of Artificial Intelligence Research, 10, 76–85, 1999.

Halpern, J. Y. Cox's theorem revisited. Journal of Artificial Intelligence Research, 11, 429–435, 1999.

Halpern, J. Y. Reasoning about Uncertainty. MIT Press, 2003.

Hannig, J. On generalized fiducial inference. Statistica Sinica, 19, 491–544, 2009.

Helland, I. S. Simple counterexamples against the conditionality principle. American Statistician, 49, 4, 351–356, 1995.

Holm, S. Implication and equivalence among statistical inference rules. In Contributions to Probability and Statistics in Honour of Gunnar Blom. University of Lund, Lund, 143–155, 1985.

Horwich, P. Probability and Evidence. Cambridge University Press. 1982.

Howie, D. Interpreting Probability: Controversies and Developments in the Early Twentieth Century. Cambridge University Press, 2002.

Huber, P. J. The use of Choquet capacities in statistics. Bulletin of the International Statistical Institute, 45, 181–191, 1973.

Ioannidis, J. P. A. Why most published research findings are false. PLOS Medicine, 2(8):e124, 2005.

Jaynes, E. T. Probability Theory: The Logic of Science. Cambridge University Press, 2003.

Jeffreys, H. Some tests of significance, treated by the theory of probability. Proceedings of the Cambridge Philosophical Society, 31, 203–222, 1935.

Jeffreys, H. Theory of Probability (3rd ed.). Oxford University Press, 1961.

Johnson, V. E. Uniformly most powerful Bayesian tests. Annals of Statistics, 41, 4, 1716–1741, 2013.

Kadane, J. B. Principles of Uncertainty. CRC Press, 2011.

Kass, R. E. and Raftery, A. E. Bayes factors. Journal of the American Statistical Association, 90, 430, 773–795, 1995.

Kalbfleisch, J. D. Sufficiency and conditionality. Biometrika, 62, 251–259, 1975.

Kemeny, J. C. and Oppenheim, P. Degrees of factual support. Philosophy of Science, 19, 307–324, 1952.

Keynes, M. A. Treatise on Probability. MacMillan, 1921.

Kolmogorov, A. N. On Tables of Random Numbers. Sankhyā Ser., A. 25, 369–375, 1963.

Kraft, C. H., Pratt, J. W. and Seidenberg, A. Intuitive probability of finite sets. Annals of Mathematical Statistics, 30, 2, 408–419, 1959.

Kyburg, H. and Smokler, D. (editors). Studies in Subjective Probability, John Wiley and Sons, 1964.

Lavine, M. and Schervish, M. J. Bayes factors: What they are and what they are not. The American Statistician, 53, 2, 119–122, 1999.

Lehmann, E. L. and Romano, J. P. Testing Statistical Hypotheses, Third Edition. Springer, 2005.

Leonard, T. and Hsu, J. S. J. Bayesian Methods. Cambridge University Press, 1999.

Li, M. and Vitanyi, P. An Introduction to Kolmogorov Complexity and Its Applications. Springer-Verlag, 1993.

Lindley, D. V. Scoring rules and the inevitability of probability. International Statistical Review, 50, 1–26, 1982.

Lindley, D. V. Understanding Uncertainty. John Wiley and Sons, 2006.

Loomis, L. H. and Sternberg, S. Advanced Calculus. Addison-Wesley, 1968.

Martin, R. and Liu, C. Inferential models: A framework for prior-free posterior probabilistic inference. Journal of the American Statistical Association, 108, 301–313, 2013.

Mayo, D. On the Birnbaum argument for the strong likelihood principle (with discussion). Statistical Science, 29, 2, 227–266, 2014.

Mayo, D. and Spanos, A. Severe testing as a basic concept in a Neyman–Pearson philosophy of induction. British Journal for the Philosophy of Science, 57, 2, 323–357, 2006.

Meng, X.-Li. Posterior predictive p-values. Annals of Statistics, 22, 3, 1142–1160, 1994.

Morgan, J. P., Chaganty, N. R., Dahiya, R. C. and Dovial, M. J. Let's make a deal: The player's dilemma. The American Statistician, 45, 4, 284–287, 1991.

Mortimer, H. The Logic of Induction. Prentice Hall, 1988.

Muthukumarana, S. and Evans, M. Bayesian inference in two-arm trials using relative belief ratios. arXiv:1401.4215, 2014.

Narens, L. Theories of Probability: An Examination of Logical and Qualitative Foundations. World Scientific, 2007.

Nozick, R. Philosophical Explanations. Clarendon. 1983.

Paris, J. B. The Uncertain Reasoner's Companion: A Mathematical Perspective. Cambridge University Press, 2009.

Pawitan, Y. In All Likelihood: Statistical Modelling and Inference Using Likelihood. Oxford University Press, 2013.

Perlman, M. D. and Chaudhuri, S. Reversing the Stein effect. Statistical Science, 27, 1, 135–143, 2012.

Plante, A. An inclusion-consistent solution to the problem of absurd confidence statements: 1. Consistent exact confidence-interval estimation. Canadian Journal of Statistics,19, 4, 389–397, 1991.

Plante, A. An inclusion-consistent solution to the problem of absurd confidence statements: 2. Consistent nonexact confidence interval estimation. Canadian Journal of Statistics, 22, 1, 1–13, 1994.

Popper, K. R. The Logic of Scientific Discovery. Routledge Classics, 1959.

Popper, K. R. Realism and the Aim of Science. Rowman and Littlefield, 1983.

Poundstone, W. Priceless: The Myth of Fair Value (and How to Take Advantage of It). Hill and Wang, 2010.

Predd, J. B., Serlinger, R., Lieb, E. H., Osherson, D. N., Poor, H. V. and Kulkani, S. R. Probabilistic coherence and proper scoring rules. IEEE Transactions on Information Theory, 55, 19, 4786–4792, 2009.

Press, S. J. Subjective and Objective Bayesian Statistics: Principles, Models and Applications. John Wiley and Sons, 2003.

Ramsey, F. P. Truth and probability. Appeared originally in The Foundations of Mathematics and other Logical Essays, Ch. VII, p.156-198, edited by R. B. Braithwaite. London: Kegan, Paul, Trench, Trubner & Co, 1931 and also in Kyburg and Smokler (1964).

Reid, N. Aspects of likelihood inference. Bernoulli, 19, 4, 1404–1418, 2013.

Rescher, N. A theory of evidence. Philosophy of Science, 25, 1, 83–94, 1958.

Rios Insua, D. and Ruggeri, E., Editors. Robust Bayesian Analysis, Springer-Verlag, 2000.

Rips, L. J. Two kinds of reasoning. Psychological Science, 12, 129–134, 2001.

Robert, C. P. A note on the Jeffreys-Lindley paradox. Statistica Sinica, 3, 601–608, 1993.

Robert, C. P. (1996) Intrinsic losses. Theory and Decision, 40, 191–214, 1996.

Robert, C. P. The Bayesian Choice, Second Edition. Springer, 2001.

Robins, J. M., van der Vaart, A. and Ventura, V. Asymptotic distribution of p values in composite null models. Journal of the American Statistical Association, 95, 452, 1143–1156, 2000.

Robins, J. and Wasserman, L. Conditioning, likelihood and coherence: A review of some foundational concepts. Journal of the American Statistical Association, 95, 452, 1340–1346, 2000.

Royall, R. Statistical Evidence: A Likelihood Paradigm. Chapman and Hall, 1997.

Royall, R. On the probability of observing misleading statistical evidence (with discussion). Journal of the American Statistical Association, 95, 451, 760–780, 2000.

Rubin, D. Bayesianly justifiable and relevant frequency calculations for the applied statistician. Annals of Statistics, 12, 1151–1172, 1984.

Ruhla, C. The Physics of Chance. Oxford University Press, 1989.

Rudin, W. Real and Complex Analysis, Second Edition. McGraw-Hill, 1974.

Savage, L. J. Elicitation of personal probabilities and expectations. Journal of the American Statistical Association, 66, 336, 783–801, 1971.

Savage, L. J. The Foundations of Statistics, Second Revised Edition (original published in 1954). Dover Publications Inc., 1972.

Schervish, M. P values: What they are and what they are not. The American Statistician, 50, 3, 203–206, 1996.

Scott, D. Measurement models and linear inequalities. Journal of Mathematical Psychology, 1, 233–247, 1964.

Shafer, G. and Vovk, V. Probability and Finance, It's Only a Game. John Wiley & Sons, 2001.

Shalloway, D. The evidentiary credible region. Bayesian Analysis, 9, 909–922, 2014.

Shang, J., Ng, H. K., Sehrawat, A., Li, X. and Englert, B.-G. Optimal error regions for quantum state estimation. New Journal of Physics, 15, 123026, 2013.

Stone, M. Strong inconsistency from uniform priors, with comments. Journal of the American Statistical Association, 71, 353, 114–116, 1976.

Strug, L. J., Hodge, S. E., Chiang, T., Pal, D. K., Corey, P. N. and Rohde, C. A pure likelihood approach to the analysis of genetic association data: An alternative to Bayesian and frequentist analysis. European Journal of Human Genetics. 18(8), 933–941, 2010.

Thompson, B. The Nature of Statistical Evidence. Springer, 2007.

Thompson, W. C. and Schumann, E. L. Interpretation of statistical evidence in criminal trials. The prosecutor's fallacy and the defense attorney's fallacy. Law and Human Behavior, 11, 3, 167–187, 1987.

Tjur, T. Conditional Probability Distributions. Lecture Notes 2, Institute of Mathematical Statistics, University of Copenhagen, 1974.

Vieland, V. J. Evidence, temperature, and the laws of thermodynamics. Human Hereditary, 78, 153–163, 2014.

Vieland, V. J., Das, J., Hodge, S. E. and Seok, S.-C. Measurement of statistical evidence on an absolute scale following thermodynamic principles. Theory in Biosciences, 132, 3, 181–194, 2013.

von Mises, R. Mathematical Theory of Probability and Statistics. Academic Press, 1964.

Walley, P. Statistical Reasoning with Imprecise Probabilities. Chapman and Hall, 1991.

Wasserman, L. A robust Bayesian interpretation of likelihood regions. Annals of Statistics, 17, 1387–1393, 1989.

Welch, B. L. On confidence limits and sufficiency, with particular reference to parameters of location. Annals of Mathematical Statistics, 10, 1, 58–69, 1939.

Wrinch, D. and Jeffreys, H. On certain fundamental principles of scientific inquiry. Philosophical Magazine, 6, 42, 369–390, 1921.

Zellner, A. An Introduction to Bayesian Inference in Econometrics. John Wiley and Sons, 1971.

Ziliak, S. T. and McCloskey, D. N. The Cult of Statistical Significance: How the Standard Error Costs Us Jobs, Justice and Lives. University of Michigan Press, 2009.

Index

acceptance region, 144
action, 44
agreeing, 35
Allais paradox, 15
ancillary statistic, 61
asymptotically weakly informative, 202

Basu's Theorem, 66
Bayes factor, 83
Bayes rule, 15, 82
Bayesian p-value, 78
Bayesian frequentism, 89
Bayesian inference, 13
Bayesian unbiased estimator, 149
Bayesian unbiasedness of credible regions, 143
Bayesian updating, 76
Black–Scholes formulas, 42
Borel paradox, 30
Brier score, 44

call option, 41
called off gamble, 39
census, 2
coherent, 37
coherent forecast, 43
coin tossing, 13
completeness, 66
composite hypothesis, 55
conditional, 4
conditional posterior, 74
conditional prevision, 39
conditional pricing function, 39
conditional prior, 74
conditional prior density, 74
conditional prior predictive density, 74
conditional relative frequency distribution, 3

conditional relative frequency function, 3
conditionality principle, 61
conditonal posterior density, 74
confidence region, 70
conjugate prior, 171
consistency, 135
contingent gambles, 39
converges nicely, 215
corroboration, 107
countable additivity axiom, 28
coverage probability, 70
credible region, 79, 122

decision function, 44
degree of evidential support, 108
degree of factual support, 110
discrepancy statistic, 179
double use of the data, 178
Dutch book, 37

empirical Bayes, 88
estimation, 19

false value, 142
falsifiability, 12
fiducial inference, 90
financial derivative, 41
finite additivity, 28
Fisher's information function, 176
form of the relationship, 4

gamble, 37
Gateaux derivative, 156
generating a hypothesis via a parameter, 119
Good's information function, 108, 110

Haldane's prior, 175

229

Printed in the United States
by Baker & Taylor Publisher Services